Protracted refugee situations

Protracted refugee situations: Political, human rights and security implications

Edited by Gil Loescher, James Milner, Edward Newman and Gary Troeller

United Nations University Press
TOKYO • NEW YORK • PARIS

© United Nations University, 2008

The views expressed in this publication are those of the authors and do not necessarily reflect the views of the United Nations University.

United Nations University Press
United Nations University, 53-70, Jingumae 5-chome,
Shibuya-ku, Tokyo 150-8925, Japan
Tel: +81-3-5467-1212   Fax: +81-3-3406-7345
E-mail: sales@hq.unu.edu   general enquiries: press@hq.unu.edu
http://www.unu.edu

United Nations University Office at the United Nations, New York
2 United Nations Plaza, Room DC2-2062, New York, NY 10017, USA
Tel: +1-212-963-6387   Fax: +1-212-371-9454
E-mail: unuona@ony.unu.edu

United Nations University Press is the publishing division of the United Nations University.

Cover design by Mea Rhee

Cover photograph by Tim Dirven / Panos Pictures

Printed in Hong Kong

ISBN 978-92-808-1158-2

Library of Congress Cataloging-in-Publication Data

Protracted refugee situations : political, human rights and security implications / edited by Gil Loescher ... [et al.].
  p. cm.
 Includes bibliographical references and index.
 ISBN 978-9280811582 (pbk.)
 1. Refugees. 2. Refugees—Civil rights. 3. Human rights. I. Loescher, Gil.
HV640.P715   2008
362.87—dc22   2008038715

Contents

Tables and figures viii

Contributors ix

Acknowledgements xiii

Part I: Themes and challenges 1

1 Introduction 3
   Gil Loescher, James Milner, Edward Newman and Gary Troeller

2 Understanding the problem of protracted refugee situations ... 20
   Gil Loescher and James Milner

3 Asylum trends in industrialized countries and their impact on protracted refugee situations 43
   Gary Troeller

4 Protracted refugee situations, conflict and security: The need for better diagnosis and prescription 69
   Eric Morris and Stephen John Stedman

v

5 Protracted refugee situations, human rights and civil society ... 85
   Elizabeth Ferris

6 Development actors and protracted refugee situations:
   Progress, challenges, opportunities 108
   Mark Mattner

7 A surrogate state? The role of UNHCR in protracted refugee
   situations 123
   Amy Slaughter and Jeff Crisp

8 A realistic, segmented and reinvigorated UNHCR approach
   to resolving protracted refugee situations 141
   Arafat Jamal

9 Historical lessons for overcoming protracted refugee
   situations 162
   Alexander Betts

Part II: Case studies 187

10 Palestinian refugees 189
   Michael Dumper

11 Somali refugees: Protracted exile and shifting security
   frontiers 214
   Peter Kagwanja and Monica Juma

12 Sudanese refugees in Uganda and Kenya 248
   Tania Kaiser

13 Bhutanese refugees in Nepal 277
   Mahendra P. Lama

14 Burmese refugees in South and Southeast Asia:
   A comparative regional analysis 303
   Gil Loescher and James Milner

15 Afghan refugees in Iran and Pakistan 333
   Ewen Macleod

Part III: Policy conclusions and recommendations 351

16 A framework for responding to protracted refugee situations .. 353
   Gil Loescher and James Milner

17 Resolving protracted refugee situations: Conclusion and policy
   implications 377
   Edward Newman and Gary G. Troeller

Index 386

Tables and figures

Tables

2.1	Major protracted refugee situations, 1 January 2005	22
9.1	Main characteristics of the two archetypal models for CPAs	176
16.1	A framework for formulating and implementing comprehensive solutions for PRS	365

Figures

10.1	Map of UNRWA's area of operations	210
14.1	Map of Myanmar	329

Contributors

Alexander Betts is Hedley Bull Fellow in International Relations at the University of Oxford, where he is Director of the Global Migration Governance Project. He is the co-author of *UNHCR: The Politics and Practice of Refugee Protection into the Twenty-first Century* (Routledge, 2008) and of the forthcoming *North-South Impasse: the International Politics of Refugee Protection*.

Jeff Crisp is Head of Policy Development and Evaluation at UNHCR, the UN's refugee agency, and was previously Director of Policy and Research at the Global Commission on International Migration. He has first-hand experience of refugee situations throughout the world and has published extensively on refugee and migration issues, as well as African affairs.

Michael Dumper is Professor of Middle East Politics at Exeter University and specializes in the Arab–Israeli conflict and in the urban politics of the Middle East. He is author of *The Politics of Jerusalem since 1967* (Colombia, 1997) and *The Future of Palestinian Refugees: Towards Peace and Equity* (Lynne Rienner, 2007). In 2007 he was awarded a five-year ESRC grant for a project entitled 'Conflict in Cities and the Contested State'.

Elizabeth Ferris is Senior Fellow and Co-Director of the Brookings-Bern Project on Internal Displacement at the Brookings Institution. She focuses on the international community's response to humanitarian crises, with a particular emphasis on the human rights of internally displaced persons.

Arafat Jamal heads the UNHCR Regional Resettlement Hub for the Middle East and North Africa, based in Beirut. With UNHCR, he has worked on IDP and returnee operations (Afghanistan), and emergency operations (Democratic Republic of Congo, Turkmenistan and Guinea); he has also served as an Operational Policy Officer and a Special Adviser in the Executive Office. He has produced a number of papers and evaluation reports on protracted refugee situations, emergency and IDP responses and other topics. He holds degrees from the Universities of Cornell and Oxford.

Monica Juma is a Senior Analyst in the Peace and Security Programme of SaferAfrica, a Pretoria-based think-tank that works on various aspects of peace and security, including post-conflict reconstruction, governance of natural resources, early warning and response and terrorism. She is also the editor of *Pax Africa*, SaferAfrica's quarterly bulletin on the state of peace and security in Africa, and an Associate of the Centre for Human Rights, University of Pretoria, and the Africa Programme of the United Nations-affiliated University for Peace (Costa Rica). Prior to joining SaferAfrica, Dr. Juma served as a Research Associate at the International Peace Academy, New York, where she focused on crisis and conflicts in Africa.

Peter Kagwanja is the Director of the Democracy and Governance Programme at the Human Sciences Research Council (HSRC), South Africa, where he also heads the Africa Division, exploring a spectrum of themes on African governance, peace and security. He is also the President of the Nairobi-based Africa Policy Institute and a Research Fellow at the Centre for International Political Studies, University of Pretoria. Prior to joining the HSRC, he served as the Director of the International Crisis Group Southern Africa Project.

Tania Kaiser is a Lecturer in Refugee Studies, Department of Development Studies, at the School of Oriental and African Studies, University of London. Her research interests include the socio-political dimensions of forced migration, protracted refugee situations, refugee security and livelihoods, humanitarian protection, the anthropology of place, space and material culture, gender and conflict. She has degrees in Literature and Anthropology from the Universities of Bristol and Oxford.

Mahendra P. Lama is Vice-Chancellor of Sikkhim University. He was formerly Chairman of the Centre for South, Central, Southeast Asian and Southwest Pacific Studies, School of International Studies, at the Jawaharlal Nehru University.

Gil Loescher is visiting Professor, Refugee Studies Centre, and Senior Research Associate, Centre for International Studies, at the University of Oxford, and Emeritus Professor of Political Science at the University of Notre Dame. He is the author or co-author of several books on refugees and international relations, including *The UNHCR and World Politics* (Oxford University Press, 2001) and *UNHCR: The Politics and Practice of Refugee Protection into the Twenty-first Century* (Routledge, 2008). He is Co-Director of 'The

CONTRIBUTORS xi

PRS Project: towards solutions for protracted refugee situations' based at the University of Oxford.

Ewen Macleod is Senior Policy Advisor to the Afghanistan Comprehensive Solutions Unit, UNHCR.

Mark Mattner is a Trudeau Scholar and PhD Candidate in Political Science at McGill University. His research focuses on local governance and resource exploration in fragile states. Mark has worked on a range of reintegration, conflict and development issues with UNHCR and the World Bank.

James Milner is Assistant Professor of Political Science at Carleton University. He has previously worked with UNHCR at Headquarters and in the field. He is author of *Refugees, the State, and the Politics of Asylum in Africa* (Palgrave Macmillan, 2009), and co-author of *UNHCR: The Politics and Practice of Refugee Protection into the Twenty-first Century* (Routledge, 2008). He is Co-Director of 'The PRS Project: towards solutions for protracted refugee situations' based at the University of Oxford.

Eric Morris is Practitioner-in-Residence of the Ford Dorsey International Policy Studies program at Stanford University. He has served in a number of senior management positions in UNHCR, as well as serving in UN peacekeeping missions and UN system coordination positions. He received his PhD from Cornell University.

Edward Newman is a Senior Lecturer in the Department of Political Science and International Studies at the University of Birmingham. Prior to that, he was Director of Studies on Conflict and Security in the Peace and Governance Programme of the United Nations University. He has published a number of books and articles, including *A Crisis of Global Institutions? Multilateralism and International Security* (Routledge, 2007).

Amy Slaughter is the director of operations for Mapendo International, a refugee NGO with headquarters in Cambridge, Massachusetts and programs in sub-Saharan Africa. Prior to that, she directed the US refugee resettlement processing program in Vienna, Austria, run by the Hebrew Immigrant Aid Society (HIAS). In addition to work with various NGOs engaged in refugee resettlement both in the US and overseas, she has served as a consultant for UNHCR and FilmAid International. Her field postings include the Former Yugoslavia and Ghana, with assessment missions to Benin, Burkina Faso, Kenya, Guinea, Chad and Thailand. Amy has a master's degree in human rights from Columbia University.

Stephen John Stedman is Professor of Political Science and Senior Fellow at The Center for International Security and Cooperation (CISAC) at Stanford University. He served as the research director of the UN High-Level Panel on Threats, Challenges and Change, created to analyse global security threats and propose far-reaching reforms to the international system. He also worked at the UN as a Special Advisor with the rank of Assistant Secretary-General, to help gain worldwide support in implementing the panel's recommendations.

Gary Troeller's recent appointments include Co-Chair, Inter-university Committee on International Migration (Harvard, MIT, Boston, Brandeis, Tufts, The Fletcher School of Law and Diplomacy and Wellesley); Research Fellow in Human Rights and Justice, Center for International Studies, and Visiting Lecturer, Political Science Department, MIT; and Advisor, UN Commission on International Migration. A former senior official with UNHCR, his publications in several languages include books, articles and essays on human rights, Middle East affairs, UN issues and international relations. He holds a PhD from the University of Cambridge and has been a Research Associate at St. Antony's College, University of Oxford.

Acknowledgements

This volume is the principal outcome of a United Nations University (UNU) research project on 'The Politics, Human Rights and Security Implications of Protracted Refugee Situations', co-directed by Gil Loescher, Edward Newman and Gary Troeller.[1] The project was generously funded by the Alchemy Foundation and the UNU, and the project directors would like to express their gratitude to these organizations whose support enabled the work to be undertaken. The co-directors would particularly like to thank the members of the UNU Peace and Governance Programme, within which this project was organized, and UNU Press. In the course of the project, a workshop was held at St. Antony's College, University of Oxford, and the project co-directors would like to thank all the participants who attended and made a contribution to our discussions. Thanks are also expressed to St. Antony's College for the use of their facilities during the workshop. In addition, the UNU representative and the office of UNHCR in New York provided very helpful assistance during project discussions and presentations in that city. Finally, all four editors would like to thank the contributors to this volume for their excellent work and dedication.

1. Earlier project results have been published as Alexander Betts, 'Conference Report: The Politics, Human Rights and Security Implications of Protracted Refugee Situations, 19–20 September, St. Antony's College, University of Oxford', *Journal of Refugee Studies* 19, no. 4, 2006, pp. 509–514; and Gil Loescher, James Milner, Edward Newman and Gary Troeller, 'Protracted Refugee Situations and Peacebuilding', UNU Policy Brief, no. 1, 2007, which was also published in *Conflict, Security and Development* 7, no. 3, 2007.

Part I
Themes and challenges

1

Introduction

Gil Loescher, James Milner, Edward Newman and Gary Troeller

Since the early 1990s, the international community's engagement with refugees has focused largely on mass influx situations and refugee emergencies, delivering humanitarian assistance to refugees and war-affected populations, and encouraging large-scale repatriation programmes in high-profile regions. In stark contrast, over two-thirds of refugees in the world today are not in emergency situations, but instead trapped in protracted refugee situations (PRS). Millions of refugees struggle to survive in camps and urban communities in remote and insecure parts of the world, and the vast majority of these refugees have been in exile for many years. Such situations constitute a growing challenge for the international refugee protection regime and the international community. While global refugee populations are at their lowest now for many years, the number of protracted refugee situations and their duration continue to increase. There are now well over 30 protracted refugee situations in the world, and the average duration of these refugee situations has nearly doubled over the past decade.

The overwhelming majority of these situations are found in some of the world's poorest and most unstable regions, and originate from some of the world's most fragile states, including Afghanistan, Burundi, Liberia, Myanmar, Sierra Leone, Somalia and Sudan. Refugees trapped in these situations often face significant restrictions on a wide range of rights, while the continuation of these chronic refugee problems frequently gives rise to a number of political and security concerns for host states and states in the region. In this way, protracted refugee situations

---

Protracted refugee situations: Political, human rights and security implications,
Loescher, Milner, Newman and Troeller (eds),
United Nations University Press, 2008, ISBN 978-92-808-1158-2

represent a significant challenge to both human rights and security and, in turn, pose a challenge to refugee and security studies.

Despite the growing significance of the problem, protracted refugee situations have yet to feature prominently on the international political agenda or in mainstream security studies. Humanitarian agencies, such as the United Nations High Commissioner for Refugees (UNHCR), have been left to cope with caring for these forgotten populations and attempt to mitigate the negative implications of prolonged exile. These actions do not, however, constitute a durable solution for protracted refugee situations. Such a response also fails to address the security implications associated with prolonged exile, with the potential consequence of undermining stability in the regions where PRS are found and peacebuilding efforts in the countries of origin.

Protracted refugee situations hold significant implications for asylum debates, international peace and security, peacebuilding and security studies. The existence of protracted refugee situations is most directly a symptom of conflict and persecution: push factors associated with armed violence and state failure, which force large numbers of people to flee their homes. This is compounded by the challenges inherent in stabilizing conflict-prone regions and societies which have experienced violent conflict. Many such situations are essentially ignored by the international community. Frequently when ceasefires and peace agreements are achieved, they are unsuccessful or give way to renewed, and often escalated, violence. Progress is often incremental, in some cases spanning decades. Many peace processes become interminably protracted: lengthy and circular negotiations in which concessions are rare, and, even if fragile agreements have been reached, they have stumbled at the implementation phase. As the UN Secretary-General observed: 'Our record of success in mediating and implementing peace agreements is sadly blemished by some devastating failures. Indeed, several of the most violent and tragic episodes of the 1990s occurred after the negotiation of peace agreements'.[1] Some estimates have suggested that as many as half of the ceasefires and peace agreements established in conflict-prone societies fail, resulting in renewed armed violence. Protracted situations of violence, which thwart efforts at stabilization (or go largely ignored), continue to obstruct the return of forcibly displaced people. Protracted refugee situations are therefore indicative of broader challenges regarding civil war and peacebuilding.

However, protracted refugee situations also reflect pathologies inherent in attitudes towards asylum in policy circles, in both the developed and developing worlds. Refugees, asylum seekers and displaced people – especially in situations of mass influx – are universally regarded with negativity as a strain upon resources and a potential threat to stability, identity and social cohesion. Protracted refugee situations stretch the original assumptions which underpinned the international legal regime on refugee protection. They are also indicative of the marginalization of refugee communities in policy circles and, above all, the reluctance on the part of governments to undertake serious remedial action, especially if that might include local integration. Protracted refugees situations are, therefore, the most acute test of refugee and asylum policy, and one that is indicative of broader challenges in this field.

Protracted refugee situations also demand new analytical thinking – as well as new policy – in the area of conflict and security. Conventional policy analysis and scholarship in the area of national and international security privilege the defence of territory and the state against external military threats. These external military threats are generally embodied in adversarial states. According to this, forced human displacement is a *consequence* of armed conflict, to be approached as an essentially secondary (humanitarian) challenge. However, there is ample evidence that protracted refugee situations are a source – as well as a consequence – of instability and conflict. Many regional conflicts demonstrate that protracted refugee situations are a driving force of ongoing grievances, instability and insurgency. In some cases, such as the conflicts in Rwanda and the Democratic Republic of the Congo, protracted refugee situations may have been the principal source or catalyst for conflict, rather than a mere consequence. Displaced communities sometimes contain combatants and militants able to exploit an environment of grievance and aimlessness amongst young men in order to build fighting forces or, on very rare occasions, groups prepared to engage in terrorism. On other occasions, conspicuous refugee communities – especially when concentrated in border regions – can upset local balances and generate local antagonism. PRS are indicative of the complex nature of contemporary conflict, which defies conventional state-centric modelling. All refugee situations are, above all, humanitarian emergencies and human rights must remain the overriding rationale for generating durable solutions. The security challenges of protracted refugee situations must not form a pretext for even greater cantonment and warehousing of refugees. Nevertheless, the security implications of leaving PRS unresolved suggest that greater efforts are essential.

Security studies is characterized by a debate between conventional military approaches and non-traditional (including critical) approaches, which seek to deepen and broaden security discourse. Some non-traditional approaches suggest that security policy and security analysis, if they are to be effective and legitimate, must focus on the individual as the primary beneficiary. Protracted refugee situations – especially in developing regions of the world where there is conflict – are highly relevant to this debate, even though they receive scant attention in non-traditional security studies. PRS are symptomatic of the reality of conflict and insecurity in much of the world: weak and failed states, civil war and persecution.

Non-traditional security studies scholarship, whilst acknowledging the nature of contemporary conflict and insecurity, has been wary of 'securitizing' forced migration as a part of the solution because of the fear that this will bolster military, exclusionary approaches to addressing the challenge. Indeed, some analysts have argued that these challenges are more humanely addressed within the realm of 'normal' politics.[2] However, the security implications of protracted refugee situations suggest that a purely humanitarian rationale in attempting to achieve durable solutions may not bring the necessary resources and attention to bear on the challenges. The security consequences – when thinking about conflicts in the Great Lakes region, Afghanistan and Burma, amongst other areas – and the regional and sometimes even global repercussions of PRS are now undeniable. Protracted refugee situations must be considered at the centre of a broadening security discourse that embraces a range of actors and challenges, including social, economic and human rights issues. Simultaneously, it is necessary to be aware of the 'normative dilemma of speaking and writing security'.[3]

One of the starting points for this volume, therefore, is that a principal challenge in approaching PRS, from both a theoretical and a policy perspective, is the need for a balance between securitization and human rights. The negative security implications of PRS must be understood and acknowledged – and policy approaches designed in light of this – but the protection of human rights must remain the overriding guiding principle. There need not be an inherent tension between recognizing, and acting upon, the security implications of PRS and their humanitarian protection. Nevertheless, as this volume demonstrates, PRS raise a range of sensitive conceptual and policy debates which are not easily resolved.

Structure of the volume

The rest of Part I includes chapters on the definition, causes and consequences of protracted refugee situations. These include the link between the securitization of asylum and migration in industrialized states and the containment and encampment of refugees in developing countries; the role of UNHCR and host states in response to PRS; the prospects for achieving durable solutions for chronic refugee populations; and the record of past programmes employing comprehensive solutions to difficult and complex protracted refugee situations. It also focuses on the perspectives and roles of actors from the humanitarian, development and security communities in addressing the problem of PRS. Part II examines several contemporary case studies of chronic refugee situations including the Palestinians in the Middle East, Somalis in Kenya and the Horn of Africa, Sudanese in Uganda and Kenya, Afghans in Pakistan and Iran, Bhutanese in Nepal, and Burmese in Southeast and South Asia. Part III concludes the book with two chapters offering conclusions and policy implications.

In chapter 2, Gil Loescher and James Milner discuss how we should understand PRS and what further conceptual and empirical questions remain to be answered. The chapter highlights the limitations of current definitions of protracted refugee situations, and argues that there is a need to further disaggregate and nuance the notion of PRS. The authors also argue that there may be many displaced people who fall 'below the radar' of policymakers, such as urban refugees, and that they should also be accounted for. Loescher and Milner illustrate the growing significance of PRS and emphasize that, contrary to popular perception, PRS are not static but often involve fluctuations in numbers and other changes within the population. While protracted refugee situations in host countries are usually viewed on a 'country-by-country' basis, the authors point to the fact that many of the largest PRS, such as the South Sudanese, Afghan and Burmese, exist in several host countries across entire regions, which suggests that solutions should sometimes be sought on a regional basis. The chapter also maintains that the underlying causes of PRS are rooted in 'impasses', themselves closely related to other issues such as security, human rights, democracy and peacebuilding. The key to finding solutions to PRS, therefore, lies in linking the refugee issue with these other issues and overcoming the impasses that give rise to the particular PRS. Moreover, the authors argue that the current impasse in finding solutions to these long-standing refugee problems is also caused by a lack of strategic, political and financial engagement with this problem among the principal donor countries.

Chapter 3 examines the links between asylum trends in industrialized countries and their impact on protracted refugee situations. Gary Troeller outlines recent developments in the industrialized Western states which have simultaneously undermined the international protection regime and reinforced the containment of protracted refugee populations in the developing world. These developments, in both the North and the South, are intrinsically linked and must be firmly borne in mind in attempting to formulate realistic policy recommendations and tools to resolve protracted refugee situations. These developments in turn point to the likelihood that any resolution of long-standing refugee problems will be concentrated in regions of origin. Moreover, for solutions in the region to be realized, the all too elusive political will on the part of all concerned must be found. Industrialized countries will have to muster sufficient resources to play a catalytic role, and all actors relevant to development and peacebuilding will need to be actively involved.

Protracted refugee situations are often associated with the phenomenon of failed and fragile states, highlighting the limitations of a purely humanitarian approach to resolving long-standing refugee situations. Moreover, these situations pose particular challenges to the human rights of refugees, especially vulnerable groups of refugees. They also pose political, development and security challenges to host states and states in the region. Given the political causes of protracted refugee situations, an effective response to this global problem must include engagement from a broader range of humanitarian, security and development actors, as chapters in this section will argue. Chapter 4 highlights the need for security planners to address the issue of PRS. Stephen John Stedman and Eric Morris argue that there remains a yawning divide between the refugee and security fields. They maintain that what is important to refugee scholars is generally not important to security scholars and that, perhaps to a lesser degree, the obverse holds as well. Moreover, there is a gap within policy planning between the humanitarian domain and the political domain. Finally, there is a disconnect between, on the one hand, those who analyse and advocate and, on the other, practitioners who make and implement conflict management policy. This results in disagreement regarding policies aimed at ending intractable conflicts. The authors maintain that security researchers and policy planners view refugees – when they do consider them at all – as a by-product of violent intractable conflicts. Security analysts tend not to think about how refugee populations are independent actors and causes of conflict, and believe that by resolving conflicts refugee crises will end and refugees will return home. In a world of scarce resources, therefore, security planners advocate putting resources into negotiation and implementing peace agreements, believing that peace settlements end the refugee crises.

Stedman and Morris argue that there is a need for refugees to be considered as an independent variable in conflict; in this way, conflict literature and forced migration scholarship need to be more fully integrated. Understanding the relationship between conflict and refugees is particularly important for the role of global and regional bodies, such as the UN Peacebuilding Commission.

Chapter 5 examines protracted refugee situations through the lens of human rights, with a particular focus on civil society's engagement with long-term refugee situations. Elizabeth Ferris discusses the links between protracted refugee situations and human rights, and the ways in which human rights actors, particularly humanitarian and human rights non-governmental organizations (NGOs), have responded to these situations. She then explores the relationships between human rights/civil society actors as well as peace and security and development actors, noting the radically different normative and political frameworks they each work under. The chapter concludes with a discussion of the roles these actors could play in implementing comprehensive solutions to PRS.

Ferris argues that the task of developing and implementing comprehensive solutions to protracted refugee situations will require the contributions of human rights actors and civil society. In particular, much more commitment by the UN and NGOs to collaborative action is needed. For example, the UN's human rights machinery should do more to highlight the human rights dimensions of protracted refugee situations, including through the special procedures, and by contributing to the development of solutions. International human rights NGOs should develop an advocacy strategy with the UN Office on Human Rights to press for more attention to protracted refugee situations. Similarly, national NGOs could press for national human rights institutions to play a more assertive role vis-à-vis protracted refugee situations in their countries, including monitoring implementation of solutions. Finally, at the regional and global levels, UN agencies and NGO/Red Cross/Red Crescent staff could work together to develop programmes which support comprehensive solutions in their areas of operation. Ferris notes, however, that this would require not only increased consultation between actors in accord with a common framework, but also a willingness by all actors to relinquish some of their tenaciously defended independence of action.

Chapter 6 addresses the link between development and humanitarian relief, and the role of development actors in addressing protracted refugee situations, focusing particularly on the role of the World Bank. The link between development and displacement has long been recognized. For example, conflict prevention and mitigation are crucial elements of the poverty reduction strategies of the World Bank. As Mark Mattner notes, 80% of the world's 20 poorest countries have suffered a major war in the past 15 years. Moreover, on average, countries coming out of war face a 44% chance of relapsing into conflict in the first five years of peace. Even with rapid progress after peace, it can take a generation or more just to return to pre-war living standards. Development itself has been shown to have a significant impact on the likelihood of conflict, as well as its duration. In addition, the recognition of the connection between conflict, poverty and displacement is now more widely accepted, and development actors have deepened the scope of their involvement in conflict-affected countries. International responses to protracted refugee situations, however, still tend to focus primarily on humanitarian assistance. In practice, protracted refugee situations are often seen as aberrations of development progress and are largely ignored by development actors. Mattner notes, however, that development actors such as the World Bank can make a positive contribution through sustained engagement with the socioeconomic roots of the crises, which are at the heart of protracted refugee situations. In situations where violent conflict has come to an end, furthermore, they can assist the sustainable reintegration of returnees through targeted development programmes.

The following three chapters all focus on various aspects of the international refugee regime, exploring in particular the record, challenges and prospects of the UNHCR. Chapter 7 argues that humanitarian agencies in general, and UNHCR in particular, have assumed responsibility for PRS in order to fill gaps in the international refugee regime that were not envisaged at the time of its establishment. Amy Slaughter and Jeff Crisp suggest that the UN's refugee agency has been limited in its ability to address the problem of protracted refugee situations. They link this to the intractable nature of contemporary armed conflicts and the policies pursued by other actors, the priorities chosen by UNHCR, and the limited amount of attention which it devoted to this issue during the 1990s. The chapter concludes by examining the organization's more recent and current efforts to tackle the issue of protracted refugee situations, and identifies some of the key principles on which such efforts might most effectively be based. Within the context of collaboration among UN agencies and NGOs, the authors suggest that UNHCR would be able to 'do more by doing less'. It could take on a more focused 'catalytic' role of facilitation and leadership as part of a broader 'clustered' approach within the international community.

Chapter 8 calls for a realistic, segmented and reinvigorated UNHCR approach to resolving protracted refugee situations. While recognizing that PRS are caused by political factors and must be resolved by political actors, Arafat Jamal argues that UNHCR has a responsibility to safeguard the rights of refugees and alleviate the plight of refugees in limbo. This role should be bold – it must accept the obligations imposed by its perceived centrality in such situations; and modest – it should attempt to responsibly devolve functions to the host state and other actors. Jamal outlines how UNHCR must take the lead in ensuring that refugees are able to enjoy secure conditions of asylum, along with their due rights and freedoms, and also develop their human capabilities, no matter what the long-term prognosis for a lasting solution is. He argues further that UNHCR could do so by segmenting the population and focusing on specific responses to receptive sub-groups, and by elaborating longer-term visions that are both principled and specific to each given refugee situation.

At the same time, Jamal argues that UNHCR must devolve certain activities to others and act as a catalyst to achieve more effective interagency cooperation. He points out that this approach should reward flexibility and imagination, and should be constantly revised and calibrated in light of regular evaluations and measurements of progress. While this chapter focuses on ad hoc, modest and segmented approaches, rather than comprehensive ones, Jamal does not dismiss the importance of politically grounded approaches. Moreover, when it seems that a political solution could be found, UNHCR should make more use of the United Nations system, including the moral authority of the Secretary-General or the use of such bodies as the UN Peacebuilding Commission, to move refugee issues higher up the international agenda and push for solid political support to resolve them. Until political solutions are obtained, however, Jamal argues that the UNHCR should act with responsibility, accountability, imagination and 'constructive impatience' to bring about immediate changes in the condition of refugees stranded in chronic and unresolved situations. An implication of the papers of Crisp and Jamal – who were amongst the first to highlight PRS as a problem and undertook some of the initial research on this issue – is that UNHCR, through its well-intentioned desire to assist and protect long-term refugees, has in a sense inadvertently become a part of the problem. By assisting refugees in these situations and institutionalizing aid delivery via long-term care and maintenance programmes, it has allowed political actors – and especially governments – to ignore PRS challenges or at least avoid making the decisions which are necessary to achieve durable solutions. This has become known as 'administering human misery': providing the minimum human needs necessary for survival while keeping the challenge off the political agenda.

Chapter 9 addresses historical lessons which exist for resolving protracted refugee situations. Drawing upon material from UNHCR's archives and interviews with stakeholders in the various initiatives, the chapter outlines what those lessons are and how they might be applied to address contemporary protracted refugee situations. Alexander Betts argues that the most successful examples have been multilateral approaches aimed at durable solutions for refugees within a given regional context – so-called comprehensive plans of action (CPAs). Such approaches had a number of characteristics. They were *comprehensive* in terms of drawing on a range of durable solutions simultaneously; *cooperative* in terms of involving additional burden- (or responsibility-) sharing between countries of origin and asylum, and third countries acting as donors or resettlement countries; and *collaborative* in terms of working across UN agencies and with NGOs. The lessons from these historical experiences suggest that future multilateral efforts to address PRS are likely to be more successful if they are based on the 'political engagement' model. That is, an approach based on sustained UN-facilitated political dialogue, culminating in political agreement between a range of governmental and non-governmental stakeholders, and including but not being confined to addressing the refugee issue.

Part II of the volume considers a range of experiences of prolonged displacement in order to derive lessons from specific cases, whether historical or contemporary. While individual cases are to a large extent unique and have specific differentiating features, it is nevertheless important to observe patterns – in terms of the nature and dynamics of PRS, and also attempts to address them – across different cases through a comparative approach. Comparative study helps to contextualize individual cases of protracted refugee situations and draw out the key elements that need to be addressed. It highlights more clearly the points of similarity and difference in each PRS with other refugee cases, and in this way highlights some of the gaps in the construction of solutions.

Chapter 10 discusses the case of the Palestinian refugees, one of the longest-standing and numerically largest refugee situations in the world. Michael Dumper argues that the Palestinian case appears to some extent unique, or at least very different from many of the other refugee cases discussed in this book. The most striking difference or unique aspect of the Palestinian refugee situation is its sheer longevity, which produces specific dynamics of exile. Over time, a degree of political and economic integration has been permitted, especially in Syria and Jordan (but not in Lebanon). Yet there has also been a strong growth in nationalist feeling and Palestinian self-identity during their long years of exile. This long duration has meant that the number of Palestinian refugees has multiplied numerous times over the decades. It is estimated that there are more than 7 million Palestinian refugees and displaced persons out of approximately 9.3 million Palestinians worldwide. Thus, this is not only the largest refugee population in the world, but the proportion of refugees to the total Palestinian population is significantly higher than in most other refugee situations.

Dumper also points out that the legal framework for refugee status and protection for Palestinians is quite exceptional. Most Palestinian refugees are registered with the United Nations Relief and Works Agency (UNRWA) and not UNHCR, leading to a highly separate culture and ethos, and a close association between UNRWA and the sense of 'refugee identity' felt by Palestinians. The case of the Palestinian refugees is also more complex and politically charged than in many other refugee cases. The Palestinian refugee case turns the principle of non-*refoulement* on its head. The issue is not whether the conditions are safe for repatriation, as in many other refugee cases, but whether they will ever be allowed to return to their original homes, given the fact that this would undermine the *raison d'être* of the Jewish and Zionist state of Israel. Moreover, any likely future repatriation programme will be to a new state of Palestine, which is not from where the majority of refugees have come.

Chapter 11 examines the case of Somalia, where there is a close link between the existence of a fragile state and the persistence of a long-standing protracted refugee situation. The chapter focuses on the fundamental connection between shifting security dynamics in the Horn of Africa, especially after 9/11, and the prolonged Somali refugee crisis. Sixteen years after the collapse of Somalia, hundreds of thousands of Somali refugees live in protracted exile throughout the Horn of Africa and Kenya. Peter Kagwanja and Monica Juma analyse the causes of the Somali refugee problem, outlining the response of the Kenyan government and how this had an impact on the protection, assistance and experiences of Somali refugees. Finally, the chapter examines the measures taken at the national, regional and international levels to strengthen the protection of Somali refugees and to expand space for durable solutions: repatriation, local integration and resettlement. Kagwanja and Juma argue for the stabilization of Somalia as the best option for ending the country's protracted refugee crisis. They urge a careful mix of 'hard' power options (military/peacekeeping) and 'soft' power such as diplomacy and dialogue to stabilize the fragile state of Somalia.

Chapter 12 examines the long and brutal exile of Southern Sudanese refugees, and the prospects for a comprehensive solution to their plight following the signing of the Comprehensive Peace Agreement in January 2005. In considering the causes and consequences of the prolonged exile of Sudanese in the region, Tania Kaiser examines the important differences in the responses of Kenya and Uganda. Kaiser argues that important similarities and differences exist between the responses of the two primary host states, with the encampment of refugees being preferred in both instances. More significant for Kaiser, however, are the differences between the policy responses, with regional politics, internal security and domestic politics contributing substantially to the different hosting policies developed by the two states. Exploring the causes of flight from Sudan and the characteristics of the Sudanese refugee populations in Kenya and Uganda, Kaiser considers the impact of prolonged exile on both refugees and states. This analysis highlights the importance of considering the range of security concerns at play in protracted refugee situations, including the physical security of refugees, security in refugee-populated areas, and perceptions of national and regional security. Kaiser then examines how this range of interests and concerns informed the responses to Sudanese refugees, namely the hosting of refugees in camps rather than settlements. This analysis is especially useful given the long-standing debate on the merits of camps versus settlements for the well-being of refugees, and the recent prominence given to the Self-Reliance Strategy (SRS) extended to refugees in Uganda. Kaiser ends by arguing that, while greater thinking is needed on the role that development approaches can play in laying the foundation for durable solutions for refugees, the most significant obstacles to such an approach remain the unwillingness of donors to support a more development-oriented approach to PRS, and the reluctance of host states to permit such an approach.

Chapter 13 deals with the case of Nepali-speaking Bhutanese, the Lhotsampa refugees, who have been in long-term encampment in eastern Nepal for the past 17 years. The Lhotsampa were expelled from Bhutan as a consequence of ethnic discrimination and the threat of force by government authorities and their nationality status has been the subject of long-standing bilateral negotiations between Nepal and Bhutan ever since their expulsion. Mahendra P. Lama provides a critical history of the nationality issues and the behaviour of the Bhutanese government leading up to the forced exile of the Lhotsampa. He then examines refugee management in the camps, before providing an overview of the recent developments in the Nepal–Bhutan negotiations over the nationality question and repatriation, as well as recent initiatives by the international community to resolve this protracted refugee situation.

Lama argues that, in addition to opening up resettlement opportunities, the only viable way to resolve the Lhotsampa refugee situation is to achieve an agreement between Bhutan and Nepal to verify once and for all the status of the refugees, thereby permitting those who want to return home to do so. If this is not possible, the author proposes the appointment of an independent commission under the South Asian Association for Regional Cooperation (SAARC) to examine the issues of identification, determination and repatriation, with a view to making recommendations that the two governments will be obliged to undertake. Moreover, the article calls for more active involvement from a number of key stakeholders who have been on the sidelines to date, including India as the region's leading state and hegemonic power, UNHCR and the King of Bhutan. A key consideration for Lama is that, after a long period of difficult exile, the Lhotsampa be offered the opportunity to choose the durable solution that they want.

Chapter 14 examines refugees from Myanmar (Burma), one of the world's most intractable protracted refugee situations. Gil Loescher and James Milner examine these situations from both a host state and a regional perspective. The ongoing conflict in Myanmar has created at least four separate but related protracted refugee situations in Bangladesh, India, Malaysia and Thailand. While each of these situations has individual dynamics and characteristics, little understanding of the regional dynamics and connections between these refugee situations exists. At the same time, the prolonged presence of these refugee populations has an impact on bilateral and regional relations. Given the particular regional and geo-strategic location of these refugee populations – on the axis between South and Southeast Asia and at the centre of regional competition between India and China – Loescher and Milner argue that situating the related protracted Burmese refugee situations within a broader comparative and regional context will prove more useful in the formulation of a comprehensive solution.

This chapter provides an overview of the root causes of conflict in Myanmar and the patterns of displacement, and traces the significant refugee flows to neighbouring states. The authors also outline the political and strategic impact of refugees, both for individual host states and at a regional level. In light of these concerns, Loescher and Milner then examine how the main host states have responded. Based on this comparative analysis, the authors consider what lessons and solutions can be generated, both for specific short-term challenges and for the refugee situations themselves. Based on preliminary fieldwork in the region and interviews with stakeholders engaged in negotiation with the regime in Rangoon, the authors conclude that in the long run a regional response, both to the situation in Myanmar and to the associated refugee populations, will likely be more successful than the current international response, which principally relies on US and European trade sanctions against Myanmar.

Chapter 15 examines the protracted Afghan refugee situation. While millions of refugees have repatriated since the late 1990s, millions more still remain in prolonged exile in Pakistan and Iran. Ewen Macleod argues that the Afghan refugee problem can no longer be understood or effectively addressed through refugee policy frameworks and humanitarian arrangements alone. Indeed, dependence on the traditional refugee solutions may only contribute to deepening the intractability of many of the complex political, economic and social issues now confronting policymakers and practitioners. Macleod maintains that the abundance of evidence and experiences in the host states in recent years suggest that the pursuit of classical refugee solutions and approaches is compromised by the range and scale of post-conflict challenges inside Afghanistan, by contemporary population movements, by poverty and exclusion, and by past and present policies and practices. Without greater political convergence on an achievable and pragmatic set of solutions, progress will remain in doubt. But, as the author argues, neither the current policy environment in Iran and Pakistan nor the situation inside Afghanistan is favourable. New approaches that go beyond the standard refugee paradigm are essential to future prospects for finding solutions.

Part III of the volume concerns policy conclusions and recommendations. Building on the conceptual understandings of the causes, consequences and possible responses presented in the volume, chapter 16 presents a possible framework for responding to protracted refugee situations. In presenting the framework, Loescher and Milner argue that PRS pose a challenge to refugees, the agencies that care for them, and a wide range of other actors. Given the diversity of these concerns, the chapter argues that responses to PRS must address the current challenges and work towards longer-term and comprehensive solutions, engaging the full spectrum of peace and security, development and humanitarian actors. Underlying their argument is the importance of shifting from the current management-driven, 'care and maintenance' approach to PRS, towards a more 'solutions oriented' approach which is based upon real engagement amongst peace and security, development and humanitarian actors. With this in mind, Loescher and Milner outline how these three sets of actors can cooperate in the short, medium and long term to develop and implement comprehensive solutions to protracted refugee situations. Finally, chapter 17 offers a range of broader conclusions and policy implications.

Core policy and conceptual issues

At the core of this book is a desire to develop a better understanding of the causes, consequences and implications of protracted refugee situations. From the beginning of the project, we have sought to encourage research that has conceptual, policy and practical relevance. In so doing, the project sought to develop a better understanding of the circumstances which give rise to protracted refugee situations in the context of contemporary patterns of violent conflict, the nature of PRS in various parts of the world, the consequences of prolonged exile, and a more conceptually rooted and empirically informed understanding of how these situations may be resolved. In this way, the authors of the chapters contained in this volume were set a similar task: to reflect on the challenge of protracted refugee situations as it relates to their area of work and research, and to provide insights that have conceptual, policy and practical relevance. As a consequence, this volume is the first major effort to draw together conceptual and empirical research on protracted refugee situations.

The chapters of this volume together relate to a number of core conceptual and policy questions that need to be addressed if our ability to resolve PRS is to be enhanced. One of the most pressing conceptual questions is how we define a protracted refugee situation. Are quantitative and qualitative measures necessary or appropriate in determining what constitutes a protracted refugee situation? Or do such definitions favour certain situations over others? Likewise, what are the causes of protracted refugee situations? In chapter 2, Loescher and Milner argue that they are simply the result of impasses in the country of origin, in the country of asylum, and in the response of the international community to particular cases. Is this sufficient, or are there deeper, more systemic causes? While an understanding of the causes of PRS will contribute to the longer-term objective of finding solutions, it is also important to understand the full range of consequences of PRS, for refugees, host states and states in the region. What are the political, human rights and security implications of PRS? How does the prolonged presence of refugees relate to other policy objectives, including development? Given these links, how can the engagement of major donor and resettlement countries be sustained?

It is also important to understand why protracted refugee situations are growing in significance and representing a higher proportion of the world's refugee population. As outlined above, PRS now account for two-thirds of the world's refugees, and the average duration of prolonged displacement has increased. Are refugee situations becoming harder to resolve, or is the international community left with a particular set of more difficult situations to resolve? How important are the links between the rise of protracted refugee situations and the rise of so-called 'failed' and fragile states? How can a better understanding of the political, security and human rights context of PRS contribute to their resolution?

Finally, it is important to ask if there are different types of protracted refugee situations. This is a question that has both conceptual and policy significance. Is a typology useful, necessary or relevant? Do different kinds of situations in countries of origin result in different kinds of protracted refugee situations? Are there some situations that remain fundamentally different from the others? To what extent can lessons from historical cases be implemented in contemporary cases? Does each PRS require an individual and separate response, or is there a common approach that can be employed in all cases?

In addition to these conceptual questions, the chapters in this volume pose a number of policy questions. Paramount among these questions is the role that UNHCR should play in responding to PRS. A number of authors stress that the inaction of other actors in the international system has left humanitarian actors, especially UNHCR, to assume the burden of managing protracted refugee situations. What implications does this have for solutions? Can UNHCR continue to act independently to find solutions for refugees, or should UNHCR play a more specialized catalytic and facilitating role? As most chapters in this volume argue that a broader range of actors are required to resolve protracted refugee situations, what role can and should be played by other humanitarian actors, both within the UN system and within the global refugee regime? Likewise, what role can and should be played by other actors within the UN system, especially peace and security and development actors?

If a range of actors are required to formulate and implement solutions to protracted refugee situations, how can these actors be more effectively coordinated? Do new structures, such as the UN Peacebuilding Commission, provide new opportunities for structured cooperation? How can the history of competition between these various actors be overcome? Have these tensions been resolved in particular cases? If these tensions can be overcome, how should the actions of various actors be sequenced? Are certain activities essential prerequisites for a solution, or will each situation call for a unique response? Should responses to PRS be designed and implemented according to countries of origin, to host countries, or on a regional basis?

Finally, it is important to consider what we mean by a solution for a protracted refugee situation. Does a solution mean that refugees are no longer in camps, or does it mean that they have achieved a legal status which no longer requires international protection? Are the three durable solutions – repatriation, local integration, and resettlement – sufficient to resolve today's PRS? Are there other solutions that are being pursued in contemporary cases? What are the limits to these solutions? How can these solutions be reinforced to make solutions for PRS more realistic?

The conclusion of this volume draws on the insights of the preceding chapters to offer some answers to these conceptual and policy questions. From the outset, this project has had two objectives. First, it was to contribute to the academic and policy understanding of the origins, nature and significance of protracted refugee situations. Second, by examining specific cases and considering attempts to find comprehensive solutions, the project sought to make specific policy recommendations for the resolution of long-standing refugee populations. The chapters of this volume make an important contribution to our understanding of the causes, consequences and possible responses to the growing challenge of protracted refugee situations. The cases also illustrate that protracted refugee situations involve a wide range of local, national, regional and international actors, and relate to a wide range of issue areas. Above all, this volume demands that we – as scholars and policy practitioners – go beyond the administration of misery.

Notes

1. *In Larger Freedom: Towards Development, Security and Human Rights for All: Report of the Secretary-General to the General Assembly*, A/59/2005, paragraph 114.
2. For example, Astri Suhrke, 'Human Security and the Protection of Refugees', in Edward Newman and Joanne van Selm, *Refugees and Forced Displacement: International Security, Human Vulnerability, and the State*, Tokyo: UNU Press, 2003; Ole Waever, Barry Buzan, Morten Kelstrup and Pierre Lemaitre, *Identity, Migration, and the New Security Order in Europe*, London: Pinter, 1993; M. Ibrahim, 'The Securitization of Migration: A Racial Discourse', *International Migration* 43, no. 5, 2005.
3. Jef Huysmans, 'Defining Social Constructivism in Security Studies. The Normative Dilemma of Writing Security', *Alternatives* 27, 2002.

2

Understanding the problem of protracted refugee situations[1]

Gil Loescher and James Milner

Some two-thirds of refugees in the world today are trapped in protracted refugee situations. The overwhelming majority of these situations are to be found in some of the world's poorest and most unstable regions, and are most frequently neglected by a range of regional and international actors. Refugees trapped in these forgotten situations often face significant restrictions on a wide range of rights, while the continuation of chronic refugee problems frequently gives rise to a number of political and security concerns for host states and states in the region. As UNHCR has argued, 'the consequences of having so many human beings in a static state include wasted lives, squandered resources and increased threats to security'.[2] Taken independently, each of these challenges is of mounting concern. Taken collectively, and given the interaction between the security, human rights and development concerns, the full significance of protracted refugee situations becomes more apparent.

Notwithstanding the growing significance of the problem, protracted refugee situations have yet to feature prominently on the international agenda, and there has, until recently, been limited attention to the problem of prolonged exile by the research and policy community. Given the fact that prolonged exile is the reality for the majority of the world's refugees, it is important to develop a more rigorous understanding of the nature and causes of the problem, and to identify practical solutions. To this end, the purpose of this chapter is to further develop a conceptual understanding of protracted refugee situations by examining its definition, causes and consequences.

Protracted refugee situations: Political, human rights and security implications,
Loescher, Milner, Newman and Troeller (eds),
United Nations University Press, 2008, ISBN 978-92-808-1158-2

Definitions

Any examination of long-standing refugee populations should begin with a definition of the nature and causes of protracted refugee situations. Such a definition has remained elusive in recent years and has arguably frustrated efforts to formulate effective policy responses. UNHCR defines a protracted refugee situation as: 'one in which refugees find themselves in a long-lasting and intractable state of limbo. Their lives may not be at risk, but their basic rights and essential economic, social and psychological needs remain unfulfilled after years in exile. A refugee in this situation is often unable to break free from enforced reliance on external assistance'.[3] In identifying the major protracted refugee situations in the world, UNHCR uses the 'crude measure of refugee populations of 25,000 persons or more who have been in exile for five or more years in developing countries'.[4] These figures exclude Palestinian refugees who fall under the mandate of the UN Relief and Works Agency for Palestine Refugees in the Near East (UNRWA).

Applying this definition to UNHCR refugee statistics from the end of 2004, there were 33 protracted refugee situations, totalling 5,691,000 refugees, at the start of 2005, as detailed in Table 2.1.

There are, however, a number of important limitations to this definition. First, this definition reinforces the popular image of protracted refugee situations as static, unchanging and passive populations and groups of refugees that are warehoused in identified camps. Given UNHCR's humanitarian mandate, and given the prevalence of encampment policies in the developing world, it should not be surprising that such situations have been the focus of UNHCR's engagement in the issue of protracted refugee situations. The UNHCR definition does not, however, fully encompass the realities of protracted refugee situations. Far from being passive, recent cases, such as that of Liberian refugees in Guinea and Somali refugees in Kenya, illustrate how refugee populations have been engaged in identifying their own solutions, either through political and military activities in their countries of origin, or through seeking means for onward migration to the West. In addition, evidence from Africa and Asia demonstrates that, while total population numbers in protracted refugee situations remain relatively stable over time, there are, in fact, often significant changes within the membership of that population. For example, while the total number of refugees in Tanzania remained relatively stable between 1997 and 2003, at just over 500,000, the refugee population itself was relatively fluid during that period, with an average of some 55,000 refugees repatriating each year, compared with an average of some 80,000 new arrivals.[5] Similar dynamics can be found in many of the world's protracted refugee situations.

Table 2.1 Major protracted refugee situations, 1 January 2005

Country of asylum	Origin	Refugee numbers as of end-2004
Algeria	Western Sahara	165,000
Armenia	Azerbaijan	235,000
Burundi	Dem. Rep. of Congo	48,000
Cameroon	Chad	39,000
China	Vietnam	299,000
Congo	Dem. Rep. of Congo	59,000
Côte d'Ivoire	Liberia	70,000
Dem. Rep. of Congo	Angola	98,000
Dem. Rep. of Congo	Sudan	45,000
Egypt	Occupied Palestinian Territory	70,000
Ethiopia	Sudan	90,000
Guinea	Liberia	127,000
India	China	94,000
India	Sri Lanka	57,000
Islamic Rep. of Iran	Afghanistan	953,000
Islamic Rep. of Iran	Iraq	93,000
Kenya	Somalia	154,000
Kenya	Sudan	68,000
Nepal	Bhutan	105,000
Pakistan	Afghanistan (UNHCR estimate)	960,000
Rwanda	Dem. Rep. of Congo	45,000
Saudi Arabia	Occupied Palestinian Territory	240,000
Serbia and Montenegro	Bosnia and Herzegovina	95,000
Serbia and Montenegro	Croatia	180,000
Sudan	Eritrea	111,000
Thailand	Myanmar	121,000
Uganda	Sudan	215,000
United Rep. of Tanzania	Burundi	444,000
United Rep. of Tanzania	Dem. Rep. of Congo	153,000
Uzbekistan	Tajikistan	39,000
Yemen	Somalia	64,000
Zambia	Angola	89,000
Zambia	Dem. Rep. of Congo	66,000

Likewise, UNHCR statistics reveal only part of a much larger story. As Crisp has argued, UNHCR refugee statistics can be the result of a particular politicized dynamic, often reflecting a process of negotiation between the Office and the host government, and typically include only those refugees under the mandate of UNHCR.[6] In many instances, host governments may limit the number of new arrivals that can enter refugee camps and settlements, thereby limiting the number of refugees under UNHCR's mandate. For example, at the end of 2005, there were some 21,000 Rohingya refugees in camps near Cox's Bazar, Bangladesh, following the repatriation of over 200,000 Rohingya in the mid-1990s. While these refugees are included in UNHCR statistics (but not in the global total of protracted refugee situations, as outlined above), as many as 250,000 other Rohingya, including all recent arrivals, are prohibited by the Bangladeshi government from being registered as refugees and entering the camps, thereby falling outside the mandate of UNHCR. Likewise, many refugee statistics do not fully include urban refugee populations, many of whom live a clandestine life in host states which require all refugees to reside in camps. If these refugees who are currently excluded from UNHCR refugee totals were to be included, the global population of protracted refugee situations would be significantly larger.

A more effective understanding of protracted refugee situations would include not only the humanitarian elements proposed by UNHCR, but also a wider understanding of the political and strategic dimensions of protracted refugee situations, and the role of broader political, strategic and economic actors. These aspects of the challenge of protracted refugee situations have not been given sufficient attention in recent policy and research discussions, and consequently need to be especially highlighted. Second, a definition should reflect the fact that protracted refugee situations also include chronic, unresolved and recurring refugee problems, not only static refugee populations. Third, an effective definition must recognize that countries of origin and host countries are both implicated in protracted refugee situations. Fourth, it should be recognized that a common characteristic of protracted refugee situation is their frequent neglect, by international and regional actors, by the global media, and by a range of other non-humanitarian actors. Taking such an understanding as a point of departure more effectively places the issue of protracted refugee situations within its proper context, thereby providing a sounder basis for policy and for practical and research engagement.

Protracted refugee situations involve large refugee populations that are long-standing, chronic or recurring, and for which there are no immediate prospects for a solution. They are not static populations, but may involve periods of increase and decrease in the total population as well as changes within it. They are typically, but not necessarily, concentrated in a specific geographical area, but may include camp-based and urban refugee populations, in addition to displaced populations currently not included in UNHCR's refugee statistics. The nature of a protracted refugee situation will be the result of both conditions in the country of origin of the refugees, and the responses and conditions in the host country. Refugees of the same nationality in different host countries will result in different protracted refugee situations. For example, the nature of the protracted refugee situation of Sudanese refugees in Uganda is different from the nature of the protracted refugee situations in any of the other seven African countries hosting Sudanese refugees. In this way, one country may produce several protracted refugee situations.

At the same time, however, there are often important connections between protracted refugee situations within a given region. While a single country of origin may produce several protracted refugee situations, as in the case of the Sudanese, this is not to say that these separate refugee situations may not have various political, strategic, economic or social links. Likewise, it is possible, and even likely, that there will be movement between related protracted refugee situations within a given region. For example, as our chapter on the regional dimensions of the Burmese protracted refugee situations argues, there are strong regional dynamics associated with the prolonged presence of refugees from Myanmar (Burma) in South and Southeast Asia. These dynamics relate to both regional discussions and responses to the causes and consequences of displacement, but also the movement of individual refugees between different protracted refugee situations. As such, while emphasis should be placed on the dynamics of the individual protracted refugee situations, it is also important to be mindful of the broader regional context within which they exist.

Given the political causes and consequences of protracted refugee situations, there is also an important distinction to be made between a 'protracted refugee situation' and a 'protracted refugee population'. Reference only to the refugee population itself frequently leads to a consideration of the problem of prolonged displacement in isolation from the broader factors that are central to explaining the causes of chronic or recurring exile. Conversely, applying the broader term 'protracted refugee situation' allows for a fuller consideration of the context within which the refugee population is situated. This distinction has important conceptual and policy benefits, as the chapters of this volume make clear.

Likewise, the political causes of protracted refugee situations make it difficult to place quantifiable parameters on a definition to say what size is necessary for a refugee population to constitute a protracted situation, or for how many years such a population must be in exile. Politically, the identification of a protracted refugee situation is, to a certain extent, the result of perception. If a refugee population is seen to have been in existence for a significant period of time without the prospect of solutions, then it may be termed a protracted refugee situation.

While useful in the discussions of overall trends and distribution, the crude measure of 25,000 refugees in exile for five years should not be used as a basis for excluding other groups that are perceived to be protracted. For example, there remained 19,000 Burundians in the DRC, 16,000 Somalis in Ethiopia, 19,000 Mauritanians in Senegal, 15,000 Ethiopians in Sudan and 19,000 Rwandans in Uganda at the end of 2005, more than a decade after fleeing their homes. Long-staying urban refugee caseloads are not typically included in an understanding of protracted refugee situations, yet tens of thousands live clandestinely in urban areas, avoiding contact with authorities and existing without legal status. At the end of 2005, there were almost 40,000 Congolese urban refugees in Burundi, over 36,000 Somali urban refugees in Yemen, almost 15,000 Sudanese urban refugees in Egypt, nearly 10,000 Afghan urban refugees in India, and over 5,000 Liberian urban refugees in Côte d'Ivoire, to name only some of the largest groups. In addition, there are hundreds of thousands of Palestinian refugees who remain in a state of limbo throughout the Middle East, several decades after their displacement.

Trends and dimensions of the problem

In the early 1990s, a number of long-standing refugee populations, which had been displaced as a result of Cold War conflicts in the developing world, went home. For example, in Southern Africa, huge numbers of Mozambicans, Namibians and others were repatriated. In Indo-China, the Cambodians in exile in Thailand returned home, and Vietnamese and Laotians were resettled to third countries. With the conclusion of conflicts in Central America, the vast majority of displaced Nicaraguans, Guatemalans and Salvadoreans returned to their home countries. In 1993, in the midst of the resolution of these conflicts, there remained 27 major protracted refugee situations, with a total population of 7.9 million refugees.[7]

While these Cold War conflicts were being resolved, and as refugee populations were being repatriated, new intra-state conflicts emerged and resulted in massive new flows during the 1990s. Conflict and state collapse in Somalia, the Great Lakes, Liberia and Sierra Leone generated millions of refugees. Millions more refugees were displaced as a consequence of ethnic and civil conflict in Iraq, the Balkans, the Caucasus and Central Asia. The global refugee population mushroomed in the early 1990s, and the pressing need was to respond to the challenges of mass influx situations and refugee emergencies in many regions of the world simultaneously.

More than a decade later, despite the decline in the overall number of intra-state conflicts, many of these post–Cold War conflicts and refugee situations remain unresolved. As a result, the number of protracted refugee situations is greater now than at the end of the Cold War. As noted above, at the end of 2004, using UNHCR's conservative figures, there were 33 major protracted refugee situations, with a total refugee population of 5.69 million. While there are fewer UNHCR-recognized refugees in protracted situations today, the number of situations has increased. Likewise, as the global refugee population has decreased in recent years, the proportion of refugees in situations of prolonged exile has increased. More importantly, UNHCR recognizes that refugees are spending longer periods of time in exile. It is estimated that 'the average duration of major refugee situations, protracted or not, has increased: from 9 years in 1993 to 17 years in 2003'.[8] With a global refugee population of over 16.3 million at the end of 1993, 48% of the world's refugees were in protracted situations. More than a decade later, with a global refugee population of 9.2 million at the end of 2004, over 64% of the world's refugees were in protracted refugee situations.

This trend and the distribution of protracted refugee situations clearly illustrate the growing significance of protracted refugee situations, not only for the global refugee regime but also for broader actors within the international system. As illustrated by the UNHCR statistics in Table 2.1, above, these situations are to be found in some of the most volatile regions in the world. East and West Africa, South Asia, Southeast Asia, the Caucasus, Central Asia, and the Middle East (CASWANAME) all host protracted refugee situations. Sub-Saharan Africa hosts the largest number of protracted refugee situations, with the largest host countries on the continent including Tanzania, Kenya, Uganda and Zambia. In contrast, the geographical area of Central Asia, South West Asia, North Africa and the Middle East hosts fewer major protracted situations, but accounts for a significant number of the world's refugees in prolonged exile, with the almost 2 million Afghan refugees remaining in Pakistan and Iran alone. While the Afghan refugees are the largest protracted refugee situation under the mandate of UNHCR, the scale of this situation pales in comparison to the more than 3 million Palestinian refugees under the mandate of UNRWA.

Causes of PRS

Protracted refugee populations originate from the very states whose instability lies at the heart of chronic regional insecurity. The bulk of refugees in these regions – Somalis, Sudanese, Burundians, Liberians, Iraqis, Afghans and Burmese – come from countries where conflict and persecution have persisted for years. In this way, the rising significance of protracted refugee situations is closely associated with the growing phenomenon of so-called 'failed and fragile states' since the end of the Cold War. While there is increasing recognition that international security planners must pay closer attention to these countries of origin, it is important to also recognize that resolving refugee situations must be a central part of any solution to long-standing regional conflicts, especially given the porous nature of these countries' borders and the tendency for conflict in these regions to engulf their neighbours. In this way, it is essential to recognize that protracted refugee situations are closely linked to the phenomenon of failed and fragile states, have political causes, and therefore require more than simply humanitarian solutions.

As argued by UNHCR, 'protracted refugee situations stem from political impasses. They are not inevitable, but are rather the result of political action and inaction, both in the country of origin (the persecution or violence that led to flight) and in the country of asylum. They endure because of ongoing problems in the countries of origin, and stagnate and become protracted as a result of responses to refugee inflows, typically involving restrictions on refugee movement and employment possibilities, and confinement to camps'.[9]

This analysis illustrates how factors relating to the prevailing situations in the country of origin and the policy responses of the country of asylum contribute significantly to the causes of protracted refugee situations. In fact, protracted refugee situations are also caused by both a lack of engagement on the part of various peace and security actors to address the conflict or human rights violations in the country of origin and a lack of donor government involvement with the host country. Failure to address the situation in the country of origin prevents the refugees from returning home. Failure to engage with the host country reinforces the perception of the refugees as a burden and a security concern, which leads to encampment and a lack of local solutions. As a result of these failures, humanitarian agencies, such as UNHCR, are left to compensate for the inaction or failures of those actors responsible for maintaining international peace and security.

For example, the protracted presence of Somali refugees in East Africa and the Horn is the result of both the consequences of failed intervention in Somalia in the early 1990s and the inability or unwillingness of the international community to engage in rebuilding a failed state. Hundreds of thousands of Somali refugees have consequently been in exile in the region for over a decade, with humanitarian agencies like UNHCR and the World Food Programme (WFP) responsible for their care and maintenance as a result of increasingly restrictive host state policy.

In a similar way, failures on the part of the international community and regional actors to consolidate peace can lead to resurgence of conflict and displacement, leading to a recurrence of protracted refugee situations. For example, the return of Liberians from neighbouring West African states in the aftermath of the 1997 elections in Liberia was not sustainable. A renewal of conflict in late 1999 and early 2000 led not only to a suspension of repatriation of Liberian refugees from Guinea,

Côte d'Ivoire and other states in the region, but also to a massive new refugee exodus. Following the departure into exile of Charles Taylor in 2003 and the election of Ellen Johnson-Sirleaf as President in November 2005, there has been a renewed emphasis on return for the hundreds of thousands of Liberian refugees in the region. In July 2006, UNHCR reported that it had helped some 73,000 Liberian refugees repatriate from neighbouring countries since 2004, and anticipated that the repatriation programme would continue through 2007.[10] In addition, the lessons of the late 1990s do not appear to have been learned. Donor support continued to be unpredictable, with only 28% of the 2006 Liberia Consolidated Appeal having been met by mid-June 2006. As cautioned by the UN Office for the Coordination of Humanitarian Affairs (UNOCHA):

> Liberia is at a critical juncture. In order to build upon the hard-won peace and political progress, international support, both financial and political, will be vital to stabilize the population by addressing the continuing urgent humanitarian needs of the population to ensure a rapid and sustainable recovery.[11]

As illustrated by these examples, among the primary causes of protracted refugee situations are the failure to engage with countries of origin, and the failure to engage in effective and sustainable peacebuilding. These examples also demonstrate how humanitarian programmes have to be underpinned by long-lasting political and security measures if they are to result in lasting solutions for refugees. Assistance to protracted refugee populations through humanitarian agencies is no substitute for sustained political and strategic action. More generally, the international donor community cannot expect the humanitarian agencies to fully respond to and resolve protracted refugee situations without the sustained engagement of the peace and security and development agencies, as discussed in the concluding chapters of this book.

Declining donor engagement in programmes to support long-standing refugee populations in host countries has also contributed to the rise in protracted refugee situations. A marked decrease in financial contributions to assistance and protection programmes for chronic refugee groups has not only had security implications, as refugees and local populations have come into competition for scarce resources, but also reinforced the perception of refugees as a burden on host states. Host states are now more likely to argue that the presence of refugees results in additional burdens on the environment, local services, infrastructure and the local economy, and that the international donor community is less willing to share this burden. As a result, host countries are less willing to engage in local solutions to protracted refugee situations.

This trend first emerged in the mid-1990s, when UNHCR experienced budget shortfalls of tens of millions of dollars. These shortfalls were most acutely felt in Africa, where contributions to both development assistance and humanitarian programmes fell throughout the 1990s. Of greater concern was an apparent bias in the allocation of UNHCR's funding to refugees in Europe over refugees in Africa. In 1999, it was reported that UNHCR spent about 11 cents per refugee per day in Africa, compared to an average of US$1.23 per refugee per day in the Balkans.[12]

These concerns continued in 2000 and 2001, with most programmes in Africa having to cut 10% to 20% of their budgets. The case of Tanzania provides one example of the implications of these budget cuts. Since 2000, UNHCR has consistently reported that its programmes in Tanzania have been 'adversely affected by the unpredictability of funding and budget cuts'.[13] In 2001, UNHCR was forced to reduce its budget in Tanzania by some 20%, resulting in the scaling-back of a number of activities.[14] In 2002, it was reported that UNHCR was forced to cut US$1 million in both the months of June and November out of a total budget of approximately US$28 million. In 2003, UNHCR reported that it 'struggled to maintain a minimum level of health care, shelter and food assistance to the refugees in the face of reduced budgets'.[15] In 2005, UNHCR reported that 'not all refugees' needs were met, a consequence of UNHCR's overall funding shortage'.[16]

Similar shortages over the past decade have also affected food distribution in the camps. Dwindling support for the WFP in Tanzania has led to a reduction in the amount of food distributed to refugees on numerous occasions in recent years. WFP was forced to significantly reduce food distribution to refugees in November 2002 and again in February 2003, resulting in a distribution of only 50% of the normal ration, itself only 80% of the international minimum standard.[17] At the end of 2004, UNHCR and WFP were still requesting additional funds to address chronic food shortages.[18]

Sensitive to these recurring shortfalls in donor support, and in response to a range of other pressures, the Tanzanian government has frequently stated that it is only willing to continue hosting refugees if the international community is willing to provide the necessary support. As Tanzanian President Benjamin Mkapa told a meeting of foreign diplomats in Dar es Salaam in 2001, Tanzania's 'sympathy in assisting refugees should be supported by the international community because it was its responsibility'.[19] This is particularly striking, given that Tanzania was once in the vanguard of local settlement for refugees, distinguishing itself as only one of two African countries to grant mass naturalization to refugees. In stark contrast, Tanzanian regulations now prohibit refugees from travelling more than 4 km from the camps, a policy which greatly reduces refugees' ability to engage in wage-earning employment.

In this way, one of the additional causes of protracted refugee situations can be found in the combined effect of inaction or unsustained international action, both in the country of origin and in the country of asylum. These chronic and seemingly irresolvable problems occur because of ongoing political, ethnic and religious conflict in the countries of refugee origin, then stagnate and become protracted as a consequence of restrictions, intolerance and confinement to camps in host countries. Consequently, a truly comprehensive solution to protracted refugee situations must include sustained political, diplomatic, economic and humanitarian engagement in both the country of origin and the various countries of asylum.

Consequences of PRS

Human rights implications

An increasing number of host states respond to protracted refugee situations by pursuing policies of containing refugees in isolated and insecure refugee camps, typically in border regions and far from the governing regime. Many host governments now require the vast majority of refugees to live in designated camps, and place significant restrictions on refugees seeking to leave the camps, either for employment or for educational purposes. This trend, recently termed the 'warehousing' of refugees,[20] has significant human rights and economic implications. As highlighted by the recent work of the US Committee for Refugees and Immigrants, levels of sexual and physical violence in refugee camps remain a cause of significant concern. UNHCR has argued: 'Most refugees in such situations live in camps where idleness, despair and, in a few cases, even violence prevails. Women and children, who form the majority of the refugee community, are often the most vulnerable, falling victim to exploitation and abuse'.[21]

More generally, the prolonged encampment of refugee populations has led to the violation of a number of rights contained in the 1951 Convention relating to the Status of Refugees (1951 Convention), including freedom of movement and the right to seek wage-earning employment. Restrictions on employment and on the right to move beyond the confines of the camps deprive long-staying refugees of the freedom to pursue normal lives and to become productive members of their new societies. Professional certificates and diplomas are often not recognized by host governments, and educational, health care and other national and local social services are limited. Faced with these restrictions, refugees become dependent on subsistence-level assistance, or less, and lead lives of poverty, frustration and unrealized potential.

UNHCR has noted that 'the prolongation of refugees' dependence on external assistance also squanders precious resources of host countries, donors and refugees ... Limited funds and waning donor commitment lead to stop-gap solutions ... Spending on care and maintenance ... is a recurring expense, and not an investment in the future'.[22] Containing refugees in camps also prevents them from contributing to regional development and state-building.[23] In cases where refugees have been allowed to engage in the local economy, it has been found that refugees can 'have a positive impact on the [local] economy by contributing to agricultural production, providing cheap labour and increasing local vendors' income from the sale of essential foodstuffs'.[24] When prohibited from working outside the camps, refugees cannot make such contributions.

Prolonged exile, especially in confined camps, further compounds the vulnerability of certain categories of refugees. It is important to include this dynamic in a consideration of the human rights consequences of protracted refugee situations. The chapter by Elizabeth Ferris in this volume highlights the extent and significance of the human rights violations suffered by long-staying refugees in host countries. While considerable research has been conducted by other researchers on the specific protection challenges of these groups,[25] and while the challenges faced by these groups are numerous, some particular aspects of their vulnerability relating to the protractedness of their exile need to be highlighted:

Refugee women: Prolonged exile, especially when combined with encampment, can have important implications for gender relations, particularly for women.[26] Significant increases in levels of domestic violence and sexual violence have been recorded in situations such as those in Thailand and Kenya, where employment opportunities are restricted and freedom of movement is curtailed, and where refugees are almost exclusively dependent on international assistance to survive over long periods of time. Likewise, prolonged exile and encampment can often lead to a breakdown of family structures, placing additional burdens on refugee women. Somewhat paradoxically, however, it has also been noted that the responsibilities and leadership opportunities for refugee women sometimes increase in such circumstances, and that engagement with international humanitarian actors can result in greater education on the rights of women, often challenging the status that women previously held in their societies.

Refugee children: Refugee children also face particular challenges in protracted refugee situations, especially those who are born and raised in exile.[27] Funding for education in many situations of prolonged exile has proven to be susceptible to frequent cuts. There are also important issues relating to the curriculum taught in refugee camps, and the decision to follow the curriculum of either the country of origin or the country of asylum. In addition, opportunities to attend secondary or post-secondary education are often denied to refugee students, often leading to higher drop-out rates when coupled with a lack of future employment opportunities. Second, a lack of opportunities within the camps, coupled with the control that a number of armed groups have over refugee populations, leads to instances of the recruitment of child soldiers from the camps, as found in cases as diverse as Guinea, Tanzania and Thailand. These challenges are frequently combined with more general concerns about delinquency and substance abuse in situations where refugee youths lack opportunities and see no prospect of a solution to their plight.

Medically vulnerable refugees: Declining donor engagement in assistance programmes for protracted refugee situations, as outlined above, often places disproportionate burdens on medically vulnerable refugees. Programmes for social support, counselling and rehabilitation are often among the first victims of a funding cut, while medically vulnerable refugees may be left to fend for themselves in the camps if a traditional care-giver opts for a clandestine life outside the camp. Notwithstanding recognition of their special needs in a number of refugee resettlement programmes, medically vulnerable refugees are often passed over during resettlement missions. At the same time, the challenges of both repatriation and local integration are typically greater for these refugees.

Urban refugees and migrants: Refugees living outside camps face a precarious existence, especially in host countries that forbid refugees from living in urban areas. Many of these refugees lack documentation and formal legal status, living on the margins of society. As a result, they are especially vulnerable to extortion and exploitation. They often work for wages far below the local level, are subjected to widespread discrimination, and can be evicted from their homes and even expelled from their host country without recourse. Urban refugees without legal status typically do not have access to the education or health systems of the host country, and, unlike camp-based refugees, seldom receive assistance from international or national agencies. In most instances, urban refugees and undocumented migrants do not benefit from international protection or assistance. While important attention has recently been paid to the successful coping mechanisms and livelihood strategies of these and other refugees,[28] some human rights organizations have also documented the range of abuses urban refugees frequently experience.[29]

'Residual caseloads': In a number of instances around the world where the majority of a refugee population has returned home or been resettled abroad, a smaller portion of the refugee population remains in the host country without a solution. Those in this so-called 'residual caseload' feel they cannot return home because of a fear of persecution or because they have been denied resettlement opportunities abroad. In these situations, residual groups are especially vulnerable, as the host government and the international community typically overlook their needs, believing that a solution to the refugee situation has already been found. There is typically pressure from host states and the international donor community to close camps and cut off assistance. These groups can be the most difficult protracted refugee situations to resolve, as they do not attract international attention, face significant pressures from host states, and are generally too small to be considered of importance by the international community.

Political and security implications

Unresolved refugee situations represent a significant political phenomenon as well as a humanitarian and human rights problem. Protracted refugee situations often lead to a number of political and security concerns for host countries, the countries of origin, regional actors and the international community. One of the most significant political implications of long-standing refugee populations is the strain that they often place on diplomatic relations between host states and the refugees' country of origin. The prolonged presence of Burundian refugees in Tanzania, coupled with allegations that anti-government rebels were based within the refugee camps, led to strained relations between the two African neighbours in 2000–2002, including the shelling of Tanzanian territory by the Burundian army. The prolonged presence of Burmese refugees on the Thai border has been a source of tension between the governments in Bangkok and Rangoon. In a similar way, the elusiveness of a solution for the Bhutanese refugees in Nepal has been a source of regional tensions, drawing in not only the host state and the country of origin, but also regional powers such as India.

Host states and states in regions of refugee origin frequently argue that protracted refugee situations result in a wide range of direct and indirect security concerns.[30] The 'direct threats' faced by the host state, posed by the spill-over of conflict and the presence of 'refugee warriors', are by far the strongest link between refugees and conflict. Here, there are no intervening variables between forced migration and violence as the migrants themselves are actively engaged in armed campaigns, typically, but not exclusively, against the country of origin. Such campaigns have the potential of regionalizing the conflict and dragging the host state into what was previously an intra-state conflict. Such communities played a significant role in the regionalization of conflict in Africa and Asia during the Cold War. With the end of the Cold War, the logic has changed, but the relevance of refugee warriors remains. This relevance was brought home with particular force in the maelstrom of violence that gripped the Great Lakes region of Central Africa between 1994 and 1996.

The outbreak of conflict and genocide in the Great Lakes region of Central Africa in the early 1990s serves as a clear example of the potential implications of not finding solutions for long-standing refugee populations. Tutsi refugees who fled Rwanda between 1959 and 1962 and their descendants filled the ranks of the Rwandan Patriotic Front (RPF), which invaded Rwanda from Uganda in October 1990. Many of these refugees had been living in the sub-region for over 30 years. In the aftermath of the Rwandan genocide, it was widely recognized that the failure of the international community to find a lasting solution for the Rwandan refugees from the 1960s was a key factor that set in motion the series of events that led to the genocide in 1994. According to UNHCR, 'the failure to address the problems of the Rwandan refugees in the 1960s contributed substantially to the cataclysmic violence of the 1990s'.[31] More than 10 years after the 1994 genocide, it would appear as though this lesson has yet to be learned, as dozens of protracted refugee situations remain unresolved in highly volatile and conflict-prone regions.

This lesson has not, however, been lost on a number of states that host prolonged refugee populations. In the wake of events in Central Africa, many host states, especially in Africa, increasingly view long-standing refugee populations as a security concern and synonymous with the spill-over of conflict and the spread of small arms. Refugee populations are increasingly being viewed by host states not as victims of persecution and conflict, but as a potential source of regional instability on a scale similar to that witnessed in Central Africa in the 1990s.

The direct causes of insecurity to both host states and regional and extra-regional actors stemming from chronic refugee populations are further understood within the context of so-called failed states, as in Somalia, and the rise of warlordism, as in the case of Liberia, noted above. In such situations, refugee camps are used as a base for guerrilla, insurgent or terrorist activities. Armed groups hide behind the humanitarian character of refugee camps and settlements, and use these camps as an opportunity to recruit among the disaffected displaced populations. In such situations, there is the risk that humanitarian aid, including food, medical assistance and other support mechanisms, might be expropriated to support armed elements. From their camps, some refugees continue their activities and networks that supported armed conflicts in their home country.

Similar security concerns may arise within urban refugee populations where gangs and criminal networks can emerge within displaced and disenfranchised populations. These groups take advantage of the transnational nature of refugee populations, remittances from abroad and the marginal existence of urban refugees to further their goals. In both the urban and camp context, refugee movements have proven at times to provide a cover for illicit activities, ranging from prostitution and people smuggling to trading in small arms, narcotics and diamonds. For example, such activities have occurred in the past among the long-standing Burmese refugee population in Thailand, and also among the Liberian and Sierra Leonean refugees throughout West Africa.[32]

The security consequences of such activities for host states and regional actors are real. They include cross-border attacks on both host states and countries of origin, and attacks on humanitarian personnel, refugees and civilian populations. Direct security concerns can also lead to serious bilateral and regional political and diplomatic tensions. Cross-border flows are perceived by host states as an impediment to their national sovereignty, especially given the tenuous control that many central governments in the developing world have over their border regions. Finally, the activities of armed elements among refugee populations not only violate refugee protection and human rights principles, but can constitute threats to international peace and security. For example, the training and arming (including by the United States and others) of the Taliban in the refugee camps in Pakistan in previous decades underscore the potential threat to regional and international security posed by refugee warriors.

In East Africa, both Kenya and Tanzania have raised significant concerns about the direct security threat posed by long-standing refugee populations fleeing from neighbouring countries at war. In particular, Kenya feels vulnerable to the spill-over of conflict from neighbouring states and from terrorist activities. Kenya's porous borders and its position as a regional diplomatic and commercial centre made it a target of international terrorist attacks in 1998 and 2002. Kenya is also concerned about the flow of small arms into its territory, and especially into its urban areas, primarily from Somalia. As a result of the link between Islamic fundamentalism, the lack of central authority in Somalia and a long history of irredentism within its own ethnic Somali population, the government in Nairobi now views Somali refugees on its territory almost exclusively through a security prism.

The presence of armed elements in western Tanzania and allegations that the refugee camps serve as a political and military base for Burundian rebel groups have been the source of significant security concerns for the government in Dar es Salaam. Tensions deriving from these allegations have led to open hostilities between Tanzania and Burundi, including the exchange of mortar-fire across the border. Concerns have also been raised by politicians and police about the perceived rise in gun-crime in urban areas resulting from the flow of small arms from Burundi. Consequently, the Tanzanian government has increased restrictions on Burundian refugees, is pushing for early repatriation, and has also adopted the official policy that refugees should be restricted to safe havens in their country of origin.

More difficult to identify, but just as potentially destabilizing as direct threats, refugee movements may pose 'indirect threats' to the host state. Indirect threats may arise when the presence of refugees exacerbates previously existing inter-communal tensions in the host country, shifts the balance of power between communities, or causes grievances among local populations. At the root of such security concerns is the failure of international solidarity and burden-sharing with host countries. Local and national grievances are particularly heightened when refugees compete with local populations for resources, jobs and social services, including health care, education and housing.[33] Refugees are sometimes seen as a privileged group in terms of services and welfare provisions, or as the cause of low wages in the local economy and inflation in local markets. Refugees are also frequently scapegoats for breakdowns in law and order in both rural and urban refugee-populated areas.

Furthermore, it has been argued that, 'in countries which are divided into antagonistic racial, ethnic, religious or other groupings, a major influx can place precariously balanced multi-ethnic societies under great strain and may even threaten the political balance of power'.[34] In this way, the presence of refugees has been demonstrated to accelerate 'existing internal conflicts in the host country'.[35]

For example, this concern was made most explicitly clear in Macedonia's reluctance to accept Kosovar Albanian refugees in March 1999, citing the concern that the mass of Kosovar Albanian refugees 'threatened to destabilize Macedonia's ethnic balance'.[36] Other examples include the arrival of Iraqi Kurds in Turkey, of Afghan Sunni Muslims in Shia-dominated Pakistan, or of Pashtun Afghans in Beluchi-dominated Beluchistan.[37]

However, not all refugees are seen as threats. The question of which refugees are seen as threats, and why, may be partially explained by understanding the perception of refugees as members of the local political community, or as outsiders. Indeed, 'in the Third World, the remarkable receptivity provided to millions of Afghans in Pakistan and Iran, to ethnic kin from Bulgaria in Turkey, to Ethiopians in the Sudan, to Ogadeni Ethiopians in Somalia, to southern Sudanese in Uganda, to Issaq Somali in Djibouti and to Mozambicans in Malawi has been facilitated by the ethnic and linguistic characteristics they share with their hosts'.[38] In this sense, the importance of affinity and shared group identity cannot be overstated. If a host community perceives the incoming refugee as 'one of us', then positive and generous conceptions of distributive justice will apply.

Conversely, if the refugees are seen as members of an 'out-group', they are likely to receive a hostile reception. Indeed, refugees, 'as an out-group, can be blamed for all untoward activities'.[39] While levels of crime may rise by no more than expected with a comparable rise in population, refugees increasingly are seen as the cause. One researcher argues in the African context that the 'presence of massive numbers of refugees' can 'create feelings of resentment and suspicion, as the refugee population increasingly, and often wrongly, gets blamed for the economic conditions that may arise within the domestic population'.[40] This can lead to a point where 'poverty, unemployment, scarcity of resources, and even crime and disease, are suddenly attributed to the presence of these refugees and other foreigners'.[41]

The indirect security concern that long-staying refugees can pose to host states is a key concept that has been lacking in both the research and policy considerations of refugee movements. In these cases, refugees alone are a necessary but not a sufficient cause of host state insecurity. It is not the refugee that is a threat to the host state, but the context within which the refugees exist that results in the securitization of the asylum question for many states. Lacking policy alternatives, many host governments now present refugee populations as security threats to justify actions that would not otherwise be permissible, especially when the state is confronted with the pressures of externally imposed democratization and economic liberalization. More generally, the presence of refugees can exacerbate previously existing tensions and can change the balance of power between groups in the country of asylum. For this reason, refugees play a significant but indirect role in the causes of insecurity and violence, but with consequences potentially of the same scale as the direct threats.

This dynamic has been evident in the dramatic restrictions on asylum that have been imposed by host states in Africa since the mid-1990s.[42] Numerous reports have pointed to the significance of the absence of meaningful burden-sharing and the growing xenophobia in many African countries as the key factors motivating restrictive asylum policies.[43] There is clear evidence to suggest that, as international assistance to refugees is cut, refugees are forced to seek alternative means to survive. This frequently places refugees in conflict with local populations and can even lead them into illegal activities.

Rather ironically, xenophobic sentiments among African populations against refugees 'have emerged at a time when most of Africa is democratizing and governments are compelled to take into account public opinion in formulating various policies. The result has been the adoption of anti-refugee platforms by political parties which result in anti-refugee policies and actions by governments'.[44] Just as Troeller's chapter outlines how politicians in Western Europe faced increasing pressures to restrict entry as asylum became a significant issue in domestic politics, 'the rise of multiparty democracy in Africa ... has arguably diminished the autonomy of state elites in determining the security agenda'.[45] This dynamic further highlights the importance of addressing the prolonged presence of refugees, not in isolation but within the broader context of domestic and international politics.

Remaining questions

As outlined in this chapter, the overwhelming majority of the world's refugees are in protracted refugee situations, which are proving increasingly difficult to resolve and affect most regions in the world. These regions are also typically associated with the phenomenon of failed and fragile states, highlighting the limitations of a purely humanitarian approach to resolving long-standing refugee situations. Moreover, these situations pose particular challenges to the human rights and livelihoods of refugees, especially a number of particularly vulnerable groups of refugees. They also pose political and security challenges to host states and states in the region. It is essential that our understanding of protracted refugee situations incorporates a deeper understanding of the political and strategic dynamics that not only give rise to the problem, but frustrate solutions. Given the political causes of protracted refugee situations, an effective response to this global problem must include engagement by a broader range of humanitarian, security and development actors, as subsequent chapters in this volume will argue.

The consideration of protracted refugee situations within this broader context not only provides a useful basis for more critically understanding the political, human rights and security implications of the phenomenon, but also highlights a number of questions that require both empirical exploration and conceptual development. These questions may be broadly clustered into two groups: the comparability of protracted refugee situations, and our ability to determine when a protracted refugee situation has been resolved.

First, given the diversity of individual protracted refugee situations, can we examine protracted refugee situations from a comparative perspective, or are all protracted refugee situations unique? This volume considers the question of protracted refugee situations both thematically and empirically, through case studies of several of the world's most prominent and chronic refugee situations. In some cases, such as that of the Palestinians or the Afghans, questions are raised as to the overriding uniqueness of the situations, and the extent to which these situations are 'a case apart'. Conversely, the thematic chapters of this volume implicitly argue that it is both possible and beneficial to consider prolonged exile as a general phenomenon, with central shared characteristics. These dynamics point to the potential benefits of disaggregating the notion of protracted refugee situations. It has been suggested that a typology of protracted refugee situations may prove beneficial in developing the conceptual clarity of the issue and highlighting the different solutions that different types of situations may require.[46] Additional research in this area would likely make an important contribution to both the study of protracted refugee situations and the formulation of more effective responses.

Additional research is also required to more fully develop our understanding of the important links between protracted refugee situations and protracted situations of internal displacement. In the vast majority of situations, protracted refugee situations in the region are mirrored by prolonged internal displacement in the country of origin. While there is increasing understanding of the relationship between the protection needs of both refugees and internally displaced persons (IDPs), and the tensions that arise when ascribing international responsibility for this protection, similar questions have yet to be fully resolved with regards to the solutions to such situations.[47] In a growing number of cases, including Afghanistan, Liberia and elsewhere, refugees repatriating after prolonged exile have returned to their country of origin only to become IDPs. This dynamic raises important questions about the meaning of solutions for protracted refugee situations and our understanding of when refugee situations may be said to have been resolved. More generally, research is also required on the links between situations of internal displacement and prolonged exile within a region, and on the political, human rights and security implications of these links.

While the chapters of this volume make important contributions to these and other debates on the causes, consequences and possible responses to protracted refugee situations, additional research is certainly required. As outlined in the Introduction, protracted refugee situations have only recently become a defined area of academic and policy research. While early progress has been made on a number of conceptual, empirical and policy issues, a greater number of issues remain unexamined and unresolved. In this sense, there is an important role for the refugee studies research community, to both identify the challenge of protracted refugee situations as a necessarily distinct area of enquiry, and cooperate with other communities to enhance our understanding of the problem. Sustained collaboration between the research, policy and practitioner communities will not only provide a better understanding of the problem of protracted refugee situations, but ultimately contribute to their resolution.

Notes

1. This chapter draws from earlier works by the authors. See Gil Loescher and James Milner, *Protracted Refugee Situations: Domestic and International Security Implications*, Adelphi Paper 375, Abingdon: Routledge for the International Institute for Strategic Studies, July 2005; 'Protracted Refugee Situations: The Search for Practical Solutions' (chapter 5), in UNHCR, *The State of the World's Refugees: Human Displacement in the New Millennium*, Oxford: Oxford University Press, 2006, pp. 105–128.
2. UNHCR, Executive Committee of the High Commissioner's Programme, 'Protracted Refugee Situations', Standing Committee, 30th Meeting, UN Doc. EC/54/SC/CRP.14, 10 June 2004, p. 2.
3. UNHCR, 'Protracted Refugee Situations', p. 1.
4. Ibid., p. 2.
5. See UNHCR, *UNHCR Statistical Yearbook 2003*, Geneva: UNHCR, 2003, p. 358.
6. See Jeff Crisp, '"Who Has Counted the Refugees?": UNHCR and the Politics of Numbers', *New Issues in Refugee Research*, Working Paper no. 12, Geneva: UNHCR, June 1999.
7. See UNHCR, 'Protracted Refugee Situations'.
8. Ibid., p. 2.
9. Ibid., p. 1.
10. UNHCR, 'Almost 300 Liberian Refugees Arrive Home on Chartered Vessel', UNHCR News Story, 31 July 2006 (available online at: http://www.unhcr.org/news/NEWS/44ce319d4.html).
11. UNOCHA, *Consolidated Appeals Process (CAP): Mid-Year Review of the Appeal 2006 for Liberia*, Geneva: UN Office for the Coordination of Humanitarian Affairs, 18 July 2006.
12. See John Vidal, 'Blacks Need, but Only Whites Receive: Race Appears to Be Skewing the West's Approach to Aid', *The Guardian UK*, 12 August 1999.
13. UNHCR, *UNHCR Global Report 2000*, Geneva: UNHCR, 2000, p. 121.
14. UNHCR, *UNHCR Global Report 2001*, Geneva: UNHCR, 2001, p. 137.
15. UNHCR, *UNHCR Global Report 2003*, Geneva: UNHCR, 2003, p. 165.
16. UNHCR, *UNHCR Global Report 2005*, Geneva: UNHCR, 2005, p. 141.
17. See UNHCR, 'Press Release: WFP and UNHCR Call for Urgent Aid for Refugees in Africa', 14 February 2003.

18. UNHCR, *UNHCR Global Report 2005*, Geneva: UNHCR, 2005, p. 141.
19. IRIN, 'Tanzania: Mkapa Calls for Assistance for Refugees', 10 January 2001.
20. See Merrill Smith, 'Warehousing Refugees: A Denial of Rights, a Waste of Humanity', *World Refugee Survey 2004*, Washington, DC: US Committee for Refugees, 2004.
21. UNHCR, Africa Bureau, 'Addressing Protracted Refugee Situations', Paper prepared for the Informal Consultations on New Approaches and Partnerships for Protection and Solutions in Africa, Geneva, December 2001, p. 1.
22. UNHCR, 'Protracted Refugee Situations', p. 3.
23. See Karen Jacobsen, 'Can Refugees Benefit the State? Refugee Resources and African Statebuilding', *Journal of Modern African Studies* 40, no. 4, 2002.
24. UNHCR, Executive Committee of the High Commissioner's Programme (ExCom), 'Economic and Social Impact of Massive Refugee Populations on Host Developing Countries, as Well as Other Countries', EC/54/SC/CRP.5, 18 February 2004 (available online at http://www.unhcr.org/excom/EXCOM/403dcdc64.pdf), p. 3.
25. For an overview of this area of research, see http://www.forcedmigration.org/browse/thematic/.
26. See Women's Commission for Refugee Women and Children, *Displaced Women and Girls at Risk: Risk Factors, Protection Solutions and Resource Tools*, New York: Women's Commission, February 2006.
27. See, for example, UNHCR, *Through the Eyes of a Child: Refugee Children Speak about Violence*, Pretoria: UNHCR, 2008.
28. See Karen Jacobsen, *The Economic Life of Refugees*, Bloomfield, CT: Kumarian Press, 2005.
29. See, for example, Alison Parker, *Hidden in Plain View: Refugees Living without Protection in Nairobi and Kampala*, New York: Human Rights Watch, 2002.
30. See James Milner, *Refugees, the State, and the Politics of Asylum in Africa*, Basingstoke: Palgrave Macmillan, forthcoming.
31. UNHCR, *The State of the World's Refugees: Fifty Years of Humanitarian Action*, Oxford: Oxford University Press, 2000, p. 49.
32. See Karen Jacobson, 'The Forgotten Solution: Local Integration for Refugees in Developing Countries', *New Issues in Refugee Research*, Working Paper no. 45, Geneva: UNHCR, 2001.
33. See UNHCR, ExCom, *Economic and Social Impact of Massive Refugee Populations on Host Developing*.
34. See Gil Loescher, *Refugee Movements and International Security*, Adelphi Paper 268, London: Brasseys for the International Institute for Strategic Studies, 1992.
35. Myron Weiner, 'Security, Stability and International Migration', in Myron Weiner, ed., *International Migration and Security*, Boulder, CO: Westview Press, 1993, p. 16.
36. Macedonian Deputy Foreign Minister, speaking at the Emergency Meeting on the Kosovo Refugee Crisis, Geneva, 6 April 1999.
37. Finn Stepputat, *Refugees, Security and Development. Current Experience and Strategies of Protection and Assistance in 'the Region of Origin'*, Working Paper no. 2004/11, Copenhagen: Danish Institute for International Studies, 2004, p. 4.
38. Loescher, *Refugee Movements*, p. 42.
39. Tiyanjana Maluwa, 'The Refugee Problem and the Quest for Peace and Security in Southern Africa', *International Journal of Refugee Law* 7, no. 4, 1995, p. 657.
40. Ibid.
41. Ibid.
42. See Milner, *Refugees, the State and the Politics of Asylum in Africa*.
43. Jeff Crisp, 'Africa's Refugees: Patterns, Problems and Policy Challenges', *New Issues in Refugee Research*, Working Paper no. 28, Geneva: UNHCR, August 2000; Bonaventure

Rutinwa, 'The End of Asylum? The Changing Nature of Refugee Policies in Africa', *New Issues in Refugee Research*, Working Paper no. 5, Geneva: UNHCR, May 1999.
44. Rutinwa, 'The End of Asylum?', p. 2.
45. Matthew J. Gibney, 'Security and the Ethics of Asylum after 11 September', *Forced Migration Review*, no. 13, June 2002.
46. Alexander Betts, 'Conference Report: The Politics, Human Rights and Security Implications of Protracted Refugee Situations', *Journal of Refugee Studies* 19, no. 4, 2006.
47. See Gil Loescher, Alexander Betts and James Milner, *UNHCR: The Politics and Practice of Refugee Protection into the Twenty-First Century*, New York: Routledge, 2008.

3

Asylum trends in industrialized countries and their impact on protracted refugee situations

Gary Troeller

The factors contributing to protracted refugee situations are to be found at a variety of levels: local, regional and international. Other chapters in this volume focus on a number factors at issue. While recognizing that causes are various and often intertwined, this chapter focuses on worrying developments in industrialized countries, which reinforce 'warehousing' of refugee populations in distant lands.

The issues of immigration and, by extension, asylum – the two inextricably linked, the distinction between them often blurred and both now complicated by security concerns – remain prominent on the political agenda. Their prominence, however, is more due to negative populist reaction to foreigners than to understanding and sympathy for victims of forced displacement and the dynamics of globalization. Tenable policy prescriptions to address long-standing refugee situations require an understanding of the cumulative factors and perceptions influencing the public and policymakers in industrialized states, whose role is key to resolving the issue.

This chapter outlines in broad brush strokes the main developments in industrialized countries, which have undermined the international protection regime in the recent past and do not augur well for responsibility-sharing in the international community. By their restrictive example, in an increasingly interdependent world, these developments reinforce warehousing or containment of refugees in poorer countries and must be firmly borne in mind in attempting to formulate realistic policy recommendations and tools to resolve protracted refugee situations. These

Protracted refugee situations: Political, human rights and security implications,
Loescher, Milner, Newman and Troeller (eds),
United Nations University Press, 2008, ISBN 978-92-808-1158-2

developments in turn point to the likelihood that any resolution of long-standing refugee problems will be concentrated in regions of origin.

The following sections in this chapter highlight the pressure of, and increasing public reaction against, the growing number of asylum seekers and immigrants over much of the last two decades and the blurring of the distinction between the two. The chapter then deals with the increasing politicization of the asylum–migration nexus, institutional responses in industrialized countries, increasing preoccupation with security and UNHCR's response to these cumulative challenges. Finally it considers the interrelationship between identity politics, multiculturalism versus integration and recent unrest in many countries around the world, from riots among France's Muslim community in November–December 2005 to the furies unleashed in many countries by the Danish cartoon controversy, further impacting the migration and asylum regime and the political will to tackle refugee problems in general and protracted refugee situations in particular. While these factors are interrelated, reinforce one another and overlap, they are treated separately here for ease of presentation. Throughout, the chapter refers to immigration problems as well as asylum issues as the nexus has become so close, irrespective of real distinctions, and both issues so politicized that governments and the general public link the two. This linkage, in turn, influences policy responses to asylum and refugee matters.

The pressure of numbers

For most of the Cold War period, asylum seekers and refugees were largely welcome in Western industrialized countries. Their numbers were relatively limited and, rightly or wrongly in the simplified optic of the times, they were perceived as products of the superpower rivalry and proxy conflicts, and were treated as pawns, if not strategic trophies, by receiving countries. In the Manichean atmosphere of the time, seeking asylum from one or the other superpower blocs was seen as an indictment of an evil empire and an affirmation of the virtues of the other. This relatively warm welcome began to recede in the early 1980s, not least given the growing number of asylum seekers arriving spontaneously in Western countries as a result of the Islamic revolution in Iran, the 10-year-long Iran–Iraq War, civil conflict in eastern Turkey and the separatist movements in Sri Lanka, to mention some of the major refugee producing situations.[1]

Between 1970 and 1980, the number of refugees and asylum seekers under UNHCR's protection increased from 2 million to 10 million. By 1990, the number under the refugee agency's umbrella had increased to 15 million. Between 1983 and 1990, the number of spontaneous arrivals in Western Europe increased from 70,000 to 200,000. Presaging the situation today characterized by 'cultural clashes', these newcomers did not fit the stereotype of earlier Cold War arrivals. They came from all over the world, not mainly from Eastern Europe, and their arrival, facilitated by cheaper air travel, coincided with the economic recession of the 1980s and declining demands for immigrant labour.

With the end of the Cold War and the collapse of the Soviet Union, perceptions of an end to conflict and the peaceful spread of democracy, as espoused by Francis Fukuyama, were quickly dispelled by a period of state fragmentation and further forced displacement. In the 1990s, 50 states underwent significant transformations, and the terms 'ethnic conflict' and 'failed states' gained prominence in political discourse. Pressures on Western asylum systems, and spiralling costs of maintaining these systems, increased dramatically.

Numbers seeking asylum reached a high point in 1992 when Europe registered 700,000 asylum applications. Well televised humanitarian disasters in the former Yugoslavia and Kosovo in Europe's backyard, along with widely reported major disasters in Iraq, Somalia, Rwanda and elsewhere uprooting millions, coupled with movements of immigrants, gave rise to the impression that rich, industrialized countries were confronted by an uncontrollable flood of foreigners.

The situation was further complicated by the forces of globalization and the absence of a European Union (EU) immigration policy, and, with few exceptions, any tradition of immigration within EU countries. Between 1980 and 2000 the number of migrants in developed countries increased from 48 million to 110 million.[2] In the absence of an immigration channel in Europe, poor people from third world countries seeking to better their economic prospects entered the asylum channel, fuelling the perception that most asylum seekers were economic migrants in disguise. The situation was not helped by the sentiment that most newcomers were a burden on, if not intent on exploiting, generous Western welfare systems rather than making a contribution to the economy, notwithstanding the fact that, with respect to victims of forced displacement, most refugees remained in poor third world countries. And in the context of the oft-mentioned asylum–migration nexus, it is worth noting that despite the hyperbolic alarmism in the 1990s, seven of eight immigrants in industrialized countries entered legally through very stringent procedures.[3]

Government responses in the 1990s

As a result of this 'flood' of arrivals, in 1993 governments sought to stem the flow. Germany, which had received over 400,000 asylum applications in 1992, changed its constitutional right to asylum and introduced far-reaching exceptions for those originating from safe countries of origin or entering from safe third countries. At the regional level, the meeting of European Union immigration ministers in London in 1992, preoccupied with control rather than asylum rights, endorsed a number of restrictive immigration and asylum measures, ranging from expanded visa requirements, through a return to safe hosts or first countries of asylum, to the expeditious handling of manifestly unfounded asylum claims (London Resolutions).[4] These measures were applied by a growing number of EU members and other industrialized countries, and complemented by carrier sanctions (fines against airlines carrying undocumented individuals) and interception at sea and/or via out-posted immigration officers in major airports ('Operation Shortstop' as Canada called it). Also in 1992, under the Treaty of Maastricht, the EU initiated a process towards the development of a common asylum policy among its then 15 members to harmonize asylum practices throughout the EU (summarized below). In the absence of means of legal entry, asylum seekers and immigrants increasingly turned to illicit modes of entry via smugglers and traffickers, which further diminished public confidence in the integrity of those wishing to enter and the systems designed to handle their applications. The debate on asylum and immigration became increasingly polarized – not to say a 'political football' – and subject to febrile treatment by political parties and the media.

In 1998, during Austria's EU Presidency, an Austrian government paper advocating a strategy proposal calling for a defence perimeter around Europe to keep out asylum seekers and immigrants was leaked. At the same time, Vienna called for amending or replacing the 1951 Convention, which, in the tangled net of asylum and immigration control hysteria, was mistaken for a failed immigration control mechanism rather than the principal international human rights instrument requiring signatories to adhere to their obligations to deal with forced displacement. Austria's call for amending or scrapping the 1951 Convention was echoed in the United Kingdom, where the Prime Minister and the Home Secretary characterized the instrument as a relic from a different time which had become irrelevant in handling contemporary flows.[5]

Across the Atlantic, the United States, while ever sympathetic to Cuban asylum seekers, did not react with the same sympathy to Haitians attempting to seek asylum. Perceiving a rising tide of Haitians fleeing chaos and repression at home as a threat to US security, President Bill Clinton declared an emergency situation, threatened military intervention in Haiti to restore democracy, and between 1992 and 1994 interdicted almost 60,000 asylum seekers at sea and instituted offshore processing on a US naval ship and subsequently in Guantanamo, Cuba.[6]

On the other side of the world, Australia intercepted the Norwegian freighter *Tampa*, bound for Australia, which had rescued 438 Afghan asylum seekers attempting to reach Australia and whose boat was sinking. Canberra refused to allow the asylum seekers to land and the Australian navy transported the group to the island of Nauru to have their claims processed there, in a notorious policy action that has come to be known as 'The Pacific Solution'.[7]

The EU harmonization process

The European Commission has sought to deal with the rising tide of immigrants and asylum seekers on a regional basis. As its name suggests, the EU harmonization process was, and is, an attempt to introduce a uniform asylum and migration system and practice into the EU.[8] Following the 1992 Maastricht Treaty's establishment of a rule-making body within the ambit of the Union, the Treaty of Amsterdam, which entered into force in 1999, asylum and migration were moved into the normal law-making structures of EU supranational competence. Articles 62 and 63 detailed *in extenso* largely *minimum* asylum, immigration and border control measures to be adopted as binding community-wide instruments within a five-year period. The asylum measures in their final form covered:
- *Dublin Regulation (II)*, focusing on responsibility for considering an asylum application for a third country national submitted in one of the EU member states.
- *Reception Directive*, dealing with minimum standards in the reception of asylum seekers.
- *Qualification Directive*, covering minimum standards regarding the qualification and status of third country nationals and stateless persons as refugees or as persons who otherwise need international protection.
- *Procedures Directive*, covering minimum standards on granting or withdrawing refugee status.
- *Temporary Protection Directive*, dealing with minimum standards for providing temporary protection to displaced people from third countries who cannot return to their country of origin.
- *European Refugee Fund*, establishing an equitable balance between EU members receiving and supporting refugees and displaced persons.

In 1999, under the Finnish EU Presidency, heads of state held a summit in Tampere, Finland. They called for a common EU asylum policy, pledged to safeguard the right to seek asylum in proper balance with immigration controls and, in the Summit's Conclusions, committed governments to a 'full and inclusive application of the Convention'. Despite the lofty rhetoric of Tampere and UNHCR's continued efforts working with the EU Commission throughout the harmonization process to safeguard protection principles, the results of this EU exercise have been mixed. The Dublin Regulation (II) risks putting the burden on southern European states bordering the Mediterranean, or new member states on the EU's eastern frontiers with fragile and under-resourced asylum systems. Both sub-regions are on the front-lines of arrival routes. Moreover, in operation, Dublin II functions slowly and at times unfairly. Months can go by without a decision on an asylum case while states wrangle over who is responsible, asylum seekers are held in detention, or a state can deny responsibility for processing a claim on technicalities. The Reception Directive does not always accord vulnerable people proper treatment as children are sometimes held in detention with unrelated adults, while other asylum seekers can be found sleeping on the streets.

The Qualification Directive, while incorporating some welcome provisions such as recognizing the importance of non-state agents of persecution and gender-based persecution, as well as persecution of children, is also not without problems. Nationals of EU countries may not seek asylum. The Directive also specifies that, if an asylum seeker can find protection under the auspices of an international organization in her/his country of origin, he/she cannot meet refugee criteria. Both these provisions have no legitimacy under international law.

The Procedures Directive outlines rights and duties of applicants during the asylum procedure. This Directive encompasses positive elements regarding interviews and safeguards for unaccompanied minors and specifies that asylum seekers should not be detained. The Directive, however, allows for so many exceptions to these provisions so as to potentially water down the provisions to an unacceptable level of minimalism. It also provides for national lists of 'safe countries of origin' below the level of EU criteria, and for a 'safe third country' and 'super-safe third country' rule which, in case of the latter, would automatically exclude access to prospective asylum seekers coming through certain countries. The Directive also does not provide for the suspensive effect of appeal – an asylum seeker appealing a case can be deported before his case is heard and posits restrictions on interviews for cases in the accelerated appeal process. All these directives must be transposed into national legislation and the likelihood is that countries will follow a minimalist approach.

Despite calls for harmonization of practices and advances – for example, in respect of the granting of refugee status – vast differences remain in awarding Convention status. The following figures are indicative. In 2005, Spain's Convention recognition rate was 11.3%, while Finland's was 1.8% and Greece's 0.8%. In 2005, Canada granted Convention status to 50.4%.[9] Of course there can be differences in the origin and bona fides of those seeking asylum in various countries, but it makes little sense that, for example, in the case of Chechens, recognition has ranged from 0% to 100%.[10] Discrimination against certain minorities and differing interpretations of criteria can play a role.[11]

On another level, given the emphasis the Nordics place on burden of proof and credibility (e.g. possession of documents, veracity and consistency in presentation), Nordic governments have long granted subsidiary status to a much higher number of applicants and been extremely restrictive in the granting of Convention Status. This practice, while affording often generous protection, can give rise to the impression, domestically and internationally, that there are few real refugees and most of those granted protection are the objects of discretionary charity. These factors not only point to disparities in harmonization, but, as more and more countries apply refugee recognition criteria restrictively, hitherto generous countries, not wishing to receive the overflow of redirected asylum seekers looking for a better chance for recognition elsewhere, are pushed to become more restrictive themselves.

One cannot underestimate the 'export value' and 'demonstration effect' of the growing tendency towards restrictive practices, both within the EU and in other regions, given global communications in an increasingly interconnected world. Central Asian governments, noting practices in the European Union, have expressed interest in developing their own 'safe country lists'.[12] Closer to home, this writer recalls a conversation with a senior justice official of a Baltic state on the subject of that country's very low Convention recognition rate. The official argued that, as a relatively poor country, his government could not be expected to do more than the minimum on the granting of Convention recognition status, citing the practice of his richer Nordic EU neighbours. As regards humanitarian status, unlike his wealthy neighbours across the Baltic Sea, his government did not have adequate resources for its own pensioners, let alone foreigners. In other words, 'charity' begins at home. It requires no great leap in imagination to appreciate that news travels fast in an interdependent world, and even poorer countries in the developing world invoke the same rationale with respect to their own refugee populations, including long-term concentrations.

The cumulative effect of the harmonization process to date, and its likely evolution, gives credence to the perception of 'fortress Europe'. The tendency towards pushing the problem of asylum, and asylum processing, beyond the borders of industrialized countries is evidenced by the growing interest in offshore processing or 'contracting out' status determination. As mentioned above, the United States did this with politically inconvenient asylum seekers such as Haitians, whom it intercepted at sea and processed on US Navy ships in the Caribbean or in

Guantanamo in the early 1990s. Moreover, Australia pursued this model most notoriously with the interception of the *Tampa* and the processing of its 400 asylum seekers in Nauru under the so-called 'Pacific Solution'.

In the EU, Brussels has considered following the Australian approach. The Netherlands, Denmark and the United Kingdom proposed in their 'New Vision for Refugees' the removal of certain groups of asylum seekers to processing centres outside Europe or on its periphery. The United Kingdom carried these proposals farthest in its 2003 paper 'New International Approaches to Asylum Processing and Protection'. Essentially, the United Kingdom proposed regional processing centres (RPCs) in the region of origin, the return of asylum seekers in the United Kingdom or a cooperating country to an RPC, international measures ranging from economic assistance to military intervention to forestall refugee flows, and managed resettlement programmes to promote burden-sharing among industrialized countries. The United Kingdom later suggested a distinction between RPCs and Transit Processing Centres closer to Europe. Denmark, Italy, Spain and the Netherlands supported the plan, whereas Sweden, Germany and France opposed it. One year later, Germany proposed off-shore processing in Libya, and Italy carried out deportations to Libya, a country that was not a signatory to the 1951 Convention.[13]

Election results

It would be difficult to think of an election in the European Union over the past few years where immigration and asylum politics have not been a key issue. The following are representative. In 2001, in Denmark's national elections, the Liberal Government lost to a coalition of the Moderate party joined by the far-right Danish People's Party, Dansk Folkeparti, which won 20% of the vote running on a virulently anti-immigrant, anti-refugee platform. The importance of the Danish People's Party influence in the ruling coalition is underscored by Denmark, the first signatory of the 1951 Convention and traditionally among the most liberal of states, now having the dubious distinction of having introduced some of the most restrictive asylum legislation in the industrialized world. Denmark's actions have contributed to deterring prospective asylum seekers. Between 2000 and 2003, Denmark's share of Scandinavian asylum applications fell from 31% to 9%, while during the same period Sweden's rose from 41% to 60% and Norway's from 28% to 31%. Three years after his election, when asked where he thought his party had been most successful, Danish Prime Minister Anders Fogh Rasmussen replied:

'I would think 80–85% of the population backs the government's policy on foreigners'.[14]

In 2002 in the Netherlands' national election, Pim Fortuyn, leader of the party List Pim Fortuyn, exploiting popular dissatisfaction with immigration, Islam and elitist political correctness but assassinated in the midst of the election, ran on a largely anti-immigrant/asylum seeker ticket and his party was successful in the elections. In the 19 October, 2003 Swiss elections, the stridently nationalist People's Party of Switzerland (Schweizerische Volks Partei, SVP), running on an anti-asylum seeker platform, used one slogan in a particularly inappropriate poster which read 'The Swiss are becoming negroes'. The party won 26.6% of the vote. The leader of the SVP won a second seat in the seven-member Federal Council, thus moving the political balance in Switzerland to the right for the first time in four decades. In the 2007 elections, the SVP, making use of another controversial campaign poster, which this time depicted three white sheep kicking a single black sheep off the Swiss flag, won 30% of the vote, the highest percentage by one party since 1919.[15]

In the United Kingdom, asylum and immigration issues have been at the top of the political agenda over the past few years. As mentioned previously, both Prime Minister Tony Blair and then Home Secretary Jack Straw have said the 1951 Convention was not working. The British Prime Minister called for measures to halve the numbers of asylum seekers coming to the United Kingdom. In the 2005 British national elections, asylum and immigration again figured prominently, particularly in the Conservative Party platform, which called for limits on asylum seekers and, as indicated above, pledged UK withdrawal from the 1951 Convention if elected.

In the French national elections of 2002, Jean-Marie Le Pen's extreme right National Front Party, running on an anti-immigrant platform, enjoyed enough electoral success, winning 17% of the vote in the first round, to force a run-off against Jacques Chirac's Gaullist party. In the French presidential election in 2007, now President Nicholas Sarkozy, the former Interior Minister, known for his tough stance on law and order, siphoned off support from Le Pen's far right constituency by focusing *inter alia* on immigration, employment and law and order, capitalizing on France's discomfiture with the 2005 autumn riots by disadvantaged immigrants that resulted in a state of emergency (see below). As a further signal of the mainstreaming of some of the tenets of the extreme right, anti-immigrant National Front Party leader, Jean-Marie Le Pen, Sarkozy adopted Le Pen's notorious slogan: 'France, love it or leave it', and called for the establishment of a Ministry of Multiculturalism and Identity. He also introduced a bill imposing tougher conditions on unskilled immigrants, limiting family reunion (recently to be reinforced by proposed DNA testing), and an abrogation of the right of illegal immigrants to receive residency papers after 10 years. 'Le Pen's influence on politics goes beyond language. An analysis in the newspaper *Le Monde* last month listed a string of Le Pen's proposals that have now been addressed by the (former) government of Prime Minister Dominique de Villepin, including some of those mentioned by Sarkozy.'[16] Le Pen's party appeared to have benefited from the immigrant riots of November–December 2005. His party reported that membership had increased by about 20%, to almost 90,000, from October 2005.[17]

In other countries such as Norway, where electoral concern about the rising tide of asylum seekers has been visible, if much more civil and muted, mainstream parties such as the Conservative Christian Democrats, which came to power in 2002, found it politically difficult not to co-opt some of the far-right Progressive Party's populist appeals against asylum seekers in its own campaign. While Norway remains the most generous contributor to humanitarian organizations per capita in the world, some Norwegian officials have indicated that they regard the co-opting of some of the tenets of the far-right Progressive Party into the mainstream platforms of major parties a more insidious, long-term development.[18]

Farther afield, in Australia, immigration has been a top election issue since the mid-1990s. While far-right, anti-immigrant, populist parties such as Pauline Hanson's One Nation failed to gain seats in national elections in 1998, Hanson further prompted her rival, Prime Minister John Howard, to mainstream migration and asylum control as key components of Australian policy throughout his successive terms in office. In the United States, immigration has complicated the bi-partisan divide between the Republican and the Democrat parties since the presidential election in 2000. The issue has split both parties over mounting concerns with border security (see below), criminalization of illegal immigrants, and guest worker programmes. The Hispanic community responded to the attempt to criminalize and deport 'illegals' by mustering well over one million predominantly Hispanic immigrants who took to the streets in protest marches in key US cities on 1 May 2006, calling for an amnesty for 12 million illegal immigrants. They also sought to demonstrate their economic importance, as workers and consumers, as hundreds of thousands of poor Hispanics took the day off work. The protestors further discomfited many US conservatives by singing the US national anthem in Spanish.[19]

For the better part of two decades, asylum seekers and refugees have been front page news on both sides of the Atlantic. In the EU, especially in the United Kingdom, the media have relentlessly portrayed asylum seekers and immigrants in a negative light, reflecting and fuelling populist reaction to newcomers. Headlines in the tabloid press such as 'Stop the Asylum Invasion', 'Poison gang are asylum seekers', 'Asylum blamed for AIDS Crisis' are typical.[20] On 15 April 2005, on the eve of the last British national election, the *Financial Times* ran a cartoon which summarized media and partisan politics. With the title 'The politics of fear have redrawn Britain's battle lines', it depicted former Conservative leader Michael Howard using the issues of crime, asylum and immigration to undermine Prime Minister Tony Blair's emphasis on the economy and public services. In Switzerland, the results of a survey conducted in 2004, and reported by Agence France Press, revealed that two-thirds of the Swiss polled thought that the influx of asylum seekers was a greater danger, compared with 51% who were concerned about terrorist activities.[21] In Denmark, in addition to headlines, several posters used by the Folkepartie in its election campaign went beyond any current civilized norms and were reminiscent of Nazi propaganda.

These developments have been inimical to the reception of asylum seekers and immigrants in industrialized countries, and have gained added impetus owing to mounting security concerns in many countries.

The immigration–asylum–security nexus

The attacks in the United States on 11 September 2001, followed by the Madrid train bombings in March 2003 and the attacks in London in July 2005, have obviously not augured well for immigrants and refugees. The advent of al-Qaeda has come to link terrorism with transnational security concerns, dependent on the migration of people, information, money and weapons. Control of borders has become a predominant political focus. Many see immigrants and asylum seekers as possible conduits for 'Islamo-fascism'.

In the United States, the Republican administration clearly linked security to immigration shortly after the events of 11 September, with the Homeland Security Presidential Directive 2, Combating Terrorism Through Immigration Policies; 'This directive, issued on 29 October 2001 locked immigration and security together bureaucratically'.[22] A few days later the USA Patriot Act was enacted into law. The Act expanded governmental authority to detain, prosecute and deport aliens suspected of terrorist activities. While the exceptionally fevered climate of the time has improved somewhat, immigration and control are still key issues, as evidenced by the growing controversy over illegal Mexican aliens in the United States, the safety of American ports and concern over porous borders. In the US House of Representatives, the Sensenbrenner

Bill sought to criminalize undocumented immigrants and deport the 12 million 'illegals' in the United States. In the south-western United States, volunteer 'vigilante-like' groups calling themselves 'Minutemen', a name replete with patriotic historical connotations from the American Revolution, and dissatisfied with US Border Patrol inefficiency, have taken to defending large stretches of the border between Arizona, New Texas and Mexico. On 13 May 2006, groups of Minutemen held rallies in Washington, DC, following Hispanic marches held earlier in May, protesting against illegal immigration.

On 15 May 2006 in a nationwide address, the first of his presidency focusing on a domestic issue, President George W. Bush promised to reinforce US Homeland Security Border Patrols with the deployment of 6,000 National Guard troops in Texas, Arizona, New Mexico and California, and 'regain full control of the border'. CNN's senior anchor Lou Dobbs' popular hour-long programme, shown in the 'prime time' slot between 6 and 7 PM, has long been devoted to the drumbeat issue of 'broken borders' and the need to control immigration. Although overshadowed by the Iraq issue, immigration and border security were issues in US congressional mid-term elections in autumn 2006. On the other side of the US northern border, since 11 September, Canada's legal provisions for deporting suspected terrorists to countries where they might face torture, in contravention of the UN Convention Against Torture, are slightly worse than in the United States.[23]

While the United States has arguably pushed furthest with the expansion of government security powers under the Patriot Act, Europe has also implemented much more stringent practices. The United Kingdom has introduced some of the most sweeping anti-terror legislation in Europe and has probably the most extensive closed circuit surveillance network in the EU. Italy, France, the Netherlands and Germany have all introduced regulations to facilitate greater surveillance by intelligence agencies, and along with Spain have increased periods of detention for terror suspects. London has increased the period of detention without charge from 48 hours in 2001 to 28 days. As Jacques Debray, a Lyon-based lawyer specialized in immigration and asylum law, noted: 'There is a clear parallel with what is going on in the United States. They detain people on administrative grounds, while we create a legal framework first. But the logic is the same: preventive detention'.[24]

Although all EU members have ratified the Convention Against Torture, the United Kingdom, the Netherlands, Austria and Germany have sought diplomatic assurances from countries with less than stellar human rights records, such as Turkey, Algeria, Jordan, Egypt and Libya, in order to deport suspected Islamic terrorists. The United Kingdom is attempting to make arrangements for deportations to Algeria without such assurances. In 2001 Sweden, with such assurances, permitted two Egyptian terror suspects to be deported on a CIA plane to Cairo and was cited by the UN Committee against Torture. The Executive Director of the Paris-based International Federation of Human Rights League, Antoine Bernard, has observed: 'The right balance between anti-terrorist concerns and human rights concerns is in full swing'.[25]

While it is clear that refugees and asylum seekers themselves are often victims of terror and not perpetrators, given the blurring of the distinction between immigrants and asylum seekers and 'the maze of fear' that has enveloped and entangled the former with prospective terrorists in the same net, the 'securitization' of immigration has had a damaging effect on victims of forced displacement. The asylum–migration nexus is now rivalled, if not enveloped, by what could be called the 'security-migration–asylum nexus'.

In the 29 April 2006 issue of *The Economist*, UNHCR complained that governments' preoccupations with security were leading to intolerance or indifference to victims of forced displacement. Citing the United States as an example, it mentioned that vague wording of the USA Patriot and Real ID Acts has led to an overly broad interpretation of providing 'material support' to terrorists. As a consequence, the resettlement of thousands of refugees, many of whom may have been victims of terror themselves, has been put on hold.[26]

However minuscule the number of terrorists may be among asylum seekers and refugees, it is fair to say that long-standing refugee situations, generating hopelessness and despair, are not immune to generating security problems. The Palestinian situation, 'refugee warriors' in the Great Lakes, the Kurdistan Workers' Party (PKK) in Europe and refugee-bred militants in the mountainous area between Afghanistan and Pakistan are examples. Myron Weiner of MIT was one of the earliest academics to make the connection between security and migration, positing five categories indicating this link: (1) when refugees and immigrants organize against their homeland regime, as Khomeini did from Paris in the 1970s; (2) when they pose a security risk to their host country; (3) when large influxes of migrants overwhelm the host country's language, customs, self-image etc.; (4) when they create a social and economic threat to natives, particularly in employment; and (5) when immigrants are used as weapons of war, as when Saddam Hussein held foreigners hostage during the first Gulf War. While categories (1), (3), (4) and (5) have long enjoyed attention, and several of these categories are in evidence as reactions to globalization, it is largely since the terror attacks on 11 September and in Madrid and London that (2) has begun to attract attention.[27] The widely reported British discovery, and thwarting, of a major plot to blow up 10 transatlantic US-bound flights in the summer of 2006 has not helped the situation.

UNHCR's response

Given the mass movements of victims of forced displacement in the 1990s, starting with Iraq, within UNHCR resettlement fell from favour as a co-equal of the two other durable solutions, local integration seemed impractical and voluntary repatriation became the preferred option.[28] Against this background, exacerbated by the collapse of the former Yugoslavia, and in an effort to keep the doors open in Europe to tens of thousands of victims of forced displacement fleeing ethnic cleansing in the Balkans, UNHCR introduced a policy of 'temporary protection', rather than traditional asylum, in the expectation that a political resolution would soon be found for the Balkan conflict. Political will proved to be an elusive commodity as the conflict in the Balkans raged for more than three years. Temporary protection did provide refuge for many, but critics argue that, in combination with UNHCR's humanitarian relief efforts, rather than providing protection in the former Yugoslavia the agency's actions struck a severe blow to the protection edifice by diluting and circumscribing the institution of asylum.[29]

Despite the Tampere declarations, mass movements of people and growing public discomfort in Europe, fuelled by alarmist reports in the media, have led several governments to publicly question the utility and relevance of the 1951 Convention, if not make an outright call for scrapping or amending it. As the attack on the asylum edifice, especially in Europe, the traditional home of international protection, gathered momentum, UNHCR decided that if it could not stop the parade it had better try to lead it in a constructive direction. Against this background, the refugee agency launched the Global Consultations in 2000 with a view to: reaffirming the integrity of the 1951 Convention; addressing contested legal issues such as gender-based persecution and non-state agents of persecution via opinions from internationally respected jurists; and examining contentious issues on which the Convention is silent, such as mass influx situations, temporary protection and internal flight, among others.

The Global Consultations process involving governments, experts and NGOs ended in 2002, with agreement on many issues. An Agenda for Protection was established, setting out six main areas for government and UNHCR collaboration. This six-point plan covered areas which the Convention did not address, such as refugee women and children, increased opportunities for durable solutions, protection concerns in mass influx situations and refugee registration. Bearing in mind the gathering pressure on the Convention in the late 1990s, the centrepiece of the Global Consultations was the first-ever meeting of state parties to the Convention in December 2001. At this meeting 150 countries attended, including 70 ministers and several heads of state. The Conference reaffirmed the enduring integrity of the 1951 Convention and undertook to come up with a more effective means of monitoring the implementation of the Treaty. Underscoring the importance of the asylum–migration nexus and its global implications, one of the principal goals of the Agenda for Protection was 'protecting refugees within broader migration movements'.[30] As was the case with the chasm between the lofty rhetoric of Tampere and the realities of asylum implementation on the ground, the translation into practice of commitments made in December 2001 has not been easy, especially in the wake of 11 September, increasing security concerns and a hardening of government attitudes.

An example of this hardening attitude presaging and reinforcing the increasing tendencies towards containment of refugees is the abovementioned attempt by the United Kingdom, supported by several other governments, to contract out asylum processing. In March 2003, shortly after the Madrid train bombings, UNHCR, in what many inside and outside the organization felt was an ill-considered move to take the lead in the latest turn of the asylum debate, offered its own version of the UK proposal, dubbed the 'Three Pronged Approach'. The UNHCR approach also suggested solutions in the region, improved domestic asylum procedures and the processing of manifestly unfounded claims in EU-operated closed detention centres – unlike the UK plan, which would have sought to do this extra-territorially outside EU borders (Moldova, Croatia and Albania were proposed). Fortunately, the Three Pronged Approach did not find much resonance among the more liberal EU members, with Sweden and Germany particularly opposed.[31] Amy Slaughter and Jeff Crisp's chapter in this volume focuses on earlier examples of UNHCR policies which have contributed to contemporary problems regarding protracted refugee situations.

Other UNHCR policy efforts, such as 'Convention Plus', endeavouring to enlist greater North–South cooperation on responsibility-sharing across a broad spectrum of areas of mutual concern, have shown more, if limited, promise. One example has been the adoption of the Multilateral Framework of Understandings on Resettlement in September 2004 and related discussions on development assistance and irregular movements with reference to Somalia and Afghanistan. However, as Alex Betts makes clear in his chapter, substantive achievements have been limited, partly owing to suspicions of this formula, given its association with asylum transit camps and contracting out protection to regions of origin. This same scepticism has cast a shadow over UNHCR's attempts at fostering protection capacity-building in regions of origin, and Refugee Aid and Development focused on developing countries has also been perceived in this context. Regarding Refugee Aid and Development, governmental reluctance to fund UNHCR appropriately in its humanitarian work, let alone subsidize its perceived move into development activities given the continuing perception of the divide between humanitarian and development activities, has further complicated the situation.[32] The 'bilateralization' or earmarking of aid through national NGOs, sidestepping the UN, has also been a factor.

As will be clear from the preceding paragraphs, the cumulative impact of numbers, public and government weariness of asylum matters and the institutionalization of restrictive measures, coupled with security preoccupations and the Agency's own problems of donor dependence, underfunding (not to mention selective earmarking) for its operations, and its sometimes ill-starred policy attempts to try to find accommodation with governments on whom it is dependent, has resulted in a reinforcement of the tendency to push the responsibility for asylum further afield. That is, back to developing countries and hence, in a manifestation of the law of unintended consequences, the above-mentioned factors have arguably contributed to containment and protracted refugee situations.

UNHCR continues to maintain that the refugee issue is a global one underpinned by the inter-connectedness of all states involved. The Agency further contends that a holistic responsibility-sharing approach is, or should be, a *sine qua non* of an effective policy. While this premise is persuasive, what is left in terms of practical solutions in light of restrictive asylum practices in industrialized countries and the example and travelworthiness of such policies to poorer countries, coupled with inadequate funding from rich countries to address the problem in countries of origin, for example local integration or peace initiatives conducive to voluntary return? Regarding solutions, resettlement, although much discussed in Western industrialized countries since the departure of former High Commissioner Sadako Ogata, is a more apparent than real possibility.

Despite the recent rhetoric about 'new' resettlement countries coming on board, the situation today has not changed that much since the early 1990s when High Commissioner Ogata gave preference to the other two durable solutions. Then, as now, the number of principal resettlement countries remains virtually the same as 15 years ago. Net quotas available to UNHCR worldwide total some 50,000. The big players such as the United States, Canada and Australia provide respectively 70,000, 10,000 and 13,000 places, but UNHCR normally receives fewer than half of these quota places. The remainder are predominantly used for special interest cases of the countries concerned and those resettled would not normally meet UNHCR's priority requirements for resettlement. The rest of the places, available primarily from the Nordics, account for at best 4,000, and some countries (e.g. Norway) reduce resettlement places when the number of asylum seekers arriving directly increases. In 2004 and 2005, UNHCR registered some 30,000 departures on resettlement under its auspices, with the number dropping to 27,700 in 2006, or 9% fewer than 2005. These figures are considerably less than the number of departures in the early 1990s under Ogata's new regime, after resettlement was de-prioritized. In 2007 HCR had a particularly good year in resettlement with some 49,600 departures, the best year in a decade and a half. This figure, however, still represents 1% of the total global population of refugees under UNHCR's care. Even in the unlikely event that current resettlement places available were doubled or tripled by the European Union, or by several countries within the European Union or elsewhere establishing annual, operational, predictable quotas, it would not make much of a dent in providing solutions for the 6.2 million refugees under UNHCR's care (this number increases to well over 7.7 million if Palestinians under UNRWA are included) who find themselves in protracted refugee situations.[33] The halcyon days of large-scale resettlement programmes like those for the Vietnamese, which saw some 2 million resettled over a 15-year period, are unlikely to be repeated. Nevertheless, some application of 'strategic resettlement' for groups such as Burmese and Bhutanese, addressed in other chapters in this volume, might offer an incentive or safety valve to poor countries hosting long-term refugee populations to proactively contribute to addressing protracted refugee situations in the region of origin.

An additional concern in viewing resettlement as a solution is the following. Those involved in resettlement matters have long had to contend with the false premise in ministries of the interior in a number of EU countries, in Australia and occasionally in other industrialized countries that the 'real refugees' are in the third world, and asylum seekers arriving directly in rich countries are undeserving, queue-jumping interlopers. This sentiment, which reinforces containment and protracted refugee situations in developing countries, is of course untrue, and undermines both the principles of the right to freedom of movement and the right to seek asylum, set out in the Universal Declaration of Human Rights and the 1951 Convention. Thus, in contemplating resettlement one has to bear in mind not only the unreality that a quantum leap in quotas could be operationally feasible, but the fact that the whole discussion often masks a tendency to become ever more restrictive on direct arrivals in industrialized countries by promising, but not delivering on, a highly selective pick-and-choose-abroad alternative.

Another factor in considering effective durable solutions and responsibility-sharing in industrialized societies is recent developments, particularly in Europe, in respect to 'identity politics' or 'societal sovereignty', which do not portend increased receptivity to foreigners.

Multiculturalism versus integration/identity politics

In considering the role of industrialized countries' involvement in dealing with protracted refugee issues, the following factors are worth noting. The European Union's receptivity to asylum and burden-sharing arrangements in the near term must be seen in the context of major recent developments. At the macro level Europe has undergone significant changes since the beginning of the new millennium. Membership in the European Union has increased by over two-thirds with the addition of 12 new members. Furthermore, the former 15 members, some with very high unemployment rates such as Germany and France, are struggling with the challenges of accommodating the free circulation of workers under EU arrangements. At the same time, the momentum towards ever more integration in the Union suffered a significant setback with the French and Dutch rejections of referenda on the EU constitution as a result of growing public dissatisfaction with Brussels making top-down decisions, most recently, *inter alia*, with the decision to start accession talks with Turkey. Electorates in both countries also registered their dissatisfaction with government elites in their own capitals and their seeming indifference to domestic concerns. In the Netherlands, fear of losing control over immigration was a prominent issue in the 'no' vote, and in France the failure of the elites to grapple with unemployment played an important role.[34]

At least in the short to middle term, the drive towards the merging, or some might say submerging, of sovereignties among the European Union's long-term members, which has been a key feature of European integration, has been put on hold. A number of countries within the European Union are asserting national prerogatives with what one might call the return of the state. At the macro-economic level, France, citing 'economic patriotism', has blocked or sought to forestall foreign takeovers of French companies (e.g. Danone by the US's Pepsico, Arcelor by Mittal Steel). Spain has followed suit by blocking an attempted German takeover of one its energy companies. Italy has blocked foreign takeovers of its banks, and Germany has done the same with its auto industry.[35] 'Identity politics' at various levels is increasingly making itself felt, both among native-born citizens and in the parallel universes of effectively segregated immigrant communities.

Stanley Hoffman has written with respect to poorer countries that globalization produces a renaissance of local cultures in reaction.[36] The same applies to some degree in richer countries confronting the velocity of change. If resistance to financial mergers by foreign companies is a concern in the higher levels of government, immigration pressures are more popularly felt throughout all levels of society. Two of Europe's main models of accommodating immigration have been shown to have great weaknesses. France's assimilationist model revealed the chasm between official policy and reality in the wake of the riots in November–December 2005 among unassimilated, poor, marginalized, immigrant communities across most of France, where unemployment among the offspring of French-born children of immigrants in the age group 19–29 is three times higher than the national average.[37] An estimated 60% of the French prison population is Muslim, which is arguably indicative of a growing Muslim underclass throughout Europe. Both factors are conducive to susceptibility to Islamic extremist impulses.[38] As riots spread across the country, the French government, confronting its worst public disturbances in 30 years, declared a three-month state of emergency. Earlier in the year, London's multicultural model revealed its weaknesses with the actions of English-born Muslims, who carried out the July 2005 terrorist bombings in the capital.

In the Netherlands, long known for its liberal policies, identity politics has also made itself felt. Tensions have risen considerably over the past few years over asylum and immigration issues. Currently about 10% of the country's population of 16 million come from immigrant families, and in the big cities the percentage rises to some 40%.[39] Disparities and tensions are also reflected in the unemployment rate, which for non-Western immigrants is 14%, in contrast to 4% among the native Dutch population. Moreover, 'criminals with foreign backgrounds make up 55% of the country's prison population'. These statistics, coupled with frictions between unintegrated immigrants and Dutch citizens, have fuelled the sentiment: 'the Netherlands is full'.[40]

The controversy over immigration flared up in 2002 with the murder by a Dutch-Moroccan Muslim radical of the Dutch film director Theo van Gogh, who had made a film critical of Islam's treatment of women. The assailant's consistently unapologetic satisfaction with his actions further outraged Dutch opinion. The furore over the decision by the former Dutch Immigration Minister to expel the Dutch-Somali member of parliament, Ayaan Hirsi Ali, who had written the script for van Gogh's film and been threatened as 'next' in the note pinned to van Gogh's body by his assailant, for misrepresentations in her asylum claim in the 1990s, 'reflects both the intensity of debate about large-scale immigration and the high profile Hirsi Ali had in the Netherlands (she has since moved to the United States)'. Hirsi Ali had police protection and lived in safe houses for several years owing to death threats given her vocal opposition to radical Islam and the plight of women in Muslim communities, even in Europe. Hers was one of many recent cases where asylum seekers were denied (or in Hirsi Ali's case was threatened with the revocation of her) Dutch nationality for giving 'false testimony or failing tests' under the aegis of former Immigration Minister Rita Verdonk, nicknamed 'Iron Rita'.[41]

In some Nordic countries, senior government officials pointing to the lack of integration of immigrant and refugee communities have privately bemoaned the practice of refugees and immigrants bringing young, often uneducated and at times under-aged, brides from their country of origin rather than marrying nationals. In another part of the spectrum, honour killings within the immediate family of young Muslim women thought to have transgressed Islamic norms have angered even traditional supporters of refugees among liberal feminist groups in some Nordic countries.[42] Honour killings among Turkish Muslim immigrant families in Germany, particularly one high-profile case in April 2006 where a young Turkish immigrant killed his sister for building an independent modern life for herself, have inflamed public opinion. Renate Kocher, head of the respected polling organization the Allensbach Institute, has observed that: 'Germans are feeling very uncertain about Islam ... This uncertainty feeds into the issue of integration and how the second and third generation born in Germany can become integrated'. This uncertainty was reflected in the results of a poll published by Allensbach in mid-May 2006 indicating that 'an increasing number of Germans believe that a clash of civilisations is taking place between Christendom and Islam and that tensions between Muslims are rising.'[43]

Immigration tensions have contributed to a rise in xenophobia. In a three-month period between February and June 2006, the following made headlines. In February 2006, in a Paris suburb, a young Jewish man was kidnapped, tortured and killed by Muslim immigrants.[44] In May 2006, in Antwerp, a Malian babysitter and the white child she was caring for were murdered and a woman of Turkish descent was seriously injured by a skinhead.[45] In mid-May in Berlin, a German parliamentarian of Turkish origin was severely beaten by right-wing extremists. Some days before, a former government spokesman, Uwe-Karsten Heye, had cautioned prospective visitors coming to Berlin for the World Cup to stay out of certain areas he called 'no-go areas' for non-whites. 'He further stated that there were several towns in the province of Brandenburg where non-white visitors may not leave with their lives.'[46] The annual report of the Federal Interior Ministry confirmed that between 2004 and 2005 '[t]he number of rightist extremists ready to use violence has increased by 400 to 10,400', and there was an increase of 27% in politically motivated crimes by far-right groups and an increase in the number of radical Muslim groups in the country from 24 to 28.[47]

The furore unleashed by the Danish cartoon controversy has also dramatized immigration tensions in Europe among certain sectors of Europe's 20 million Muslims, not to mention repercussions abroad. The cartoon controversy, in addition to questions of bad taste, raised issues of Western values of free speech and secularism versus Islamic autonomy, and for some seemed a further indictment of multiculturalism and nativist reaction to immigration. Part of the problem may be due to the implications of what the author and columnist William Pfaff has called 'ghettoization through political correctness', where people were encouraged to think of themselves as members of specific communities rather than citizens of their new country. Increasingly, the traditional left liberal emphasis on multiculturalism and racial equality, which fostered a culture of unquestioning protective political correctness, and passive discrimination, making it difficult to even raise certain societal problems arising from this 'ghettoization', is giving way to concerted attempts at integration of immigrant communities in a number of countries.[48]

Since the election of the Rasmussen government in Denmark in 2001, the Danes, among other measures, have introduced stringent language requirements linked to financial assistance and made it harder for refugees and immigrants, as well as Danes, to marry abroad and bring their wives to Denmark. In Milan, the Italian Ministry of Education rejected a plan to establish Muslim-only classes, declaring the proposal 'unconstitutional', and saying that the aim was to 'overcome any form of discrimination and increase moments of integration and dialogue between cultures'.[49] In Germany, the state of Baden-Württemberg has established questions in a citizenship test to be asked only of Muslims, dealing with the rights of women, domestic life and religion. The United Kingdom introduced a new citizenship test in late 2005, emphasizing 'Life in Britain' with induction ceremonies similar to those in the United States to promote civic appreciation and shared identity.[50] The Netherlands has instituted more intensive citizenship requirements, including proven capability to function in the new language. As mentioned above, France, following the riots of autumn 2005, has introduced more restrictive immigration measures making it more difficult for foreigners to bring in relatives and fostered measures to facilitate integration, particularly of second-generation immigrants.

In the United States, a perennial bastion of immigration and since the 1960s multiculturalism, immigration and identity along with security concerns have become major political issues. Samuel P. Huntington, author of *The Clash of Civilizations*, has challenged the prevailing orthodoxy of political correctness surrounding immigration in the United States in a book in which his concerns are mirrored in the title: *Who Are We? The Challenges to America's National Identity*.[51] Huntington's book, published in 2004, both presaged and has added to the growing debate, principally on Hispanic immigration to the United States, and not least on matters of multiculturalism, bilingualism, integration and tolerance to newcomers from native-born Americans. Issues addressed by Huntington are reflected in the growing debate around border control and how to handle an estimated 12 million illegal aliens, which has dominated media coverage over recent months. The controversy, which has multiple dimensions, has raised the issue of the value of applying US laws in a country that considers itself a 'nation of laws', as well as integration and identity as evidenced by the Senate's approval of a Republican proposal designating English as the national language. It is worth repeating that the significance of the immigration issue is underscored by the fact that President Bush, in his first nationwide television speech focusing on a domestic issue, addressed immigration, given the controversy swirling around the subject and the 2006 mid-term congressional elections. Immigration remains an issue in the 2008 US presidential election.

Conclusion

It follows from the foregoing that the inter-linked factors of security, the perception that immigration needs stronger controls, the preoccupation with and challenge of integrating existing immigrant communities already in many European countries, identity politics, rising xenophobia and racism, coupled with increasingly institutionalized restrictive asylum procedures and systems that push asylum seekers to the periphery, and the cumulative politicization of migration and asylum, have reinforced the trend of warehousing and containment in developing countries and thus contributed to protracted refugee situations.

If one can talk of the return of the state to protect national sovereignty, one could also think of growing preoccupation with 'societal sovereignty', or the identity politics of safeguarding the cultures of native-born populations in rich countries from too many foreigners as a reaction to globalization, immigration and asylum.

Spain and Malta have struggled with boatloads of undocumented African migrants trying to reach their shores. Over 32,000 attempted to reach Spain via the Canary Islands in 2006 compared with 8,000 in 2005, and the European Union has had to send emergency patrol boats and planes to stem the influx. Malta has been described as 'sinking' under the tide of illegal arrivals.[52] And the recent controversy originating in the United Kingdom over some Muslim women wearing the veil has become a media issue in Europe. In the Netherlands, the victorious Christian Democrats in the November 2006 national elections made the banning of the wearing of the *burka* part of their policy platform. As Germany assumed the presidency of the European Union in January 2007, the German Interior

Minister, Wolfgang Schaubel, condemned the wearing of the *burka* in outlining his agenda for the EU presidency.[53]

Another factor that may be at play in reinforcing the fortress mentality in industrialized countries and the tendency to push asylum to the periphery, and well beyond to distant third world countries, is, at least in the case of Europe, the reality of the 'near abroad'. Geo-politically Europe lives in close proximity to nine Muslim states bordering the Mediterranean extending from North Africa through the Middle East to Turkey on Europe's eastern rim. For years, a large percentage of Europe's asylum seekers have been Muslim men. As Thomas Friedman has written, Arab countries in this arc, most without democratic governments and traditions, have another notable push factor: unemployment. A recently published ILO report has noted that 60% of the Arab world is under the age of 25. Unemployment in North Africa and the Middle East is 13.2%, the highest in the world (higher even than in sub-Saharan Africa), and adding more than 500,000 unemployed a year.[54] Policymakers in European capitals are mindful that not a few people from this area might seek to enter Europe, either as economic migrants or, given the instability in parts of the region, as a result of forced displacement.

In the circumstances, solutions to protracted refugee situations, if they are to be found, will be focused on the 'new refugee paradigm' or 'refugee aid and development', concentrating on managing solutions in the regions of origin. To make this work, industrialized countries would have to play a significant political, diplomatic and financial enabling role. In the current political climate, marked by preoccupations with societal sovereignty and terrorism, policymakers would have to be convinced that it is in their own security and peacebuilding interests not to underplay their shared, if illusive, moral responsibility to effectively address issues of failed and failing states and their human costs in an increasingly interconnected world. Most rich countries remain reluctant to accept asylum seekers and immigrants; in the current asylum-migration climate they will have to find the political will, resources and sustained interest necessary to play an effective role in addressing protracted refugee situations. They will also have to overcome the scepticism of developing countries that their efforts are not just a way of externalizing solutions as well as asylum.

Notes

I thank Dr. Gregor Noll, Professor of Law, Lund University, Sweden, and Brian Gorlick, Senior Policy Advisor, UNHCR Liaison Office in New York, for their very useful comments on this chapter.

1. See G. Troeller, 'Refugees and Displaced Persons in Contemporary International Relations; Reconciling State and Individual Sovereignty', in E. Newman and J. van Selm, eds., *Refugees and Forced Displacement, International Security, Human Vulnerability and the State*, Tokyo: UN University Press, 2003, pp. 50–65; hereafter Troeller, *Refugees*.
2. UNHCR, *State of the World's Refugees. Human Displacement in the New Millennium*, Oxford: Oxford University Press, 2006, p. 12. Hereafter *SOWR 2006*. UN, *Migration in an Interconnected World: New Directions for Action*, Report of the Global Commission on International Migration, Switzerland: SRO Kundig, 2005; hereafter GCIM.
3. The point is made by D. Papademetriou, 'Migration', *Foreign Policy*, Winter 1997–1998, pp. 15–18.
4. See chapter 7 in UNHCR, *The State of the World's Refugees. 50 years of Humanitarian Actions*, Oxford: Oxford University Press, 2000; hereafter *SOWR 2000*.
5. Brian Gorlick, 'Misperceptions of Refugees, State Sovereignty, and the Continuing Challenge of International Protection', in Anne F. Bayefsky, ed., *Human Rights and Refugees, Internally Displaced Persons and Migrant Workers*, Amsterdam: Koninkiijke Brill, 2006, pp. 65–89; hereafter Gorlick.
6. *SOWR 2000*. See pp. 176–177 for a summary of US interdiction practices since the 1970s.
7. *SOWR 2006*, p. 41. See also David Marr and Marian Wilkinson, *Dark Victory*, Canberra: Allen & Unwin, 2005.
8. Much has been written on the EU Harmonization Process. Among the best of many competent treatments of the subject this section draws upon: UNHCR, 'Towards a Common European Asylum System: A View from UNHCR', *European Series*, Special issue 2003, pp. 1–7; C. Levy, 'The European Union after 9/11: The Demise of a Liberal Democratic Asylum Regime?', in *Government and Opposition*, no. 1, Winter 2005, pp. 26–59; hereafter 'Levy'; the excellent paper by J. Kumin, 'The European Union's Refugee Policy: A View from the UNHCR', Presentation at the Free University of Brussels, July 15, 2005, on which this section draws in particular – hereafter 'Kumin'; and Gregor Noll, *Negotiating Asylum: The EU Acquis, Extraterritorial Protection and Common Market of Deflection*, Amsterdam: Martinus Nijhoff Publishers, 2000.
9. UNHCR, *2005 Global Refugee Trends*, pp. 46–49, online at www.unhcr.org/statistics.
10. *SOWR 2006*, p. 46.
11. Kumin provides compelling examples of how asylum seekers can fall through the cracks, pp. 1–3.
12. Ibid., p. 12. The 'demonstration effect' in an increasingly 'wired world' will be clear to serious students and practitioners of refugee protection, as it should be to well-intentioned advocates. As mentioned in the most recent *State of the World's Refugees*: 'Indeed, developing countries often point to Western-countries' policies to justify their increasingly restrictive practices ... As a result the rights of ... refugees are often violated due to the indiscriminate implementation of measures aimed at combating illegal migration.' *SOWR 2006*, p. 180. This point has also been made by, among many others, UNHCR Assistant High Commissioner for International Protection, Erika Feller, who remarked in meetings with Nordic officials, in which the author participated, that restrictive European asylum practices are often quoted back to her in meetings with her counterparts in developing countries, in connection with their own adaptation of such practices. Tania Kaiser, whose chapter on Uganda appears in the volume, has mentioned similar experiences in her meetings with East African government officials.
13. *SOWR 2006*, pp. 38–39.
14. See 'Send Back Your Huddled Masses', *The Economist*, 16 December 2004, quoted in Gorlick, p. 66.

15. Alan Cowell, 'Swiss Political Axis Tilts with Victory by Rightist', *International Herald Tribune (IHT)*, 11 December 2003. See also Nick Cumming-Bruce, 'Rightists Strengthen Hold in Swiss Vote', *IHT*, 22 October 2007.
16. Katrin Bennhold, 'New Arena for Ideas of Le Pen', *IHT*, 13 January 2006, p. 3.
17. Ibid.
18. Author's discussions with senior Norwegian officials.
19. 'More Marches, a Growing Backlash', *The Economist*, 6 May 2006, pp. 29–30.
20. Jeff Crisp, 'A New Asylum Paradigm? Globalisation, Migration and the Uncertain Future of the International Refugee Regime', Evaluation and Policy Analysis Unit, UNHCR, Working Paper no. 100, December 2003, p. 1.
21. 'Swiss More Fearful of Foreign Immigrants than Terror Attacks', *Agence France Press*, 1 December 2004; cited in Gorlick.
22. J. Tirman, ed., *Maze of Fear. Security and Migration after 9/11*, Washington, DC: Social Science Research Council, p. 2; hereafter *Maze*.
23. See O. Obiora, *Deportations to Torture after 9/11. An Analytical Comparison of Canadian and US Practice*, Toronto: University of Toronto, forthcoming.
24. Katrin Bennhold, 'Europe's Terror Battle: Liberties Eroding', *IHT*, 14 April 2006, p. 4.
25. Ibid. See also Gregor Noll, 'Diplomatic Assurances and the Silence of Human Rights Law', *Melbourne Journal of International Law*, no. 7, 2006, pp. 104–126.
26. 'Refugee Policy, Less Hope, a Longer Wait', *The Economist*, 29 April 2006, p. 34.
27. M. Weiner, ed., *International Migration and Security*, Boulder, Colorado: Westview Press, 1993, cited in *Maze*, p. 5.
28. See G. Troeller, 'UNHCR Resettlement: Evolution and Future Direction', *International Journal of Refugee Law* 14, no. 1, 2002, pp. 85–95; hereafter Troeller, *IJRL*.
29. See G. Loescher, *The UNHCR and World Politics*, Oxford: Oxford University Press, 2000, and, in contrast, S. Ogata, *A Turbulent Decade. Confronting the Refugee Crises of the 1990s*, New York: Norton, 2005.
30. *SOWR 2006*, p. 57. For further elaboration of the importance of the asylum–migration nexus see UNHCR, 'Note on International Protection', Part IV, EC/57/SC/CRP.14, June 2006 and, *inter alia*, the Office's 'Discussion Paper' of 29 June 2006, which states: '... UNCHR considers that it is essential to be actively engaged in the issue of international migration if it is to effectively discharge its mandate', p. 7, as well as the 'Extracts of UNHCR's Agenda for Protection dealing with the protection of refugees within wider migration movements', which provides a list of the 10-plus international and regional migration fora in which the Office participates, in addition to its close collaboration with the International Organization for Migration. All three documents are found on the UNHCR website under 'Refugee Protection and International Migration', www.UNHCR.org.
31. Ibid.
32. Ibid. *SOWR 2006*, pp. 181–185. It should also be noted here that finance ministries in developing countries have not wished to see development aid which they regard as for their own nationals being used to benefit refugees they host.
33. Resettlement was much discussed during Ruud Lubbers' tenure as High Commissioner for Refugees. See also the 'Multilateral Framework for Understandings on Resettlement', referred to above, among UNHCR's new policy initiatives. For a historical perspective on this durable solution compare the situation in 1990 in G. Troeller's 'UNHCR Resettlement as an Instrument of Protection', *IJRL* 3, no. 3, 1991, with the article Troeller, *IJRL* referred to above. Figures given for recent resettlement departures under UNHCR auspices are taken from *2004 UNHCR Global Refugee Trends*, p. 4, *UNHCR – Refugees by Numbers 2006 Edition. Basic Facts*, p. 9, and *UNHCR 2007 Global Trends* (provisional), June 2008, p. 10, available online at www.UNHCR.org.

34. 'Dutch Nees Up', *The Economist*, 4 June 2005, p. 49.
35. Patrick Sabatier, 'The Return of Protectionism: Globalization a la Carte', *IHT*, 19 May 2006, p. 8.
36. S. Hoffman, 'The Clash of Globalizations', *Foreign Affairs*, July–August 2002, p. 108.
37. Report from the French National Statistics Office, September 2005, cited by Katrin Bennhold, 'Even Those It's Supposed to Help, It Seems, Oppose French Jobs Law', *IHT*, 15 March 2006, p. 3.
38. The point is made by the French scholar Farhad Khosokhavar in an article entitled 'Growing Muslim Prison Population Poses Huge Risks', *IHT*, 9 December 2004, p. 8.
39. Toby Sterling, 'Dutch Asylum Debate Has a Face and a Name', *IHT*, 9 March 2006, p. 3.
40. Jeffery Fleishman, 'Tolerance and Fear Collide in the Netherlands', UNHCR, *Refugees* 2, no. 135, 2004, p. 19.
41. Marlise Simons, 'Dutch Aide under Fire over Islam Critic Ruling', *IHT*, 18 May 2006, pp. 1, 4.
42. Author's discussions with senior Nordic officials.
43. Judy Dempsey, 'Turkish Leader Faces German Mood Swing', *IHT*, 26 May 2006, p. 3.
44. Doreen Carjeval, 'French Murder Suspect Smiles for the Camera', *IHT*, 1 March 2006, p. 3.
45. 'Antwerp on Edge after Killing Prompts Racial Hate Worries', *IHT*, 13–14 May 2006.
46. Judy Dempsey, 'Racial Attack on German Politician Angers Germans', *IHT*, 22 May 2006, p. 3.
47. Judy Dempsey, 'Germany Uses Schools to Combat Extremism', *IHT*, 23 May 2006, p. 3.
48. Geoffrey Wheatcroft, 'The Muddied Waters of Identity', *IHT*, 26 December 2005, p. 6.
49. Elisabetta Povoleda, 'Italy Rejects Plan for Muslim-only Class', *IHT*, 15 July 2004, p. 4.
50. Edward Rothstein, 'Putting Citizenship to the Test', *IHT*, 25–26 February 2006, p. 18.
51. S. P. Huntington, *Who Are We? The Challenges to America's National Identity*, New York: Norton, 2004.
52. Meg Bortlin, 'Making Money in Senegal off Human Cargo', *IHT*, 30 May 2006, p. 2; Dan Bilefsky, 'Malta Fears Sinking under Migrants', *IHT*, 7 June 2006, pp. 1–2; Victoria Burnett, 'As Seas Calm, Spanish Brace for More Migrant Boats', *IHT*, 15 May, 2007, p. 3.
53. Dan Bilefsky, 'German Official Condemns Burka', *IHT*, 12 January 2007, p. 3.
54. Thomas L. Friedman, 'Empty Pockets, Angry Minds', *IHT*, 23 February 2006, p. 6. Niall Ferguson makes similar points regarding the actual and potential for volatility in the region in his 'The Next War of the Worlds', *Foreign Affairs*, September/October 2006.

4

Protracted refugee situations, conflict and security: The need for better diagnosis and prescription

Eric Morris and Stephen John Stedman

This chapter differs from the other analytical chapters in this volume. It is primarily an essay which critiques some of the existing approaches to explaining the relationship between refugee movements and conflict management, and argues for better diagnosis of the causes and consequences of long-term exile. The chapter maintains that refugee and security planners will become better at prescribing solutions to protracted refugee situations when they better understand both the causes of refugee movements and extended exile and their relationship to protracted conflict and peacebuilding, a theme that is returned to in more detail in the concluding chapters of this volume.

Despite a few exceptional attempts to bridge the gap between the study of refugees and the study of international security and conflict management, there remains a deep chasm between these two topics.[1] Indeed, the international refugee regime is based on the separation of the humanitarian from the political. UNHCR's Statute states that: 'The work of the High Commissioner shall be of an entirely non-political character; it shall be humanitarian and social and shall relate, as a rule, to groups and categories of refugees'.[2] The 1967 UN General Assembly Declaration on Territorial Asylum states that the granting of asylum to persons fleeing persecution 'is a peaceful and humanitarian act and that, as such, it cannot be regarded as unfriendly by any other State'.[3]

The study of refugees has been dominated by disciplines such as anthropology, sociology and geography, and scholars in these fields tend to focus on those who are victims of human rights abuses or conflict and

Protracted refugee situations: Political, human rights and security implications,
Loescher, Milner, Newman and Troeller (eds),
United Nations University Press, 2008, ISBN 978-92-808-1158-2

who have been forcibly displaced. Researchers in conflict and security studies, on the other hand, place states at the centre of their analysis. Few refugee researchers try to understand why states respond in the way they do to refugees, nor do many attempt to understand under what circumstances states are willing to try to find solutions to protracted refugee situations. In fairness, however, the record shows that refugee scholars more frequently attempt to reach out to their security colleagues, and Gil Loescher and James Milner's work on protracted refugee situations is the latest example.[4] They argue that protracted refugee situations cause conflict directly and indirectly, and that such situations diminish human, regional and international security.

International security scholars, for the most part, are also blind to the significance of refugees for their study of conflict and conflict management. When they notice refugees at all, they usually see them as a symptom of large-scale violence and do not ascribe to them any independent causal agency, for instance in starting civil wars or contributing to their duration and intensity, much less the possibility that solutions for refugees might contribute to the lessening of violence. This is true whether or not the scholars in question base their conclusions on large-N quantitative data or small-n comparison of detailed case studies.

Among those who study war and peace quantitatively, Fearon and Laitin, Collier and Hoefler, Walter, Doyle and Sambanis, Fortna and others have little to say about how refugees may instigate war, prolong it, and pose a barrier to making peace, because few of them include variables related to refugee populations and their management.[5] The only quantitative study of civil wars that does not treat refugees as epiphenomena is a recent article by Idean Salehyan and Kristian Skrede Gleditsch that shows empirically what refugee scholars have claimed for some time, namely that refugees are independent variables; once violence and political crisis have created them, refugees continue to act in their own right, with observable effects on war and peace in their home and host countries.[6] While demonstrating that refugees from neighbouring states significantly increase the risk of civil conflict for receiving states, they also emphasize that this is not a deterministic relationship: many refugee movements do not result in violence. Salehyan and Gleditsch focus narrowly on the spread of civil war and conflict diffusion, but their analysis should open the door for the conflict and security field to take refugees more seriously in the future.

When we turn to case study analysis, we largely find the same absence. To give one example, prominent experts on conflict management have recently developed the concept of intractable conflicts – conflicts that are particularly stubborn or difficult but not impossible to manage.[7] These experts produced two volumes and neither references the concept of protracted refugee situations – situations involving 'large refugee populations that are long standing, chronic, or recurring'. Indeed, the two volumes mention refugees only once, and that is in their potential role as diaspora funders of conflict, citing the role of Sri Lankan émigrés as supporters of the Tamil Tigers. Obviously, these concepts beg to be brought together, and prompt two critical questions: Are there particular dynamics of protracted refugee situations that contribute to the intractability of conflict? And how do the particular dynamics of intractable conflicts make it difficult to manage protracted refugee situations?

When it comes to conflict prevention, refugee and security scholars should have much to say to each other, but, again, one finds little evidence of this interchange taking place. If protracted refugee situations contribute directly or indirectly to violence, it would be valuable for conflict prevention experts to ask some basic questions. Where protracted refugee situations are the product of civil wars, what lessons can be learned to prevent forced displacement that will exacerbate the conflict and render a negotiated settlement all the more difficult? Displacement and failure to realize return were at the heart of the Rwandan tragedy, not only for the Tutsi who became refugees in Uganda and elsewhere in the 1960s, but also for the hundreds of thousands of Hutu internally displaced by the civil war in the early 1990s. Where protracted refugee situations are not the product of civil wars, such as the refugees from Bhutan in Nepal, what can be done to address them so that they do not instigate war? A quick survey of recent studies of conflict prevention, however, shows few, if any, references to refugees at all.[8]

Challenge of the refugee–security nexus

As described above, security scholars, to the extent that they think about refugees at all, view them as a by-product of tough, intractable conflicts. They do not tend to think about how refugee populations are independent actors and causes of conflict. Rightly or wrongly, security scholars assume that, if you resolve the conflict, then the refugee crisis will end. In a world of scarce resources, they advocate putting resources into negotiation and implementation of peace agreements, and sometimes they are right: a peace settlement ends the refugee crisis. Loescher and Milner agree that protracted refugee situations are caused by political crises, and it will be political solutions that will end them. Yet this formulation remains at such a level of generality that no clear policy prescriptions present themselves.

The analysis of protracted refugee situations needs to be better informed by the accumulated records of international conflict management since the end of the Cold War. As the *Human Security Report 2005* documents, the world of today is much less violent than the world of 1992.[9] Since then, the numbers and intensity of civil wars have declined dramatically. The Report states that civil wars have decreased by 40% since 1992 – a dramatic historical decline. Only twice in the last 200 years has the world seen such a precipitous drop in civil wars in such a short period of time. Hand in hand with the reduction of the numbers of civil wars has been a steep reduction in the numbers of refugees in the world. At least some of the credit for this downturn should go to the myriad activities of international conflict management – mediation, peacekeeping and peacebuilding. The record is certainly a more hopeful one than anyone would have dared to predict in the early 1990s.

Most of today's protracted refugee situations are related to those remaining civil wars that have defied the post-1992 trend, and these situations are clearly a product of particularly intractable conflicts. To the extent that international actors get better at mediating and implementing peace agreements for the hard-core remaining wars, they will reduce the number of protracted refugee situations. In this respect, we wish to sound at least one tentative, optimistic note. If international and local actors succeed in implementing peace after the North–South war in the Sudan and the war in Burundi, in addition to consolidating peace in Liberia, the number of protracted refugee situations in Africa will be reduced almost by half. This is how protracted refugee situations are ultimately resolved – bringing one war after another to an end.

The challenge is to say what policymakers should do differently. What should they do with the refugees and their plight that they are not doing now, that will shorten the time needed to make peace and consolidate it in such a way that the chances that violence resumes are lessened? Few would dispute that this is a question of great normative importance, yet we lack clear and persuasive answers. Perhaps it could not be otherwise at this stage, as there is no consensus on what is really important. We have seen that what is important to refugee scholars is generally not important to security scholars and that, perhaps to a lesser degree, the obverse holds as well. There is a gap between the humanitarian domain and the political domain, with actors in the former sometimes endowing their principles with a 'sacred' aura, in opposition to the 'profane' arts of negotiation and compromise practised by actors in the latter. Finally, there is a disconnect between, on the one hand, those who analyse and advocate and, on the other, practitioners who make and implement conflict management policy – again, different perspectives on what is really important and, thus, what policies are crucial for ending intractable conflicts.

The way forward: Better diagnosis, prescription and policy formulation

A first step towards improving the policy utility of the concept of protracted refugee situations is to disaggregate it. As presently conceived, the symptom of protractedness can mask an abundance of causes and, without knowledge of those causes, prescriptions will be either too general to have much effect, or wrong. We suggest as first steps better differentiation among protracted refugee situations in terms of causes and in terms of their conditions of exile. Beyond better diagnosis, there is a need for better prescription, especially concerning the relationships between refugee management and conflict management. Finally, there is a need to link better understanding of protracted refugee situations to policy formulation and implementation.

Better diagnosis of protracted refugee situations

It is important to devise a politically informed analysis, or diagnosis, of the causes that lead to the onset and continuation of displacement. To do so one must necessarily expand the scope beyond protracted refugee situations to the broader dynamics of forced displacement, including internally displaced persons. Indeed, we believe that dealing with protracted refugee situations without considering the often closely connected issue of internal displacement in the country of origin limits understanding. What follows below relates for the most part to situations where displacement is linked to intractable conflicts, although we will return to situations where the link is not so clearly established or remains dormant.

The central message of this volume is that there are political and strategic consequences of unresolved refugee situations. The analysis contained in several chapters regarding the regional dynamics of refugee situations, in terms of diffusion of conflict and contagion, is particularly persuasive. For the most part and to their credit, refugee scholars are often more attentive to the regional dynamics of such situations than security scholars – it seems to come with the territory since the focus is on cross-border population movements.

Yet looking at the consequences or the symptoms alone does not translate easily into policy prescriptions. For example, Loescher and Milner have argued elsewhere: 'Prolonged and unresolved refugee crises almost universally result in politicization and militancy of refugee communities with predictable adverse consequences for host state and regional security'.[10] This clear argument calls for comment. First, assuming that there is indeed a likelihood of 'adverse consequence', what is striking is how limited our ability is to predict the onset and sustainability of politicization and militancy of refugee communities. In fact, the onset of such phenomena can be at the very beginning of a refugee situation and thus not a consequence of a prolonged exile. Second, the argument sounds too automatic. The suggestion that all protracted refugee situations, if left untended, will produce large-scale violence is not demonstrated. Further analysis is required to reach an understanding of under what conditions this will or will not happen, and this may make prediction a little more accurate than is the case now. In short, there is a need for better diagnosis.

Protracted refugee situations and the causes of displacement

Diagnosis hinges first on the causes and the nature of displacement. When examining the cause – the purpose of the war and the manner in which it was waged – one must first establish whether any of the parties to the conflict have the intention to displace a particular segment of the population, or whether displacement is, at least initially, a consequence of the war. The causes of displacement can be broken down into two broad categories: wars of exclusion and wars of control.

Wars of exclusion can also be characterized as wars about identity. Exclusion can occur when one group displaces another in order to bring about a congruence between political and cultural boundaries. The wars attendant upon the break-up of the former Yugoslavia represent the clearest examples. Likewise, exclusion can occur when a minority group seeking secession sees the presence of other groups (particularly the politically dominant group nationally) as an obstacle to its goal. Both cases (and there are other variations on the theme) involve issues of ethnicity, religion and cultural differences. With wars of exclusion, displacement is the very purpose of the war.

Wars of control, sometimes known as power wars, are essentially about the struggle for political domination, where displacement follows outbreaks of violence – a by-product of the war. Over time, however, displaced populations can assume political and military significance. When counter-insurgency campaigns are conducted, populations may be displaced purposefully in the interest of 'national security'. The underlying purpose is usually to deny insurgents the support of the local population, or as a means of group punishment for presumed support of insurgents. At that point, cultural differences that may not have initially been politically salient can be exploited for political and military purposes. As a result, displaced civilians become hostages of the parties to the conflict and are subject to varying degrees of manipulation, including forced military recruitment. While the initial motivation of the war may not have been mass displacement, the original intent can change over time as differences among groups become exacerbated by the violence.

The manner in which the war is waged is a critical element to be examined as well, although our consideration will be brief in that it flows in some measure from the purpose of the war. In wars of exclusion, where displacement is the very purpose of the war, protagonists make no meaningful distinction between combatants and civilians. Mass atrocities follow. Admittedly, wars of control constitute an exceedingly broad category, which calls for much greater refinement than we can offer here. Reference has already been made to counter-insurgency campaigns and their legacy in terms of intractable conflicts. The logic of these wars entails the need to maintain control of populations, not only for obvious war purposes – provision of food, manpower and other resources – but also as a basis for claiming legitimacy in an attempt to get a seat at the bargaining table. In this way are 'pseudo-states' born.[11]

Beyond protracted refugee situations linked to ongoing intractable wars, there is a small subset of such situations linked to wars that have ended, if one measures them in terms of the reduction in numbers of those killed, but where, because one side won a victory or due to instability surrounding the settlement, a portion of the population remains alienated or fearful, or both, and refuses to return. There is also a small subset of situations tied to political exclusion and an outburst of state-led violence at the time of expulsion, but not linked to ongoing violence. Not all these situations will revert to violence, but some do.

Diagnosis and the conditions of exile

We also need better diagnosis of the conditions of displacement and exile to understand why some situations will revert to violence while others do not. The length of time in displacement is an indicator of a protracted situation but has no predictive power unless the purpose and means of war, as well as other elements of the conditions of displacement, are considered.

The organization of camps and settlements for refugees and internally displaced persons is critical. In theory, a reasonably well-controlled and well-managed camp, in contrast to dispersed refugee settlements, should provide an environment where security could be maintained, assuming there is a host government able and willing to assume responsibility for security. In practice, such camps often lend themselves more easily to becoming politicized and militarized, depending on the motives and organizational capacity of those who claim to represent the displaced. A study of Hutu refugees in Tanzania in the 1980s noted a striking difference between those in camps who embraced extremist politics and those in towns who just wanted to get on with their lives.[12] Militarized camps may look remarkably similar, yet there is an important distinction intrinsic to the camp population – the degree to which the relationship between combatants and civilians is primarily coercive or in some measure consensual.[13]

The most critical element to consider in terms of the link between conditions of displacement and security threats is the situation in the country of asylum. The response of the host government to an influx of refugees will be determined by a number of political and strategic considerations. The nature of the relationship between the host country and the country of origin can be adversely affected, but not always in easily predicted ways. Several protracted situations – refugees from Bhutan in Nepal, from Myanmar in Thailand, from Burundi in Tanzania – can cause diplomatic strains and perhaps a sporadic military incident but not much beyond that. If the relationship between the two states is already adversarial, the host government can use the displaced as a weapon, or at least a point of pressure, against the country of origin. Alternatively, a convergence of interests between the two states can lead to early and unsustainable repatriation under less than voluntary circumstances, a development that Loescher and Milner correctly highlight in their chapters.

A more differentiated understanding is required of the political and security interests of the host government and other engaged states. This is a matter not simply of bilateral relations (assuming there is only one host country), but of regional states and, in some cases, global powers. The crucial question for states is, what are the political uses of refugees? This question recasts the debate over security threats presented by protracted refugee situations. In the 1980s, Thai government officials rarely missed an opportunity to emphasize the immense burden that hundreds of thousands of displaced Cambodians placed on their country or to highlight the indirect security threats that followed in terms of black markets, proliferation of small arms and conflict with Thai villages. Did that mean that Thai authorities believed the protracted refugee situation to be a direct security threat? Most certainly not – the Vietnamese Army in Cambodia was the security threat. The displaced Cambodians – living in camps under the control of the resistance – constituted a genuine security asset. Thailand, of course, enjoyed the strong support of the United States, China and the states of the Association of Southeast Asian Nations (ASEAN), which explains, in large measure, the sustainability of the endeavour to bring about a Vietnamese withdrawal. A clear understanding is required of the perception of host and other states of the political uses of refugees and of their desire and capacity to sustain the endeavour. At the same time, a rapid shift in such a perception or desire and capacity – in short, when the displaced are no longer seen as a security asset – can be destabilizing and sometimes lead to an attempted or successful armed return to the country of origin. A prominent example of this dynamic is Museveni's treatment of Rwandan Tutsi refugees, the rise of the Rwandan Patriotic Front (RPF) and the invasion of Rwanda in 1990, setting in motion a chain of events that led to the Rwandan genocide, as outlined in chapter 2.

Refugees and civil wars: Better prescription

In arguing for the need to differentiate among protracted refugee situations through more detailed diagnosis we come to two conclusions. First, it is not possible to move from identification of symptoms to policy prescriptions. Second, while all indirect security threats from protracted refugee situations are problematic in a similar way, at least superficially, direct security threats – that is, those potentially leading to large-scale violence and intractable conflict – are threatening in their own specific way. In sum, we need better understanding of the relationship between solutions for the displaced and the ending of intractable conflicts through mediation and implementation of peace agreements or, even better, through the prevention of violence altogether.

In chapter 16, Loescher and Milner correctly call for a renewed emphasis on a range of durable solutions and responses to protracted refugee crises. Leaving aside whether Western countries would embrace the option, it is certainly plausible that, if third country resettlement once again became a robust response to refugee crises, it could have an ameliorative effect in the termination of civil wars. Our only caveat concerns those conflicts where refugee diaspora organize to fund violence and are often more strident on possible compromise solutions than combatants on the ground. Local integration of refugees into neighbouring states is rarely considered an acceptable solution by host governments. This leaves the solution of repatriation to the country of origin.

Researchers do not know enough about the relationship of refugees and the ending of intractable conflicts. The one analysis that looked at peace implementation, following civil war from 1980 to 1999, came to some surprising conclusions about the role of refugees in the success and failure of peace agreements. The refugee scholar Howard Adelman expected to find strong universal evidence that successful repatriation of refugees was crucial for successful peace agreements. In some cases this turned out to be so, but he concluded: 'Evidence and arguments challenge the presumption of humanitarian scholars that successful repatriation is essential to the successful implementation of a peace agreement. At the same time, they also challenge the possibly implicit assumption of most scholars of security studies that the issue of repatriation is a marginal factor in the successful implementation of a peace agreement. Instead, repatriation may be a complicating factor, with many different dimensions, in the successful implementation of a peace agreement'.[14]

An understanding of the many different dimensions of the relationship may well require rephrasing the question in both a broader and a more specific manner. The broader question is whether repatriation is central to the peace process or not, in that the peace process puts in place the conditions that permit all groups – refugees, internally displaced persons and war-affected populations – to benefit, without discrimination, from the peace.

The issue of return is most certainly central to how wars of exclusion are ended. As displacement was the very purpose of the war, attempts in the peace process to reverse the logic of the war assume such political and strategic significance that, in the end, peace implementation becomes impossible or extremely difficult. The internationally negotiated Arusha Agreement for Rwanda did not bring about the return of the Rwandan Tutsi; this was accomplished by the RPF victory. Nor, for that matter, did international efforts bring about the return of the Rwandan Hutu from the militarized camps in eastern Zaire; this was accomplished by another RPF victory. The painfully slow process of minority returns in Bosnia after the Dayton Agreement attests to the difficulties this issue created for peace implementation.

In other cases, the issue of return does not assume such a degree of centrality. This can be for a number of reasons, one being that repatriation is marginal to the peace agreement. An important distinction needs to be made here: saying that repatriation is marginal to any number of peace agreements is not the same (or should not be the same) as saying that successful return and reintegration are marginal to the sustainability of the peace. Perhaps a rising tide of peacebuilding can lift all boats. If there is a well-founded presumption that returning refugees will be able to avail themselves of the same rights as their compatriots who remained, then it is not really that important whether refugee return is highlighted in a peace agreement. Indeed, conventional wisdom among practitioners argues against preferential treatment of returning refugees, lest this cause resentment from other war-affected groups.

It would also be beneficial to recast the question in a narrower manner. While many conditions for successful peace implementation will be relevant to refugees, it will be important to identify conditions that are return specific. These will vary from situation to situation, but two return-specific conditions are fairly constant and need to be addressed. The first is the issue of property restitution or compensation. This issue can often be foreseen and, to the degree that it is problematic, solutions should be included in a peace agreement. The second is the presence of armed groups – be they government, militia or rebel – that caused the displacement of potential returnees. Knowledge of these situations will probably be imperfect at the time of peace negotiations, but resolution of such potential conflicts will be required in the early phases of peace implementation.

Of course, attempted linkages should be made in the conflict cycle well before the implementation of a peace agreement. A particularly promising linkage, as we have argued above, is between the concept of intractable conflicts and the concept of protracted refugee situations. As the majority of protracted refugee situations are clearly linked to intractable conflict, both refugee and security scholars need to work in tandem to bring the two concepts together. Equally, refugee and security scholars should look to find operational linkages between the two. To the degree that refugee scholars and practitioners can look beyond their domain and position their concerns within the broader process of conflict management, they can make a greater contribution in improving how international actors mediate and implement peace agreements.

Protracted refugee situations that are not immediately perceived as linked to ongoing intractable conflicts present a greater challenge in engaging conflict management actors. Yet, under certain circumstances, some of these situations that have been dormant for years can lead to large-scale violence. What is needed to engage conflict management actors is a better understanding of the possible triggers that can turn such situations into war. The more proficient that international actors become at understanding and resolving protracted refugee situations where war has not begun, the better they will be at preventing new intractable wars.

Finally, in making the linkages between refugees and security, it is important not only to take preventive action at the onset of conflict but also to stem, contain and reverse forced displacement at an early stage so that it does not further exacerbate the conflict. In early 2001, the emergence of ethnic Albanian rebel groups in southern Serbia and in the former Yugoslav Republic of Macedonia was perceived as having the potential to unravel the fragile situation in the Balkans. The international response was reasonably coherent, involving NATO, the European Union, the United Nations and the Organization for Security and Co-operation in Europe (OSCE). Central to the effort to prevent the incipient conflict in southern Serbia was an emphasis on prevention of forced displacement and the quick return of those who had felt compelled to leave. The conflict simmered for six months, ceasefires were periodically broken, but in the end political concessions were granted by Belgrade, the rebels disarmed and were given amnesty, and the displaced quickly returned. In Macedonia, the fighting was much more severe and more widespread geographically. The rebels rapidly made military gains and displacement was significant. Ethnic Albanians fled to Kosovo. Ethnic Macedonians remained displaced in their own country. It was the latter group that presented the larger political problem, for nationalist leaders said there could be no settlement until the displaced ethnic Macedonians could return home. Thus, the emphasis was given to putting security conditions in place so this could happen. This was more containment and reversal instead of prevention of displacement, but inasmuch as the displacement was contained and reversed it facilitated the conclusion of a peace agreement.

Linkages to conflict management policy

There is also a need to make linkages to conflict management policy formulation and implementation. These linkages are required not only for resolution of protracted situations but also because what is to be proposed in terms of policy must be informed by better diagnosis.

Loescher and Milner correctly put the issue this way: 'So long as discussions on protracted refugee situations remain exclusively within the humanitarian community, and do not engage the broader security and development communities, their impact will be limited.'[15] It must be acknowledged, however, that the disconnect among the three communities – humanitarian, security and development – as it pertains to protracted refugee situations is just one on a lamentably long list of topics where optimal multilateral action is hampered by the divergent institutional interests of the three communities.

The challenges in making the linkages from a specific issue, such as protracted refugee situations, to policy formulation and implementation are considerable. It should be recognized that management of refugees is a sub-goal in a much larger endeavour of conflict management. Mediators and implementers of peace agreements do not begin their work by thinking about how they will solve a conflict's refugee crisis. They will first think about how a war can be brought to an end, and their attention will likely be focused on meeting the political and military needs of the combatants and providing security for the civilians caught in-between. Moreover, the challenges of managing refugees are never exactly the same in any two cases, and an insistence that they are essentially the same in every case dooms attempts to formulate policy prescriptions. Finally, different sub-goals interact – success in one area can affect success in another area; for example, to the extent that implementers of a peace agreement succeed in demobilizing soldiers and demilitarizing politics, the more likely that any repatriation of refugees will also be successful. The importance of this interaction will be missed if any given sub-goal is analysed in isolation.[16]

There is also a threshold that needs to be considered. As we have seen, chapter 2 describes two kinds of implications for international security – direct and indirect. Security scholars and practitioners will of course privilege the direct implications, for those seem to be linked directly to producing large-scale violence above and beyond the violence that produced the refugee situation in the first place. Indirect violence means smaller-scale violence or tougher lives lived by the refugees, or greater vulnerability to small-scale violence because of crime and small arms proliferation. Security scholars will tend to ignore indirect security effects unless it can be shown that there is a potential for indirect effects to produce actions that could lead to direct security effects and large-scale violence.

In looking for institutional entry points that would make a linkage between protracted refugee situations and policy formulation and implementation, Loescher and Milner in chapter 16 are right in suggesting that the new United Nations Peacebuilding Commission should look far more closely at the connections between refugee crises and conflict management. While it is still early days for the Peacebuilding Commission, the first two countries to be considered – Sierra Leone and Burundi – clearly lend themselves to making the linkages, in that both have produced significant refugee movements and both conflicts had widespread regional security implications, in the Mano River and Great Lakes regions respectively. For these connections to be sustained, though, the new Peacebuilding Support Office (PBSO), the secretariat to the Commission, must have robust participation from UNHCR and other concerned actors. It is thus gratifying to note that in May 2007 the UN High Commissioner for Refugees made a presentation to the Peacebuilding Commission. Appropriately, and to the credit of the Commission and Support Office, the Representative of the Secretary-General for Internally Displaced Persons also made a presentation during the same session.

A second potential institutional entry point for better integration of concerns of the displaced in conflict management is the new Mediation Support Office, in the United Nations' Department of Political Affairs. This office, like the PBSO, is concerned with lessons learned, best practice and informing strategies of peacebuilding and mediation. We have earlier pointed to examples of how better understanding and treatment of the displaced can make the attainment of a mediated agreement easier, and how the avoidance of the worst kinds of displacement can improve the chances that any mediated agreement will be sustained. To the extent that analysis of the displaced and greater understanding of protracted refugee situations can better inform policies earlier in the conflict cycle, the greater the likelihood that many of the worst consequences will be mitigated.

A final question that needs to be posed is whether the post 9/11 security context is more propitious for an agenda that ties refugees to security. Loescher and Milner hope that this is the case and they allude to arguments that some policymakers have made about globalization, interconnection of threats, and failed states as facilitators of terrorism and lawlessness that have made the most powerful states take notice. While one can easily find such rhetoric, we see a mixed scorecard. There is more stated desire for effective results in peacebuilding, more capacity for peacekeeping and more unity in mediation. In addition, there has been institutional innovation by governments and international organizations to improve results.

At the same time, there continues to be ineffective international leadership in mobilizing resources to better address civil wars and failing states, particularly in making and consolidating peace in individual cases like Darfur, where leadership has been largely absent. That the situation in Darfur has been allowed to deteriorate over the last four years would have been inconceivable to most observers a decade ago, after Bosnia and Rwanda.

Conclusion

While there is growing evidence from historical and contemporary cases that refugee movements can affect levels of conflict, this relationship remains poorly understood by both conflict and refugee planners. Refugee movements are all too often seen only as a by-product of conflict, with limited attention paid to the various ways they may cause conflict, prolong conflict, or frustrate efforts to resolve conflicts. This chapter has argued that both refugee and security planners should pay much greater attention to the changing dynamics of protracted refugee situations and the links between prolonged exile, conflict and peacebuilding. While protracted refugee situations are largely the result of intractable conflict, and while a resolution of some of the world's most protracted conflicts would contribute greatly to the resolution of many of the world's most protracted refugee situations, it is also important to understand how resolving protracted refugee situations could contribute to the resolution of long-standing conflicts.

To further our understanding of these relationships, greater emphasis should be placed on enhanced diagnosis, prescription and policy for responding to protracted refugee situations. First, it is important to improve our diagnosis of protracted refugee situations and the various ways in which they are related to conflict in countries of origin or in the region. In particular, it is important to develop a more disaggregated understanding of situations of prolonged exile, to help understand why certain situations become politicized and part of broader conflict dynamics. Second, there is a need for better prescription, including a more refined understanding of how solutions for refugees could enhance efforts to resolve conflict. Finally, there are important policy lessons from the field of conflict management that should be more fully incorporated into our policy responses to protracted refugee situations. Paramount among these lessons are those relating to peacebuilding. In this sense, this chapter has argued that responses to both refugees and conflict would benefit from closer dialogue and understanding. Given the range of cases in which situations of prolonged exile are more than simply a by-product of conflict but a factor contributing to its continuation, it is important that security scholars consider the role of refugee movements more carefully and more rigorously. Likewise, refugee scholars could learn important lessons from the conflict literature, especially in areas relating to peacebuilding and the regional dynamics of conflict, which would enable them to advance their understanding of the preconditions for resolving protracted refugee situations.

Notes

1. For several recent exceptions, see Sarah Kenyon Lischer, *Dangerous Sanctuaries: Refugee Camps, Civil War, and the Dilemmas of Humanitarian Aid*, Ithaca: Cornell University Press, 2005; Kelly Greenhill, 'Engineered Migration as a Weapon of War', in Michael Innes, ed., *The Clandestine Politics of Sanctuary After the Cold War*, London: Taylor and Francis, 2008; Stephen John Stedman and Fred Tanner, eds., *Refugee Manipulation: War, Politics, and the Abuse of Human Suffering*, Washington, DC: Brookings Institution, 2003.
2. UN General Assembly Resolution 428V, 14 December 1950.
3. UN General Assembly Resolution 2312 XXII, 14 December 1967.
4. See Gil Loescher and James Milner, *Protracted Refugee Situations: Domestic and International Security Implications*, Adelphi Paper 375, Abingdon: Routledge for the International Institute for Strategic Studies, July 2005.
5. See, for example, Michael Doyle and Nicholas Sambanis, *Making War and Building Peace: United Nations Peace Operations*, Princeton: Princeton University Press, 2006; Virginia Page Fortna, *Peace Time: Cease Fire Agreements and the Durability of Peace*, Princeton: Princeton University Press, 2004; Barbara Walter, *Committing to Peace*, Princeton: Princeton University Press, 2001; James D. Fearon and David D. Laitin, 'Ethnicity, Insurgency, and Civil War', *American Political Science Review* 97, no. 1, February 2003, pp. 75–90; and Paul Collier and Anke Hoeffler, 'Greed and Grievance in Civil Wars', *Oxford Economic Papers* 56, 2003, pp. 563–595.
6. Idean Salehyan and Kristian Skrede Gleditsch, 'Refugees and the Spread of Civil War', *International Organization* 60, Spring 2006, pp. 335–366.
7. Chester Crocker, Fen Osler Hampson and Pamela Aall, eds., *Grasping the Nettle: Analyzing Cases of Intractable Conflict*, Washington, DC: United States Institute of Peace, 2005; and Chester Crocker, Fen Osler Hampson and Pamela Aall, *Taming Intractable*

Conflicts: Mediation in the Hardest Cases, Washington, DC: United States Institute of Peace, 2005.
8. See, for example, Chandra Lekha Sriram and Karin Wermester, eds., *From Promise to Practice: Strengthening UN Capacities for the Prevention of Violent Conflict*, Boulder: Lynne Rienner, 2003; Fen Osler Hampson and David M. Malone, eds., *From Reaction to Conflict Prevention: Opportunities for the UN System*, Boulder: Lynne Rienner, 2002; and Albrecht Schnabel and David Carment, eds., *Conflict Prevention: From Rhetoric to Reality*, Lanham, MD: Lexington Books, 2004.
9. Human Security Centre, *Human Security Report, 2005: War and Peace in the 21st Century*, Oxford: Oxford University Press, 2006.
10. Gil Loescher and James Milner, *Protracted Refugee Situations: Domestic and International Security Implications*, Adelphi Paper 375, Abingdon: Routledge, 2006, p. 34.
11. Stephen John Stedman, 'Conclusions and Policy Recommendations', in Stedman and Tanner, *Refugee Manipulation*, pp. 169–171.
12. Liisa H. Malkki, *Purity and Exile: Violence, Memory, and National Cosmology Among Hutu Refugees in Tanzania*, Chicago: University of Chicago Press, 1995.
13. Stedman, 'Conclusions and Policy Recommendations', pp. 171–172.
14. Howard Adelman, 'Refugee Repatriation', in Stephen John Stedman, Donald Rothchild and Elizabeth Cousens, eds., *Ending Civil Wars: The Implementation of Peace Agreements*, Boulder: Lynne Rienner, 2002, p. 296.
15. Loescher and Milner, *Protracted Refugee Situations*, p. 77.
16. George Downs and Stephen John Stedman, 'Evaluation Issues in Peace Implementation', in Stephen John Stedman, Donald Rothchild and Elizabeth Cousens, eds., *Ending Civil Wars: The Implementation of Peace Agreements*, Boulder: Lynne Rienner, 2002, pp. 49–50.

5

Protracted refugee situations, human rights and civil society

Elizabeth Ferris

This chapter examines protracted refugee situations (PRS) through the lens of human rights, with a particular focus on civil society's engagement with long-term refugee situations. Following a discussion of the linkages between protracted refugee situations and human rights, attention then turns to ways in which human rights actors have responded to these situations, with an emphasis on the roles played by both humanitarian and human rights non-governmental organizations (NGOs). Relationships between human rights/civil society actors and peace and security and development actors are then explored. The chapter concludes with a discussion of the roles these actors could play in implementing comprehensive solutions to PRS.

Protracted refugee situations as a human rights issue

Most fundamentally, people flee their communities and their countries because their human rights have been violated. As a training manual published by the Office of the High Commissioner for Human Rights makes clear:

> In essence, a refugee or IDP [internally displaced person] who reaches a camp is already a person who has suffered a series of serious human rights violations. In many cases, the fact of being obliged to leave one's home itself entails violations of certain rights, such as the right to security of person, and the freedom to choose one's residence. Very often, the factors which led to the displacement

---

Protracted refugee situations: Political, human rights and security implications,
Loescher, Milner, Newman and Troeller (eds),
United Nations University Press, 2008, ISBN 978-92-808-1158-2

– discrimination, armed conflict, other forms of generalized violence, etc. – themselves involve violations of human rights.[1]

A range of human rights abuses lead people to become refugees. 'Such violations can range across the spectrum of economic, social, cultural, civil and political rights. This may include, for example, the denial or blocking of humanitarian access and assistance, or the destruction of agricultural land and the poisoning of wells, or the killing, mutilation and sexual assault of civilians and the forced recruitment of civilians, including children, to serve as soldiers. Vulnerable groups, such as female-headed households, children, members of minority groups and the uprooted – both refugees and internally displaced persons (IDPs) – can typically experience multiple forms of human rights deprivation'.[2]

In other words, refugees are people who have suffered human rights violations and who are in search of protection from further human rights violations.[3] In some cases, they find the protection they need in a country of asylum, but in too many they suffer further human rights violations while in exile.

Rights in theory

There are three basic bodies of international law which apply to refugees: international human rights law, refugee law and international humanitarian law. International human rights law affirms that all human beings have certain basic human rights, regardless of their legal status, including those who are internally displaced and those who have not been recognized as refugees by the host government. Beginning with the Universal Declaration of Human Rights, the international community has negotiated a number of international conventions, including:
- The Convention Against Torture and Other Cruel, Inhuman, or Degrading Treatment or Punishment
- The Convention relating to the Status of Stateless Persons
- The Convention on the Reduction of Statelessness
- The Convention on the Prevention and Punishment of the Crime of Genocide
- The International Covenant on Civil and Political Rights
- The International Covenant on Economic, Social and Cultural Rights
- The International Covenant on the Elimination of All Forms of Racial Discrimination
- The Convention on the Elimination of All Forms of Discrimination Against Women
- The Convention on the Rights of the Child

Some of these international human rights instruments include specific references to refugees. For example, the Convention on the Rights of the Child holds:

> The child shall be registered immediately after birth and shall have the right ... to a name, ... a nationality. ... States Parties shall take appropriate measures to ensure that a child who is seeking refugee status or who is considered a refugee ... shall ... receive appropriate protection and humanitarian assistance in the enjoyment of ... rights.[4]

In addition, there are regional instruments upholding basic human rights, as well as a large body of 'soft' international law which includes declarations, guidelines and principles. While these do not have the same weight as international conventions and law, they do represent a commitment by the international community to uphold basic human rights for all people. For example, the Declaration on the Elimination of Violence Against Women proclaimed by UN General Assembly Resolution in December 1993 states that states should:

> Develop, in a comprehensive way, preventive approaches and all those measures of a local, political, administrative and cultural nature that promote the protection of women against any form of violence, and ensure that the re-victimization of women does not occur because of laws insensitive to gender considerations, enforcement practices or other interventions.[5]

Refugees also enjoy human rights specifically linked to their refugee status, most obviously through the 1951 Convention relating to the Status of Refugees and its 1967 Protocol, including freedom from forcible return (*refoulement*) (Article 33), access to courts of law (16), education (22), public relief and assistance (23), social security (24), freedom of movement (26, 31), and identity papers (27). Regarding employment, refugees are to receive 'the most favourable treatment accorded to nationals of a foreign country ... as regards the right to engage in wage-earning employment ... Restrictive measures imposed on aliens or the employment of aliens for the protection of the national labour market shall not be applied to a refugee who ... has completed three years' residence in the country'.[6] The Convention also obliges the contracting states to facilitate the assimilation and naturalization of refugees as far as possible (Article 34).

Finally, international humanitarian law makes it clear that refugees have identifiable rights. For example, Article 44 of the Geneva Convention Relative to the Protection of Civilian Persons in Time of War states that: 'the Detaining Power shall not treat as enemy aliens exclusively on the basis of their nationality ... of an enemy State, refugees who do not, in fact, enjoy the protection of any government'. Article 73 of the related protocol states that 'persons who, before the beginning of hostilities, were considered as stateless persons or refugees ... shall be protected persons ... in all circumstances and without any adverse distinction'.[7]

Rights in practice

While international law thus provides comprehensive human rights for refugees, the reality for those refugees living in protracted situations is quite different. Many governments confine refugees to designated camps where there is neither freedom of movement nor freedom to choose one's residence. Many governments do not provide access to education or permit refugees to be employed or self-employed. There are still too many cases where children born in refugee camps do not have proper documentation or nationalities. Violence against refugees, particularly refugee women, remains widespread. The economic and social rights of refugees are usually lacking, such as the right to the highest possible standard of health and to an adequate standard of living, including adequate food, shelter and clothing.

For example, some 30 States Parties to the Refugee Convention seek to limit their obligations to allow refugees to pursue wage-earning employment by, for example:
- Requiring permits (Malawi, Sweden) and extended residence (Chile, Cyprus, Jamaica, United Kingdom)
- Subjecting refugees to alien employment quotas (France, Honduras, Madagascar)
- Privileging members of certain other nationalities above refugees (Angola, Brazil, Denmark, Guatemala, Luxembourg, Norway, Portugal, Spain, Sweden, Uganda)
- Categorically denying Article 17's rights or treating them merely as 'recommendations' (Angola, Botswana, Burundi, Ethiopia, Iran, Latvia, Liechtenstein, Mexico, Moldova, Papua New Guinea, Sierra Leone, Zambia, Zimbabwe).[8]

The lack of economic rights for refugees in protracted refugee situations has far-reaching consequences for the refugees. As Gil Loescher and James Milner have written: 'Restrictions on employment and on the right to move beyond the confines of the camps deprive long-staying refugees of the freedom to pursue normal lives and to become productive members of their new societies ... Containing refugees in camps prevents their presence from contributing to regional development and state-building'.[9] It also increases the vulnerability of refugees to other forms of exploitation.

The publication of a report in 2002 entitled 'Sexual Violence and Exploitation: The Experience of Refugee Children in Liberia, Guinea, and Sierra Leone'[10] sent shock waves through the international humanitarian community. The researchers found not only that sexual exploitation was widespread, but that it was perpetrated by aid workers, peacekeepers, and community leaders.[11] They highlighted the twin causes of poverty and abuse of power as the main reasons for sexual exploitation of children. Because the refugees do not have access to the labour market, or to the means to become self-reliant, they are forced to rely on humanitarian assistance, which does not meet their needs. Humanitarian agencies have not been able to provide the necessary assistance to allow people in protracted refugee situations to live in dignity and security.

The pattern of humanitarian assistance has led to overwhelming dependency by the refugee population. The size of the plastic sheet determines the size of the house. The food ration is for 30 days but it is calculated on kilocalories and not quantity. It finishes within 10 days, but there is not enough land to grow food. The non-food items given are not replaced and there are not enough income-generating jobs for the refugees to earn money to buy their own. Education is free but all the other related expenses are left for the parents to provide, such as books, pencils, uniforms and shoes. The parents have no income and the children have to fend for themselves. Girls' bodies are the only currency they have left. At the same time, surrounding the refugee population and controlling so much of their lives is a moneyed elite – UN and NGO workers, peacekeepers, etc. – whose resources are considerably more than what the refugees have. They can afford to exploit this extreme disparity and pay for sex when they want and with whom they want.[12]

While this study had a substantial impact, in part because of its findings that humanitarian workers were perpetrators of sexual abuse of refugee children, the fact is that sexual and gender-based violence is common in situations where the basic material needs of refugees are not met. Violations of economic and social rights thus can lead to violations of civil rights.

Refugee and displaced women are particularly at risk of sexual and gender-based violence due to the psychological trauma and stress of conflict, flight and displacement, disrupted roles within the family and community, dependence, and ignorance of their individual rights enshrined under national and international law.[13]

The right of refugees to education has a solid basis in international and regional human rights, humanitarian and refugee law.[14] Primary education should be free, compulsory and non-discriminatory. States which have ratified the Convention on the Rights of the Child must take active steps to introduce free secondary and higher education. Experts have recognized the importance of education for refugees to bring back a sense of normalcy, to build capacity, to protect and empower women and girls, and to prepare the ground for lasting peace and prosperity. As stated in the UNHCR's 2002 report 'Educating Refugees Around the World': 'To foster this potential through education is the single most effective way to enable war-affected populations to rebuild their lives, to improve overall living standards and to promote long-term peace and economic development'.[15] Moreover, education has a clearly recognized role in protection. Susan Nicolai makes the point that 'parents feel safer if children are in school rather than out. Education lessens the chance that the child will be recruited, exploited or exposed to other risks. In practical terms, education structures can play a more protective role in children's lives' through deterring a cycle of violence.[16]

Around the world, the relationship between education and self-reliance is very clear; education increases people's ability to provide for themselves and their families. For refugees, the situation is sometimes different, as refugees with high levels of education are often prevented from working in their fields and take jobs below their qualifications due to non-recognition of their credentials in the host countries. Nonetheless, refugees' exercise of their right to education remains an important factor in their ability to become economically self-sufficient and to develop their full potential.

The Global Survey on Education in Emergencies found that approximately half of refugee children and adolescents were not in school. Or, to put it another way, about half the global refugee population are unable to exercise their right to education. For those refugees who are able to attend school, the survey found problems with the quality of teaching, as many refugee teachers do not meet the minimum requirements of their governments. Moreover, the majority of those enrolled in school are enrolled in the early primary grades. While girls are almost as likely as boys to be enrolled in pre-primary and grade one, their enrolment decreases steadily after that.[17] Globally, only an estimated one out of ten adolescent refugee girls goes to class.

What are human rights actors doing about these human rights violations in protracted refugee situations?

UNHCR is responsible for providing assistance and protection – and for upholding the human rights of refugees living in camps which they administer. As there is another chapter in this volume specifically analysing UNHCR's activities, this chapter looks at other actors who are, or who may be, involved with protecting the rights of refugees in protracted refugee situations.

The Office of the High Commissioner for Human Rights (OHCHR) is mandated 'to promote and protect the enjoyment and full realization, by all people, of all rights established in the Charter of the United Nations and in international human rights laws and treaties.... The mandate includes preventing human rights violations, securing respect for all human rights, promoting international cooperation to protect human rights, co-ordinating related activities throughout the United Nations, and strengthening and streamlining the United Nations system in the field of human rights. In addition to its mandated responsibilities, the Office leads efforts to integrate a human rights approach within all work carried out by United Nations agencies.'[18] This mandate clearly applies to refugees in protracted situations.

Over the years, the (former) UN Human Rights Commission heard many statements on refugees in protracted refugee situations, and provided an opportunity for hundreds of organizations to provide information on the human rights violations of refugees.

In addition, OHCHR has a number of special procedures to focus attention on particular human rights issues. Presently there are 28 thematic and 13 country special representatives (or rapporteurs). They carry out urgent appeals, country visits, follow-ups to country visits, and, in some cases, normative work. In this regard, the UN Representative of the Secretary-General on the Human Rights of Internally Displaced Persons can be highlighted for developing the Guiding Principles on Internal Displacement. A number of the special procedures are relevant to the rights of refugees in protracted refugee situations, including those on adequate housing, arbitrary detention, the right to education, the right to food, and violence against women, its causes and consequences. Virtually all of the countries with special representatives are ones which have experienced major refugee flows (e.g. Liberia, Democratic Republic of Congo, Sudan and Somalia).

While their mandates are broad enough to include protracted refugee situations, these special procedures – with a few exceptions – have not focused on refugees. The Special Rapporteur on violence against women has highlighted the specific protection needs of refugee women in her reports. In 2002, the Working Group on Arbitrary Detention visited Australia to examine the country's use of detention of asylum seekers and recommended that the Australian government change its laws to be in conformity with international law.[19] However, the Working Group has not yet looked at restrictions on movement in protracted refugee situations. The Special Rapporteur on torture and other cruel, inhuman or degrading treatment or punishment is another possible avenue for increasing the engagement of the human rights system with protracted refugee situations. Nevertheless, when the Special Rapporteur undertook a mission to Nepal (January 2006), he did not look into the situation of Bhutanese refugees in the country.

There are possibilities for human rights actors to become more engaged with protracted refugee situations by using the special procedures.[20] Another option would be for the Human Rights Council to establish a special working group, or special representative, to examine protracted refugee situations. This would have the advantage of highlighting the constellation of human rights abuses which occur in protracted refugee situations and could be a way of pressuring governments to lift some of the restrictions on refugees.

While OHCHR has contributed to a global understanding of human rights abuses in protracted refugee situations, it has not been able to translate this commitment into action on the ground – where refugees live. This is due partly to the lack of OHCHR's capacity in the field and partly to an understanding of its mandate.

In fact, OHCHR defers to UNHCR in refugee settings. As pointed out in its training manual, 'it would not generally be the role of a UN human rights operation to visit a refugee camp managed by the UNHCR to review camp conditions. The UNHCR has the greatest experience and the most appropriate mandate to provide protection to refugees. However, the mandate and expertise of UN human rights operations can often be complementary to an HCR role, provided that there is adequate coordination'.[21] In countries where OHCHR has a field presence, staff participate in the UN or Inter-Agency Standing Committee (IASC) Country Teams and can often provide human rights perspectives on humanitarian programming. Physical protection of refugees is linked with being present where they are, and OHCHR, unlike UNHCR, is not equipped to provide that continual presence in the field. Moreover, there is a close relationship between protection and assistance which UNHCR is well placed to monitor.

In particular, human rights operations in the field may contribute to the search for solutions for refugees in protracted situations. For example, information on the state of human rights in the country of origin and monitoring of returnees can contribute to the viability of repatriation as a solution for refugees. Actions to help identify the alleged perpetrators of human rights violations can lead to justice which, in turn, can make returns more sustainable. The issue of how OHCHR and other human rights actors might more effectively engage in protracted refugee situations will be considered later in this chapter.

Another set of human rights institutions which have not been particularly active on refugee questions are the national human rights institutions (NHRIs). Since the UN General Assembly adopted the so-called 'Paris Principles' in 1993 – minimum standards concerning national human rights institutions – many countries have worked with the United Nations to establish or strengthen such bodies. There are presently over 100 NHRIs whose mandates may include: commenting on existing or draft laws, monitoring domestic human rights situations and acting on complaints or petitions from individuals or groups, advising on compliance with international standards, cooperating with regional and international bodies, and educating and informing the public about human rights.[22]

In his 2002 report, UN Secretary-General Kofi Annan placed priority on the development by the United Nations of the capacity of national institutions. 'Building strong human rights institutions at the country level is what in the long run will ensure that human rights are protected and advanced in a sustained manner. The emplacement or enhancement of a national protection system in each country, reflecting international human rights norms, should therefore be a principal objective of the Organization. These activities are especially important in countries emerging from conflict'.[23] As national-level organizations, NHRIs are well placed to monitor and advise on human rights issues emerging from protracted refugee situations, but in practice only a few NHRIs have taken up the issue. The International Conference of the National Human Rights Institutions, meeting in 2004 in Seoul, South Korea, considered the issue of 'migration in the context of terrorism', and noted that refugees have become more vulnerable in the post-9/11 world. The Asian Centre for Human Rights called on NHRIs to become more active in addressing questions of non-*refoulement*, equal access to humanitarian assistance and long-term refugee situations. The Centre argues that 'the NHRIs have a critical role to play for protection of the refugees living within its geographical jurisdiction, irrespective of whether they are under the care of the UNHCR or the government'.[24] The Australian Human Rights and Equal Opportunity Commission has been very involved in the debate over detention centres for asylum seekers. Some NHRIs have also conducted monitoring missions and become involved in advocacy campaigns on an ad hoc basis. In Jordan, for example, the Commission monitors the situation of refugees and camps and the Commissions of Venezuela and Ecuador have, on a few occasions, addressed the Colombian refugee issue. Having said that, NHRIs have rarely taken the lead in considering the human rights implications of protracted refugee situations, and in fact many are institutionally quite weak and are sometimes viewed as being too close to their governments.

One of the obstacles to both OHCHR and NHRIs becoming more engaged in addressing the human rights of refugees in protracted refugee situations is the dominant role played by UNHCR in refugee camps. When resources are scarce and staff are stretched thin, it is hard for human rights organizations to decide to devote additional resources to human rights violations which are understood to be under the mandate of UNHCR. However, as we have seen, UNHCR has not been able to assure the human rights of refugees living in camps under its jurisdiction. Moreover, none of the human rights actors has done an adequate job in assuring the human rights of refugees in protracted refugee situations. The focus on refugees living in camps is a major shortcoming. As James Milner points out, 'long-staying urban refugee caseloads are not typically included in an understanding of protracted refugee situations, yet tens of thousands live clandestinely in urban areas, avoiding contact with authorities and existing without legal status. There are almost 40,000 Congolese urban refugees in Burundi, over 36,000 Somali urban refugees in Yemen, almost 15,000 Sudanese urban refugees in Egypt, nearly 10,000 Afghan urban refugees in India, and over 5,000 Liberian urban refugees in Côte d'Ivoire, to name only some of the largest caseloads.'[25] The Asian Centre for Human Rights has looked at the situation of Burmese Chin refugees in New Delhi, noting that 'UNHCR has been cutting the number of people who can obtain subsistence allowance as it seeks to promote a disastrous "self-reliance" policy by encouraging participation in vocational training courses with a view to obtain jobs. In the absence of the right to work, the UNHCR's self-reliance policy is nothing but promotion of illegal work'.[26] A similar concern was raised in the NGO Statement on Protracted Refugee Situations at the 2004 meeting of the UNHCR Executive Committee with respect to Bhutanese refugees in Nepal. The statement noted that 'we are concerned about UNHCR's plan to phase out aid for the refugees and promote self-sufficiency when the Government of Nepal has not yet supported or acknowledged the refugees' right to work or engage in other self-reliance activities'.[27]

In their chapter in this volume, Crisp and Slaughter argue that the dominant role played by UNHCR in protracted refugee situations may actually make implementation of durable solutions more difficult. Refugees and host countries may not take steps to address or remedy the situation if UNHCR is 'coping' with the situation, and refugees may be dissuaded from conceiving of self-made solutions to their protracted exile or interim strategies for survival while awaiting a durable solution. Similarly, the dominance of UNHCR, particularly in camp settings, may have dissuaded human rights actors from closer scrutiny of the human rights dimensions of protracted refugee situations.

While there are clear links between the protection of human rights and the protection of refugees, Amnesty International has stated that 'for too long refugee protection issues have remained outside the mainstream of the UN human rights machinery. Part of the reason for this marginalization of refugee issues is the unwillingness of governments to allow international scrutiny of their policies towards refugees. In addition, the fact that refugees are the responsibility of a specific UN agency – the UN High Commissioner for Refugees (UNHCR) – has tended to keep refugee protection issues separate from the UN human rights program'.[28] Given the fact that serious human rights violations in PRS remain despite UNHCR's efforts, it would seem timely for human rights actors to become more engaged with long-term refugee situations.

What about civil society and non-governmental organizations (NGOs)?

The proliferation of NGOs makes generalizations about the NGO community very difficult. Refugee-serving NGOs include small organizations staffed by volunteers and housed in church basements, as well as organizations with annual budgets of US$2 billion per year – about twice the annual budget of UNHCR. Given the trend that high-profile emergencies attract funds, there is considerable competition between NGOs – competition for media coverage, visibility and funds from the general public, national governments and intergovernmental bodies.

At the global level, NGOs have played an important role in raising awareness of the human rights violations of refugees in protracted situations. International human rights organizations such as Amnesty International, Human Rights Watch and Refugees International have provided credible analyses and information about long-standing refugee situations, as visits to their websites demonstrate. The US Committee on Refugees and Immigrants in 2004 launched an initiative against the 'warehousing' of refugees,[29] which further raised awareness of protracted refugee situations. A number of international humanitarian NGOs have also provided invaluable insights into the reality on the ground in specific situations, including Save the Children, Oxfam and the Jesuit Refugee Service. The 2003 Assembly of the International Council of Voluntary Agencies (ICVA) focused on 'forgotten emergencies' and participants shared their frustrations at the inadequate international response to PRS.

NGO statements to meetings of the UNHCR Executive Committee (ExCom) have often focused on protracted refugee situations and have sought to draw attention to those long-standing refugee situations which are not considered emergencies. In its joint statement to the 2004 meeting of UNHCR's ExCom, the NGOs welcomed many aspects of UNHCR's document on 'Protracted Refugee Situations',[30] particularly the focus on the Convention rights of refugees, but also raised several concerns. Specifically, the NGOs noted that the international community had paid insufficient attention to refugees' rights to economic activity and choice of residence, and that refugees' rights should be upheld even as durable solutions are being sought. 'At the same time', the statement notes, 'we oppose the phasing out of assistance to refugees – even in the name of self-reliance – in situations where refugees do not yet enjoy the legal rights to work, engage in professions and enterprises, own property, move about, and choose their residence, or are not receiving adequate education or healthcare'.[31]

NGOs have often raised issues with global bodies that they have observed from their work in the field. Advocacy by NGOs was crucial to pushing UNHCR to pay attention to gender, for example, and particularly the protection needs of refugee women and girls.[32]

There has long been an assumption that there is a division between human rights and humanitarian NGOs, with human rights NGOs more active in denouncing human rights abuses while humanitarian NGOs are more circumspect in their public advocacy. The argument has been that if humanitarian NGOs are too outspoken in denouncing human rights abuses, particularly at the hands of the government, they risk being asked to leave the country. However, at least in recent years, these distinctions have become blurred. Many humanitarian organizations have been active in denouncing human rights abuses and are working increasingly closely with human rights organizations. Moreover, many large international NGOs, as well as national ones, are involved in both provision of humanitarian assistance and advocacy on human rights.

Both national and international NGOs have been active in working with UNHCR and with the various UN human rights mechanisms. They monitor governmental compliance with international treaties and submit notes to UN treaty bodies. For example, four NGOs recently submitted a note to the UN Human Rights Committee in response to a report submitted by Bosnia and Herzegovina, which raised a number of human rights issues related to the return of refugees and internally displaced persons in that country.[33]

The Committee on the Rights of the Child explicitly encourages governments to consult with NGOs in preparing their reports. In its work on education, the Women's Commission for Refugee Women and Children has pages of suggestions for individuals to become more involved in advocacy for the rights of refugees to education, including writing to the Special Representative on education and issuing alternative reports. In Colombia and Mexico, two NGO coalitions have created alternative reporting and monitoring mechanisms on the fulfilment of child rights and submitted them to the Committee on the Rights of the Child.[34]

NGOs are also active in protecting and promoting human rights on the ground. In 1998, staff of Médecins Sans Frontières (Netherlands) systematically catalogued hundreds of cases in Sierra Leone of mutilation of civilians by rebel forces. This information was shared with the UN human rights bodies as well as with the international media, and played an important role in the decision to establish a UN human rights team in Sierra Leone, and in raising awareness about human rights abuses in that country.[35] In 1996, the Norwegian Refugee Council (NRC) established a network of civil rights projects[36] providing free legal aid/information on issues related to refugees, IDPs, returnees and minority groups in Kosovo.[37] In 1995, the International Committee of the Red Cross initiated a project for the teaching of international humanitarian law in schools in seven countries of the Commonwealth of Independent States, to ensure that international humanitarian law was included in official school curricula or programmes.[38]

Many NGOs have adopted a rights-based approach to both their humanitarian and development work. This approach requires a deliberate focus on protection to underpin all aspects of NGO planning. The rights-based approach is 'a conceptual framework for the process of human development that is normatively based on international human rights standards and operationally directed to promoting and protecting human rights'.[39] With a rights-based approach, human rights are the means, the end, the mechanism of evaluation and the central focus of sustainable human development. The rights-based approach starts from the ethical position that all people are entitled to a certain standard in terms of material and spiritual well-being. Within a rights-based approach, people are engaged as claim-holders to rights rather than beneficiaries of charity and aid; they are considered active subjects rather than objects of development. Governments, intergovernmental organizations and NGOs are thus seen as duty-bearers that must ensure that the human rights of claim-holders are met and not abused.[40] Adopting a rights-based approach has meant that humanitarian and development NGOs have become more familiar with international human rights law and have developed closer relations with human rights actors. For some NGOs, it has also meant a fundamental change in their relationship to beneficiaries, who are seen no longer in terms of their 'needs' but rather as individuals and communities seeking to claim their basic human rights.

Rights-based approaches to humanitarian issues and to development offer new possibilities for protecting human rights in protracted refugee situations and for increasing accountability. Over the past decade, the issue of accountability has become a major issue for NGOs, which have sought to develop common standards based on a rights-based approach. Thus, the Sphere Project was launched in 1997 to improve the quality of assistance and enhance the accountability of the humanitarian response through an explicitly rights-based approach. The Humanitarian Charter and Minimum Standards were first published in 2000 and proposed universal standards in fields such as nutrition and water/sanitation. The 2004 edition includes incorporation of cross-cutting issues of gender, children, the elderly, the disabled, HIV/AIDS and protection.[41] In 2005, ALNAP, the Active Learning Network for Accountability and Performance, published 'Protection: An ALNAP Guide for Humanitarian Agencies',[42] which includes helpful, practical advice for organizations working to deliver humanitarian assistance to also include a protection dimension throughout their work, from setting protection objectives to monitoring protection outcomes. The guide notes that humanitarian assistance can be an entry point to protection, that protective assistance can save lives, that information serves a protection role, and that presence and accompaniment are also important in protecting the human rights of beneficiaries.[43]

These rights-based networks and standards contribute to NGO involvement on the ground in areas where refugees' rights are being violated. For example, to respond to the violation of refugees' rights to education, in 2000, the Inter-Agency Network for Education in Emergencies (INEE), made up of UN agencies and NGOs, was established. This network recently launched the global INEE Minimum Standards for Education in Emergencies, Chronic Crises and Early Reconstruction, after extensive consultations with more than 2,250 individuals from around the world.[44] These standards provide a tool for NGOs working in protracted refugee situations and a standard for determining when basic human rights are not being met.

Over the past 20 years, there has been increasing recognition of the important role which NGOs play in providing protection to refugees and of the links between protection and assistance. 'Protecting Refugees: A Field Guide for NGOs',[45] a joint production of UNHCR and NGOs, was an early effort to provide resources to NGOs that often find themselves in situations where their programmes offer protection to refugees. Although the Guide does not make specific reference to protracted refugee situations, many of the suggestions made are applicable to those situations, such as the sections on protecting refugee women, refugee children and adolescents and older refugees.

Relationships between human rights/civil society actors and peace and security and development actors

There are several very positive developments taking place within the international humanitarian community which increase the potential for collaborative efforts to address protracted refugee situations. First of all, there are increasing numbers of joint initiatives between NGOs and UN agencies, as noted in some of the examples mentioned above. Since 1992, the Inter-Agency Standing Committee has provided a forum where UN agencies working in the fields of humanitarian assistance and development can come together with representatives of NGOs, the Red Cross/ Red Crescent movement and other intergovernmental organizations, to talk about humanitarian issues. The Humanitarian Response Review, initiated in 2004 by then-Emergency Relief Coordinator Jan Egeland, called for more coordinated action within the larger humanitarian community.[46] The reform of the humanitarian system includes establishing clusters to promote coordination within specific sectors and cross-cutting issues, strengthening the humanitarian coordinator system and strengthening NGO–UN relations.

NGOs see possibilities within these reform efforts of shifting from a UN-centric paradigm, in which NGOs are viewed either as 'junior partners' or in marginal terms, to a new system which recognizes the important role which NGOs play in humanitarian response. Thus, in July 2006, a meeting between UN and non-UN organizations on enhancing the effectiveness of humanitarian action brought together some 40 directors of UN and non-UN humanitarian organizations to consider how to work together more effectively. Among other things, the meeting began to conceptualize the international humanitarian system as being made up of three equally important poles: the United Nations, NGOs and the Red Cross/Red Crescent movement.[47] Recognizing NGOs and the Red Cross/Red Crescent movement as equal partners offers new possibilities for developing genuinely collaborative initiatives in protracted refugee situations.

The fact that a rights-based approach has been adopted by both UN agencies and NGOs provides a common framework which can be used to strengthen collaboration.[48] In 2006, the IASC Task Force on Human Rights and Humanitarian Action prepared a human rights guidance note for Humanitarian Coordinators.[49] This note suggests the need to work with other human rights partners at the country level, including dedicated UN human rights operations, NHRIs, and local and international NGOs. Humanitarian Coordinators are asked to gather data on human rights and to advocate with the relevant parties for the application of humanitarian principles and human rights law. Although there is often a large gap between guidelines from headquarters and implementation on the ground, it is clear that there is a push for UN organizations working in development and humanitarian response to base their work on principles of human rights.

UN agencies themselves are recognizing the need to work more closely with each other in the area of relief-to-development. For this reason, UNHCR has built partnerships and has become a member of the UN Development Group. 'UNHCR representatives are under instruction to engage with UN Country Teams in the preparation of Common Country Assessments/United Nations Development Assistance Frameworks. Furthermore, over the past two years, UNHCR has forged strategic partnerships with the International Labour Organization and the Food and Agriculture Organization to help develop the productive capacities of refugees'.[50] UNHCR's initiative on 'Strengthening Protection Capacity' has developed innovative ways for donor governments, local governments, UN agencies and civil society to work together to increase protection in the field.

International development organizations, such as the UN Development Programme (UNDP) and the World Bank, have developed programmes in reconstruction, human rights and post-conflict settings which offer further possibilities for collaboration. UNDP's Bureau for Crisis Prevention and Recovery has looked at the relationship between natural disasters and development, including the exacerbating effects of armed conflict.[51] Recognizing the relationship between human rights and development, the UN Development Programme is collecting the experiences of UNDP country offices in supporting national human rights institutions.[52]

As of 2004, about one-third of the World Bank's Post-Conflict Fund (PCF) funding was disbursed for programmes to help IDPs and refugees. A recent review of these grants begins by noting that 'refugees and IDPs are not usually the objects of development programs; rather, they are the focus of relief and humanitarian programs. Regular World Bank projects address their needs by including them in the category of "vulnerable" groups'.[53] The review of these 17 PCF grants found that more work needs to be carried out in political and security assessments and on gender relations, and that the programmes generally perform better when they take a rights-based approach.

While there is increasing recognition of the need to strengthen collaboration within the United Nations and within the broader international community, there are serious obstacles to greater coordination. The fact remains that UN agencies and NGOs alike are eager to protect their 'territories' and turf battles are, unfortunately, common. As noted above, one of the primary obstacles to greater coordination at the field level is increasing competition for funds: competition between donors, between different UN agencies, between UN agencies and NGOs, and between different NGOs. Moreover, in many countries there is a sense of 'coordination overload', where staff feel that coordination meetings take time away from their programmatic work.[54]

Towards comprehensive solutions

The development of comprehensive solutions requires political will and commitment from governments, intergovernmental actors and civil society. It is important to note that there is always some tension between *de jure* and *de facto* durable solutions – as refugees and NGOs, perhaps more than governments, have long recognized. Refugees may become integrated into local communities where this is possible – even when this has not been formally adopted as a durable solution. Similarly, when conditions permit, refugees return home on their own, usually not waiting for the implementation of formal repatriation programmes.

Human rights actors and civil society, including both human rights and humanitarian NGOs, can support comprehensive solutions in a number of ways:

1. They can draw attention to the human rights violations occurring in protracted refugee situations and thus mobilize support for the development of comprehensive solutions. They are particularly well placed to draw attention to long-standing refugee situations which would not otherwise receive much media attention.
2. They can advocate for the implementation of relevant national, regional and international law in decisions about comprehensive solutions, for example by insisting on the voluntariness of refugees' decisions to return or by advocating for compensation provisions for returnees in peace agreements.
3. Based on their presence on the ground, they can provide critical perspectives on the development of comprehensive solutions, by identifying factors which need to be taken into consideration and providing a 'reality check' of what will be needed to successfully implement the plan.
4. They can provide information to refugee communities to support solutions, e.g. information about the country of origin to support repatriation, information about job possibilities in host countries and information about resettlement countries.
5. They can monitor implementation of the plan and alert the international community to problems which emerge. For example, they can

accompany and monitor returning refugees and identify problems with local integration which develop.
6. They can design their assistance programmes to support comprehensive solutions. For example, they can tailor vocational training programmes to accord with the job market in host countries, countries of origin and resettlement countries. In host countries, by providing services to the local communities as well as to refugee populations, they can help minimize the risk of local resentment at perceived special treatment of refugees. They can provide support to refugees resettled in third countries through language training, social assistance and job placement.

NGOs can play a role of critical engagement with proposed comprehensive solutions and can implement specific programmes which support those solutions. NGOs played a significant role in both the development and implementation of the three major, past comprehensive solutions: the Comprehensive Plan of Action for Indo-Chinese Refugees, the International Conference on Central American Refugees, and the Conference on the Commonwealth of Independent States. NGOs mobilized support for the plans at the political level, interpreted the plans to refugees with whom they worked closely, and monitored the implementation of the plans, with a particular focus on human rights violations.

One of the main obstacles to greater engagement by NGOs in comprehensive solutions is their fierce independence and reluctance to be perceived as coming under a broader UN umbrella. Thus, the question of 'integrated missions' has provoked heated debate within the NGO community, with many NGOs fearing that the dominance of political issues in such missions limits the space for humanitarian action. A decision may be made, for example, in an integrated mission to target a particular area for assistance to support peace initiatives. While this may be beneficial to the country as a whole, it is also in violation of the Code of Conduct developed by NGOs and the Red Cross/Red Crescent movement, which states that assistance will be given on the basis of need alone. In order to assure the widest possible 'buy-in' of NGOs and human rights actors, these concerns will have to be addressed; in particular, the independence of NGOs needs to be recognized.

Another obstacle to greater engagement by the NGO community in comprehensive solutions concerns the transition from relief to development. Although many national and international NGOs work with both relief and development, it is easier for many international NGOs to mobilize resources for refugee assistance than for long-term development (and obviously it is easier to mobilize resources in the initial phases of an emergency than in protracted refugee situations). In particular, national NGOs should play a more active role in development than international NGOs, but may lack the capacity for doing so. Although international NGOs are often involved in building the capacity of national NGOs, there is also an element of competition between national and international NGOs.

The IASC Working Group suggested four types of engagement between UN and non-UN humanitarian organizations at the field level, which can be adapted to the issue of NGO involvement in comprehensive solutions to PRS:

Level 1: Do no harm (to each other's programmes)
This minimal level of coordination essentially commits all parties working on comprehensive solutions to share information with each other on their plans and programmes.

Level 2: Technical coordination
This level involves agreement on technical standards in such areas as health, water, sanitation and nutrition.

Level 3: Operational coordination
This involves agreeing on common objectives in specific areas of operation, such as timing and modalities of return programmes, identification of land to be made available to refugees remaining in host countries, and planning of resettlement programmes. This involves the sharing of sensitive information and analyses, and a commitment to adapt one's programmes to support common operational objectives.

Level 4: Strategic coordination
This implies joint analysis, needs assessment, priority setting and resource allocation in support of comprehensive solutions.[55] This level implies not only a commitment of field staff to work in a coordinated fashion, but also the commitment of the headquarters of both the UN agencies and NGOs.

While there are many examples of successful coordination at the first three levels (as well as many examples where even minimal coordination has not been achieved), the fourth level of strategic coordination in humanitarian response remains an aspiration. It would, of course, be even more difficult to have a system of strategic coordination involving a wider range of actors, e.g. peacekeeping forces, World Bank and UN development agencies. And yet, such coordination could make comprehensive solutions both more likely and more durable. It is suggested here that there are no short cuts to increasing coordination among disparate actors – even when they are committed to achieving comprehensive solutions. Not only is consultation required, but individuals and organizations need to get to know each other better in order to understand the different pressures which organizations face, to appreciate the expertise and resources which they bring and to develop the necessary trust to work together. The different mandates, experiences and 'culture' of organizations are a tremendous asset in the search for comprehensive solutions, but, in order for this asset to be realized, people and organizations need to understand and trust each other. In short, this requires more meetings for individuals whose agendas are already very full. In this respect, implementation of the cluster approach – by mandating more inter-agency meetings and working groups – may create opportunities for joint action, which can build the necessary understanding and trust between different agencies to make increased coordination possible.

The task of developing and implementing comprehensive solutions to protracted refugee situations will require the contributions of human rights actors and civil society. While there are some hopeful signs of increased willingness on the part of different kinds of actors to work together, much more commitment to collaborative action is needed. The UN's human rights machinery could do much more to highlight the human rights dimensions of protracted refugee situations, including through the special procedures, and to contribute to the development of solutions. International human rights NGOs could develop an advocacy strategy with OHCHR to press for more attention to protracted refugee situations. Similarly, national NGOs could press for national human rights institutions to play a more assertive role vis-à-vis protracted refugee situations in their countries, including monitoring implementation of solutions. Both at the level of UN/IASC Country Teams and at the global level, UN agencies and NGO/Red Cross/Red Crescent staff could work together to develop programmes which support comprehensive solutions in their area of operations. This would require increased consultation between actors in accord with a common framework. More painfully, it would require a willingness by all actors to give up some of their tenaciously defended independence of action. Given the suffering and human rights abuses taking place in protracted refugee situations, this does not seem to be an unreasonable trade-off.

Notes

1. OHCHR, 'Training Manual on Human Rights Monitoring', Professional Training Series, no. 7, Geneva: OHCHR, 2004, p. 168.
2. Inter-Agency Standing Committee, 'Human Rights Guidance Note for Humanitarian Coordinators', IASC Task Force on Human Rights and Humanitarian Action, Geneva, June 2006, p. 2.

3. Although this chapter focuses on protracted refugee situations, it is important to note that those who are internally displaced also often experience serious human rights violations, exacerbated by the fact that they do not enjoy the rights provided under the 1951 Refugee Convention.
4. Convention on the Rights of the Child, Articles 7 and 22, excerpts from 'The People's Movement for Human Rights Education', http://www.pdhre.org/rights/refugees.html.
5. Article 4 (f).
6. Convention relating to the Status of Refugees, 1951, Article 17.
7. Protocol Additional to the Geneva Conventions of 12 August 1949, and Relating to the Protection of Victims of International Armed Conflicts, Article 73.
8. UNHCR, Declaration under Section B of Article 1 of the Convention (as of 1 October 2003), cited by Merrill Smith, 'Warehousing Refugees: A Denial of Rights, a Waste of Humanity', *World Refugee Survey 2004*, Washington, DC: US Committee for Refugees and Immigrants, p. 50.
9. Gil Loescher and James Milner, 'Protracted Refugee Situation in Thailand: Towards Solutions', presentation to the Foreign Correspondents' Club of Thailand, 1 February 2006, pp. 4–5.
10. UNHCR and Save the Children UK, 'Sexual Violence and Exploitation: The Experience of Refugee Children in Liberia, Guinea, and Sierra Leone', report of an assessment mission carried out 22 October–30 November 2001, http://www.reliefweb.int/rw/rwb.nsf/AllDocsByUNID/6010f9ed3c651c93c1256b6d00560fca.
11. Security and military forces, including UN peacekeepers, were also identified as being involved in sexual exploitation of children, as were local businessmen.
12. UNHCR and Save the Children UK, 'Sexual Violence and Exploitation', p. 56.
13. UNHCR, *Sexual and Gender-Based Violence against Refugees, Returnees and Internally Displaced Persons: Guidelines for Prevention and Response*, Geneva: UNHCR, 2003.
14. See Women's Commission for Refugee Women and Children, *Right to Education During Displacement: A Resource for Organizations Working with Refugees and Internally Displaced Persons*, New York: Women's Commission, 2006, pp. 9–10, for a comprehensive listing of applicable human rights instruments.
15. UNHCR, 'Educating Refugees Around the World', 2002, http://www.unhcr.org/cgi-bin/texis/vtx/partners/opendoc.pdf?tbl=PARTNERS&id=3fcb52bf1.
16. Susan Nicolai, 'Education That Protects', *Forced Migration Review*, no. 22, pp. 11–12.
17. Women's Commission for Refugee Women and Children, *Global Survey on Education in Emergencies*, New York: Women's Commission, 2004, pp. 9–10, and Lori Heninger, 'Who Is Doing What and Where', *Forced Migration Review*, no. 22, p. 7. Note that the survey highlights that the figures are worse for internally displaced people.
18. http://www.ohchr.org/EN/AboutUS/Pages/Mandate.aspx.
19. United Nations, Report of the Working Group on Arbitrary Detention, Visit to Australia, 24 October 2002, UN Doc. E/CN.4/2003/8/Add.2, http://www.unhchr.ch/Huridocda/Huridoca.nsf/0/6035497b015966fec1256cc200551f19?Opendocument.
20. For example, see Amnesty International, *The UN and Refugees' Human Rights: A Manual on How UN Human Rights Mechanisms Can Protect the Rights of Refugees*, London: Amnesty International, 1997.
21. OHCHR, 'Training Manual', 2004, p. 173.
22. For an effort to evaluate the effectiveness of NHRIs, see International Council on Human Rights Policy, *National Human Rights Institutions: Impact Assessment Indicators*, Draft Report for Consultation, Geneva: ICHRP, 2005.

23. United Nations, *Strengthening of the United Nations: An Agenda for Further Change. Report of the Secretary-General*, 9 September 2002, UN Doc. A/57/387.
24. Asian Centre for Human Rights Review, 8 September 2004, http://www.achrweb.org/Review/2004/37-04.htm.
25. James Milner, 'Protracted Refugee Situations: Human Rights, Political Implications and the Search for Practical Solutions', paper presented at the Canadian Council for Refugees, Fall Consultations, London, Ontario, 17–19 November 2005.
26. Asian Centre for Human Rights, Review, 8 September 2004. Also note that UNHCR's policy on urban refugees has been in the process of revision for several years.
27. NGO Statement on Protracted Refugee Situations, presented to the Standing Committee of the Executive Committee of the High Commissioner's Programme, 29 June–1 July 2004, p. 2. Available on the website of the International Council of Voluntary Agencies: http://www.icva.ch/.
28. Amnesty International, *The UN and Refugees' Human Rights*, 1997, p. 4.
29. Merrill Smith, 'Warehousing Refugees', pp. 38–56.
30. UNHCR, Executive Committee of the High Commissioner's Programme, Standing Committee, 'Protracted Refugee Situations', EC/54/SC/CRP.14.
31. NGO Statement, op. cit., p. 2.
32. For an overview of the role of NGOs in gender issues, see Alice Edwards, 'Overview of International Standards and Policy on Gender Violence and Refugees: Progress, Gaps and Continuing Challenges for NGO Advocacy and Campaigning', paper presented to the Canadian Council for Refugees, June 2006. The PDF can be accessed here: http://www.amnesty.org/en/library/info/POL33/004/2006.
33. International Committee for Human Rights, International Displacement Monitoring Centre, Minority Rights Group, Benjamin N. Cardozo School of Law Human Rights and Genocide Clinic, submission to the UN Human Rights Committee, 2006.
34. Women's Commission for Refugee Women and Children, *Right to Education During Displacement*, p. 27.
35. Inter-Agency Standing Committee, *Growing the Sheltering Tree: Protecting Rights through Humanitarian Action*, New York: UNICEF/IASC, 2002, p. 134.
36. NRC, ICLA Information Counseling and Legal Assistance, http://www.flyktninghjelpen.no/arch/_img/9069513.pdf.
37. IASC, *Growing the Sheltering Tree*, p. 37.
38. Ibid., p. 68.
39. Office of the United Nations High Commissioner for Human Rights, *Frequently Asked Questions on a Human Rights-Based Approach to Development Cooperation*, New York and Geneva: United Nations, 2006, p. 15. This publication also contains an extensive bibliography of resources on rights-based approaches.
40. See, for example, 'Facts and Issues: A Rights-Based Approach to Development', *Women's Rights and Economic Change*, no. 1, August 2003.
41. For further information on Sphere, including the handbook, see http://www.sphereproject.org.
42. ALNAP, *Protection: An ALNAP Guide for Humanitarian Agencies*, London: Overseas Development Institute, 2005.
43. Ibid., see especially section 7, pp. 79–98.
44. See http://www.ineesite.org, p. 17.
45. UNHCR, *Protecting Refugees: A Field Guide for NGOs*, Geneva: UNHCR, 1999.
46. UNOCHA, *Humanitarian Response Review*, Commissioned by the United Nations Emergency Relief Coordinator and Under-Secretary-General for Humanitarian Affairs, New York and Geneva: OCHA, 2005.

47. Jan Egeland and Elizabeth Ferris, 'Chairs' Summary', Enhancing the Effectiveness of Humanitarian Action: A Dialogue between UN and Non-UN Humanitarian Organizations, 12–13 July 2006.
48. See, for example, 'The Human Rights Based Approach to Development Cooperation – Towards a Common Understanding Among UN Agencies', adopted at Stamford, USA, May 2003.
49. IASC, 'Human Rights Guidance Note', 2006.
50. UNHCR, 'Protracted Refugee Situations', 2004, p. 6.
51. UNDP, Bureau for Crisis Prevention and Recovery, *Reducing Disaster Risk: A Challenge for Development*, New York: UNDP, 2004.
52. Danish Institute for Human Rights, 'Supporting National Human Rights Institutions – a Strategic Niche for UNDP?', Mission to Baku, Azerbaijan, April 2004.
53. Swarna Rajagopalan, 'Within and Beyond Borders: An Independent Review of Post-Conflict Fund Support to Refugees and the Internally Displaced', Social Development Papers: Conflict Prevention & Reconstruction, Paper no. 17, World Bank, October 2004, p. 1.
54. IASC Working Group, 'Background Paper 2: Enhancing UN/Non-UN Engagement at Field Level', prepared for the meeting on 'Enhancing the Effectiveness of Humanitarian Action: A Dialogue between UN and Non-UN Humanitarian Organizations', 12–13 July 2006.
55. Ibid., p. 4.

6

Development actors and protracted refugee situations: Progress, challenges, opportunities

Mark Mattner

Conflict prevention and mitigation are crucial elements of poverty reduction strategies. There is clear evidence that many of the world's poorest countries are trapped in situations where poverty causes conflict and conflict causes poverty. In the past 15 years, 80% of the world's 20 poorest countries have suffered a major war. On average, countries coming out of war face a 44% chance of relapsing in the first five years of peace. Even with rapid progress after peace, it can take a generation or more just to return to pre-war living standards. Development itself has been shown to have a significant impact on the likelihood of conflict, as well as its duration.

This linkage is well recognized.[1] At the level of policy, it is clear that the distinction between relief and development is largely artificial. There is no continuum between the two phases, but rather both processes unravel at the same time. There is also not necessarily a country-wide pattern. While some areas might be affected by war, other regions of a country might enjoy healthy economic growth. In terms of operations, development actors have deepened the scope of their involvement in conflict-affected countries. Nobody held their breath, for example, in the mid-1990s for the World Bank to be among the early supporters of reconstruction in post-Dayton Bosnia. This reflected its institutional focus on long-term economic development as well as its need to interact with a clear government counterpart. By the beginning of the crisis in Timor-Leste, by contrast, the Bank had adjusted its internal operations procedures to enable it to be among the first actors on the scene.[2]

Protracted refugee situations: Political, human rights and security implications,
Loescher, Milner, Newman and Troeller (eds),
United Nations University Press, 2008, ISBN 978-92-808-1158-2

Responses to protracted refugee situations, however, still tend to focus primarily on humanitarian assistance.[3] In practice, protracted refugee situations are often seen as aberrations of development progress and are largely ignored by development actors. This is despite the fact that they can make a positive contribution through sustained engagement with the socioeconomic roots of the crises, which are at the heart of protracted refugee situations. In situations where violent conflict has come to an end, furthermore, they can assist the sustainable reintegration of returnees by bringing to bear the full range of their portfolios.

This chapter revolves primarily around the reception of 'relief-to-development' approaches among development actors, and the headway those actors have made in addressing the socioeconomic, security and political dimensions of conflict. The chapter's main argument is that the role of development actors in conflict situations goes far beyond forging operational partnerships with relief actors. While efforts to strengthen linkages between relief and development support should be further strengthened, such partnerships offer only a starting point in efforts to bring development assistance to bear in achieving genuinely durable solutions. The role of development actors encompasses supporting overall efforts at conflict prevention, mitigation and post-conflict reconstruction. With reference to protracted refugee situations, this translates into ensuring that the specific needs of refugees are met and that they have access to the range of resources made available to local populations across development portfolios.

This chapter will offer a review of approaches to development assistance in transition environments. It will then discuss the involvement of development actors in security, human rights and peace support activities, as well as discuss development operations in conflict environments more generally. The chapter will conclude by considering institutional barriers that impede effective interventions, and will propose a set of suggestions.

Protracted refugee situations: Development challenge, opportunities and responses

Refugee presence can represent both a challenge and an opportunity for local communities. It is, however, contingent on the characteristics of specific situations. Most analytical case studies agree on this point and do not attempt to draw more general lessons. Development actors have sought to strengthen their approaches to development aid for refugees. To some extent, this has represented an effort to overcome the relief-to-development 'gap' and to increase inter-agency cooperation in transition situations. A key distinction is between host countries and countries of origin, which both experience specific sets of challenges and opportunities with respect to development aid to refugees.

With respect to host countries, a 2003 UNHCR Executive Committee report on Tanzania, Pakistan, Côte d'Ivoire and Sudan reflects the ambivalent and case-specific socioeconomic significance of refugee presence. According to UNHCR, the most serious impact of refugee presence pertains to the degradation of environmental resources and the insecurity associated with competition over socioeconomic benefits. In Pakistan, for example, refugees were said to have complicated the country's efforts at poverty alleviation. At the same time, UNHCR cites evidence that refugees in Tanzania have had a positive impact on local communities through increased overall levels of government and bilateral donor support, increased demands for certain goods and services, and expanded employment opportunities. The presence of Liberian refugees in Côte d'Ivoire in 2002 was found to have both acted as an impetus for increased agricultural productivity while increasing pressure on scarce land. Similarly, in eastern Sudan in 2002, UNHCR studies found that pressure on land had increased, partially because of the protracted nature of settlements, while there had been some improvements in the education and health sectors for the local population.[4]

There is ample further evidence on both adverse and positive effects of refugee presence on development in host areas. Robert Chambers, for example, shows that poor households in hosting areas are easily overlooked in the design of assistance approaches. They do not possess the assets or social safety nets to compensate in case of increased competition for scarce resources. Interventions should thus be designed to benefit both refugees and poor households, in a manner they can all access.[5] At the same time, local communities and host countries can benefit from the presence of refugees. This is not only rhetoric applied by agencies in search of funding, but also borne out by a great number of case studies. There tends to be an overall influx of relief resources. Karen Jacobsen, furthermore, points out that states are compelled by refugee populations to expand their effective presence into rural areas where refugees are located, thereby achieving state-building objectives which carry benefits that are potentially beyond the immediate impact of refugee situations.[6]

The livelihood strategies of refugees cannot be evaluated independently of the local context. Refugees bring with them skills, assets and networks, and acquire new ones as refugees.[7] In this context, common understandings of 'dependency' of refugees might be deeply flawed and more efforts should be made to understand the socioeconomic motivations behind the livelihood choices made by refugees. Angolan refugees in Zambia in the late 1990s, for example, had developed local livelihood systems and were not necessarily planning to return to Angola during the Lusaka peace process in 1996/1997. The conclusion would be to adopt a less stringent view on the differences between refugees and their host, conceptualizing assistance approaches through the lens of 'migration'.[8]

Determining the socioeconomic impact of refugee presence, furthermore, is a deeply political process. In their political interactions and in attempting to obtain additional resources from donor governments, host country governments on the whole tend to subscribe to the view that the presence of refugees on their soil represents a significant additional burden they are not sufficiently equipped to deal with by themselves. There have been indications of concomitant erosion in the willingness of host countries to provide refuge. Conversely, developed donor nations have expressed their hope that increased refugee assistance *in situ* will prevent large-scale movements of people to the North. Even the basic numbers are controversial, as similar dynamics are at work with respect to determining refugee statistics more generally.[9]

In this vein, approaches to supporting refugees need to be mindful of their economic impact on host communities. UNHCR has increased its focus on livelihood approaches to support refugees in host countries. This approach aims to equip refugees with the assets, skills and capabilities to make their own living. In the long term, conceptually, this is also intended to prepare them for the return to their places of origin, thereby avoiding the harmful effects of long-term dependency on the provision of relief aid in care and maintenance programmes.[10] The operational arrangements made to implement this strategy will be discussed below.

In countries of origin, the policy approaches to promoting refugee repatriation after the cessation of violent conflict are relatively less contested. This can be attributed in part to the modern state system, which demands and prescribes clear identities based on citizenship status, but leaves little room for recognizing and harnessing the day-to-day livelihood responses made at the local level by refugees and their hosts. Donors and most host countries also continue to prefer reintegration as a durable solution and tend to be eager to avoid setting a precedent for refugee settlement in host countries. In this context, refugee assistance is an established aspect of most post-conflict reconstruction packages. This is not to say, of course, that the return of refugees itself is not problematic and does not create a potential burden on the receiving economy. In fact, in terms of reintegration specifically, the mechanisms are much the same as in host countries. The distinction, however, is that return represents the most popular solution and governments in the countries of origin have little power to prevent it. In the needs assessments that typically precede these interventions, refugee assistance is a well-established focus area. It is a key part in post-conflict reconstruction packages with expected benefits in terms of stability and peacebuilding.[11]

In responding to refugee situations, including protracted ones, in both countries of origin and host countries, actors tend to be divided into humanitarian or development actors. This is despite the fact that, in conceptual terms, the distinction between relief and development activities has been seriously questioned. In particular, questions have been raised about the nature of the continuum this approach implies, and which forms the basis for distinguishing between humanitarian and development activities. In essence, relief and development can be seen as two separate sets of objectives which do not necessarily build on each other in a sequential manner. While relief is concerned with survival, development describes a set of improved socioeconomic outcomes which are sustainable and appropriate to the local context. Furthermore, relief tends to be provided without political conditions in response to short-term emergencies. While it often goes beyond the short term, such as in protracted refugee situations, it differs from development in that it does not tend to aim at state-building objectives.[12]

At present, the most immediate and visible role played by development actors in protracted refugee situations is in direct operational cooperation with humanitarian agencies. This aims to support post-conflict transition in countries of origin or development approaches for refugees in the host countries. To facilitate this cooperation, development actors have devised a variety of tools and procedures. The Post-Conflict Fund (PCF) of the World Bank, for example, has been set up to provide a flexible mechanism for the Bank to respond relatively swiftly to situations in which it would usually not get involved. In this vein, the PCF has financed a number of refugee-related projects.[13] While these projects by themselves have tended to receive positive reviews, it appears that their success has not led to a strengthening in interest in refugee issues within the World Bank. On the whole, the Bank has relied primarily on its status as a relatively disengaged funding agency, with UN agencies, mostly UNDP, providing the direct implementing expertise. Agencies such as the UNDP, on the other hand, have become significant players in implementing transition assistance, in particular through its Bureau for Crisis Prevention and Recovery (BCPR). In this case as well, however, it should be noted that there are internal divisions within UNDP between the staff and bureaux dealing with transition situations and others working on more mainstream development projects. This is despite the fact that both frequently operate in parallel in the same countries.

Development actors maintain large aid portfolios in many refugee-hosting countries and in some countries of origin. These range across many different sectors and types of activities. While working directly on refugee issues does not usually form part of 'mainstream' development business, development actors have begun to pay attention to conflict, including refugee flows and displacement, in their policy planning. This is a highly relevant development because it offers the opportunity to harness development resources at the macro level towards alleviating some of the factors which either have caused refugee flows or have contributed to freezing them into protracted states. There are now a variety of relevant operational and analytical tools used by most development actors. Almost every agency, for example, has developed approaches to conflict analysis, seeking to strengthen the sensitivity of aid portfolios to contextual factors on the ground. While this has been standard practice for humanitarian agencies for quite some time, particularly through 'do no harm frameworks', development agencies still have some way to go in completely mainstreaming these efforts.[14]

There have been a number of specific initiatives to help mainstream a focus on mitigating and preventing conflict in development operations.[15] The most noteworthy initiatives have been at the macro level. Efforts have also been made to strengthen the methodology used in, and to understand and mitigate the shortcomings of, Poverty Reduction Strategy Papers (PRSPs) in conflict countries. For both multilateral and bilateral donors, the participatory exercises which inform the PRSP drafting process now represent the most crucial policymaking process. Contingent on these strategies are access to debt relief and further aid flows. Mainstreaming conflict in PRSPs is therefore particularly noteworthy and represents an attempt at tackling drivers of conflict at the macro level.[16] A particular problem faced by conflict-affected countries is their need for targeted resources despite their frequent inability to achieve the prerequisite rating in the relevant performance-based allocation mechanisms. To address this situation, the World Bank has introduced special preferential allocations through the International Development Association (IDA) for countries emerging from conflict. Under this arrangement, countries deemed to be in post-conflict recovery are measured by a separate set of indicators and receive supplementary allocation for a maximum period of three years.[17] While this has not solved the problems related to performance-based resource allocation altogether, the policy shows that adjustments can be made if all stakeholders work together.

A related macro-level issue which has a significant bearing on refugee situations is the increasing interest in governance work in fragile states. Common characteristics of fragile states are not only their weak institutional structures but also the existence of ongoing or recent conflict and of protracted displacement. Evidence suggests that state weakness is correlated with violent conflict; this is a key element of preventing future conflicts. That development is particularly promising given development actors' poor record on conflict prevention, as opposed to reactive reconstruction. The World Bank has made available a US$25 million Fragile States Trust Fund dedicated to its reengagement in fragile states. It also finances analytical work on service delivery and state capacity strengthening in fragile states. Other donors have similar initiatives. This represents a very promising development in that it merges a predominantly mainstream work area of 'governance' with a key security concern. A key donor coordination body, the OECD's Development Assistance Committee, has been highly active in this regard.

Furthermore, community-based projects at the local level can be of particular relevance for the socioeconomic integration of refugees in their host communities and in their return destinations. Efforts are currently under way to strengthen the methodology of these projects to improve available targeting techniques to channel resources to vulnerable groups in a manner that strengthens community bonds. This can include internally displaced persons (IDPs), refugees and other vulnerable groups such as former combatants.[18]

Development activities and humanitarian, peace and security actors

As has been noted above, responses to protracted refugee situations tend to focus on humanitarian assistance.[19] This is particularly true in the host country, where the perception of the long-term perspective of development actors tends to preclude their engagement. This reluctance of development actors to become involved is partially predicated on a perception of the 'relief-to-development gap', in which the long-term objectives of development programmes are conceived as sequential to the short-term objectives of humanitarian relief activities.[20] Given that development actors are focused on broader work in conflict-affected countries, we should look at partnerships across relevant sectors. Protracted refugee situations are distinctive in this respect, because they often unfold and persist in parallel to mainstream development cooperation.

There have been numerous attempts over time to strengthen the operational collaboration between humanitarian and development agencies. Building on the Brookings Process of the late 1990s, UNHCR adopted a new *Framework for Durable Solutions* in 2003. This policy framework was designed to tighten the cooperation between relief and development actors by including long-term aspects in planning at an early stage in the process. It thus envisages close collaboration between UNHCR, UNDP and the World Bank, as well as a host of other agencies according to their comparative advantages (ILO, UNICEF, etc.). The package of specific concepts comprised an area development approach (Development Assistance for Refugees, DAR), a framework for local integration (Development through Local Integration, DLI), and an approach to reintegration in the country of origin (Repatriation, Reintegration, Rehabilitation and Reconstruction – the so-called '4Rs').[21] The Framework has been piloted in a number of programmes. In terms of the 4Rs, for example, development and humanitarian organizations have collaborated actively in refugee reintegration in Sierra Leone and Sri Lanka as well as, to some extent, in Afghanistan.[22] With respect to DLI, the Zambia Initiative envisages the participation of development actors in addressing the needs of the local population in refugee-hosting areas in Zambia. It has been supported by a number of bilateral development donors and has been fully integrated in the Zambian government's National Development Plan.[23]

Under Ruud Lubbers, High Commissioner from 2001 to 2005, UNHCR followed up this policy framework by developing its Convention Plus approach. By strengthening the notion of burden-sharing within the framework of the Refugee Convention, it sought to increase the involvement of donor countries in addressing the challenges posed by the presence of refugees in developing countries. The main strategic thrust of this framework, therefore, was less operational but aimed at ensuring longer-term stability of financial support for refugee assistance in host countries. It has been noted above that determining the actual number of refugees and the socioeconomic impact of their presence is deeply political. The Convention Plus approach should be seen in this light. By conceptually linking burden-sharing assistance to state obligations under the Refugee Convention, it has been controversial and has to date not been formally adopted.

For development organizations, however, these initiatives do not tend to represent 'mainstream' business. The World Bank, for example, is only very rarely involved in DAR/DLI/4Rs initiatives. This is primarily owing to its larger country-level projects and the perceived politicization of such initiatives. When it is involved, largely through specific trust funds or pilot initiatives as noted above, the Bank acts primarily as a funding agency and less as an involved operational partner. Similarly for UNDP, the Bureau of Crisis Prevention and Recovery represents a relatively specialized entity within the organization, and cannot be held to represent the mainstream activities of the agency at large. In this context, it should be noted that refugee-related activities and those aimed at other sectors often unfold side by side. Large aid portfolios often do not take into account the existence of issues related to conflict because they tend to be perceived as marginal to the bigger picture.

The key to strengthening the cooperation of development actors with other actors in protracted refugee situations, therefore, is to clarify the fact that security challenges represent mainstream development issues and have a potential bearing on the success of development portfolios. There are a number of ongoing initiatives where this link has been made, and can potentially be capitalized upon for further collaboration and coordination.

Disarmament, Demobilization and Reintegration (DDR) projects have become a significant budget line for development actors, including UNDP and the World Bank. The rationale of such projects is to remove one driver of conflict by reintegrating former combatants. The Multi-Country Demobilization and Reintegration Program for the Great Lakes (MDRP) alone is a US$500 million multi-donor project managed by the World Bank. It covers the Great Lakes countries as well as Angola and the Central African Republic, attempting to provide the framework conditions for peace and stability in a coordinated manner. In practice, however, DDR programmes tend to experience similar challenges as other refugee-related operations in moving beyond short-term approaches to assistance. This can be seen in the MDRP, where few participating governments have developed the national frameworks for DDR, which would enable a more effective reintegration of combatants toward the socioeconomic reintegration stage of the process. In addition, some serious questions have been raised on the validity of the DDR process, outlining doubts about the extent to which reintegration of former combatants is in fact a useful approach to peacebuilding. Addressing the needs of former combatants through specific programmes may in fact be counterproductive and serve to stigmatize rather than reintegrate them in the long run. On the funding side, cooperation and collaboration between different donors have not always been smooth. In addition to the political problems often encountered by 'technical' development agencies seeking to work in the security domain, which tends to be perceived as more political, some development agencies face problems regarding their mandate. The World Bank, for example, is restricted in its use of military infrastructure, such as transport capacity within DDR programmes.

A further area of overlap between development and security activities is Security Sector Reform (SSR). In particular, SSR is often conceived in parallel with reintegration assistance for former combatants of military personnel. Donor interest in this issue is growing rapidly and there is increasing consensus that security institutions are relevant for development, and should therefore be discussed openly. The OECD Development Assistance Committee, a key coordination mechanism for donor governments, for example, issued a set of comprehensive guidelines on SSR in 2005.[24] The operational cooperation and institutional relationships between development and security actors, however, remain sporadic and largely uncoordinated. This can be traced to actors' perceptions of their respective roles, comparative advantages and mandates. Key actors in this domain tend to be bilateral donors, which are unconstrained by institutional limits to their engagement and have the diplomatic clout to engage such sensitive issues. Conversely, given the sensitivity of issues related to the security sector, multilateral actors often act cautiously. These problems are particularly pronounced for the World Bank, which is constrained from direct engagement with armed or police forces. Under these conditions, an entry point to security sector reform can be public financial management, and the need to include security-related spending therein. First experience with this approach has been gained in Afghanistan, where a review of state finances was conducted which explicitly included the security sector.[25] This is an example of a creative solution to address an issue that is central to peacebuilding efforts through a mainstream development activity.

The reform of justice institutions, including the judiciary and police, also represents a significant field for development actors, and is closely tied to security-related activities. This includes activities aimed at strengthening human rights institutions, and frequently training in human rights. In conflict-affected environments, furthermore, related activities involve mechanisms of transitional justice. In practice, however, there is very little specific methodology on supporting justice reform in conflict-affected countries. Efforts to link transitional justice mechanisms with more long-term approaches of justice reform often fail. A recent review of justice reform initiatives in post-conflict countries has supported this conclusion. It points out that rule of law initiatives have, on the whole, had very little impact. This is despite the fact that such programmes have now become ubiquitous components of reconstruction packages. Very little is known, in fact, about how to bring about the complex and interdependent social goods which are expected to be delivered through justice reform.[26] While this is true in any developing country context, the challenge is even more urgent in the context of violent conflict and state fragility.

Potential roles for development actors in the short, medium and long term

As has been indicated repeatedly above, development, humanitarian and security concerns can no longer be discussed and addressed separately.

This is particularly relevant for protracted refugee situations, given their long-term nature and the fact that large development aid portfolios often exist in the same countries. Uganda, while not a refugee but a protracted IDP situation, represents a paradigmatic example. It has rightfully been praised for success in fostering socioeconomic development and stable governance in the south, and has become one of the favoured recipients of aid among donors. At the same time as it was a favoured aid recipient of the international community, Uganda was facing continuous armed conflict in the north, along with protracted displacement of a large majority of the entire northern population. Solutions will not be found through one-size-fits-all approaches and have to be, above all, case specific. They are bound to depend on the specific conditions and socioeconomic interactions which have developed in any given situation. This requires flexibility of all actors involved, as well as strengthened knowledge of local situations.[27]

At the most basic level, addressing protracted refugee situations requires that development agencies continue coordinating their activities among each other, as well as with humanitarian actors, governments and security actors. It has been noted above that significant attempts have been made in the past in this regard, and that there have been repeated expressions of commitment to this end. However, these efforts need to be built on and implemented in practice. Engaging conflict effectively requires a willingness to engage early and expand relevant activities during conflict and prior to the post-conflict 'reconstruction' phase. At present, development actors still tend to be reactive, assuming that the assistance will have the greatest impact following the end of conflict. Earlier engagement not only would improve the effectiveness of development efforts, but can also make a substantial contribution to conflict mitigation. It has been noted above as well that significant parts of the methodology for such engagement already exist. The question is therefore primarily one of mainstreaming, complemented by the development of new tools as required.

Coordination with donors also needs to be improved. This refers both to the predictability of funding as well as to the alignment of the foreign policy and development arms of donor governments. Donors' role in multilateral institutions and continued advocacy for attention to conflict and security issues in development fora are crucial. Development actors need to reflect on the political dimensions of their activities and the sociopolitical environment in which they operate. To achieve durable solutions to protracted refugee situations, bilateral donors, for their part, need to ensure that the foreign policy, development and humanitarian objectives pursued by their governments are coherent and closely aligned to support conflict mitigation and peacebuilding efforts. In short, the comparative advantage of development actors in protracted refugee situations lies at the macro level, where issues related to refugee situations can be reflected and addressed.

Compared with other actors, the chief comparative advantage of development actors is their ability to provide resources for addressing the long-term underlying drivers of the conflicts, which in turn have created protracted refugee situations. This macro-level perspective goes beyond, and has a potentially more durable impact than, the narrower focus on operational partnerships with humanitarian actors in the short and medium term. As a matter of course, development agencies need to put the conflict and state fragility issues which are at the root of many refugee situations at the core of their policy agenda. Refugee situations should no longer be seen as a narrow set of issues that will be dealt with by a small group of specialized agencies. Instead, the security and socio-economic ramifications of these situations make them a concern for all groups of actors, cutting straight across the divide between development, humanitarian and security actors. They also cut across the confines of sector approaches to development assistance.

Conflict, crises and displacement are deeply political phenomena and are closely interconnected. Technical approaches to development assistance easily miss this crucial dimension, which can severely hamper their effectiveness. Development actors should be encouraged to take a more explicitly political role, or at the very least make a concerted effort to strengthen and act on their cognizance of the political environment surrounding protracted refugee situations. The existing array of conflict analysis tools is a good starting point to conduct the necessary analysis, but should be complemented with more political analysis approaches. In addition to their engagement with host governments and governments in countries of origin, development actors can play a crucial role in drawing attention to the macro-level implications of refugee situations. The World Bank, for one, enjoys significant respect for its analytical capacities. With the prerequisite political will, it therefore also enjoys some significant leeway in setting agendas conducive to achieving durable solutions to refugee situations. Additionally, area development approaches can be effective tools to support the reintegration of refugees or their integration into local communities. In the design of their aid packages, bilateral and multilateral donors in particular command a significant amount of agenda-setting power. This also applies to processes at the multilateral level, such as the nascent UN Peacebuilding Commission (PBC). The PBC offers excellent opportunities to press for early and sustained engagement of development actors in conflict and post-conflict situations, including protracted refugee situations and other regional dimensions of conflict. In this, peacebuilding should not be left to security actors alone.

Development agencies have a crucial role to play in bringing to bear their expertise and knowledge of local contexts and challenges.

Working in environments of state fragility and conflict is labour intensive, and therefore requires adequate human resources addressing conflict-related issues that are related to human resources. In development agencies, internal incentives tend to be structured in a way that is unfavourable to working on fragile and conflict-affected countries. Career advancement, explicitly or implicitly, tends to be based on quantifiable success. This structure does not reward risk taking. The effect of such structures is to direct both junior-level talent and senior-level experience away from portfolios in risky settings of fragility and conflict. In addition, in a climate veering increasingly toward rewarding success in development cooperation, conflict and state fragility remain issues outside the mainstream and are generally considered too risky to demand sustained individual or institutional attention. A 2006 review of the World Bank's work in fragile states, for example, emphasized the need to strengthen staffing provision, incentives and monitoring and evaluation.[28]

In attempting to address crises and protracted refugee situations, prevention of future crises is easily overlooked. However, given the demonstrated linkages between socioeconomic development and conflict, mainstream economic activities can have a significant conflict prevention impact. This can be garnered and capitalized upon only through concerted efforts at evaluation, learning and analysis. Crucially, this is particularly important at the country and macro level, where efforts should be made to tailor approaches to the overall conflict and security situation.

It should be noted that the buy-in of governments in both countries of origin and host countries is a necessary condition for successfully achieving durable solutions. Development actors tend to enjoy more successful long-term relationships with governments that are, for the most part, characterized by technical and operational dialogue. While development activities are never non-political, the actual implementation of such programmes is characterized by technical relationships that are deliberately kept non-political. This is a unique opportunity to engage governments toward resolving protracted refugee situations.

Conclusion

This chapter has argued that debates on linking relief and development should go beyond narrow conceptions of operational partnership, and instead look at country situations in a more holistic manner. Development actors have much to contribute to such a macro-level approach. They command significant analytical capacity and often have close access to policymakers at different levels in both host countries and countries of origin. In many countries, aid portfolios represent a highly significant socioeconomic factor. In those situations, development actors also command significant political leverage. While it is appropriate for development institutions to strive to maximize the technical quality of their work, this does not necessarily need to preclude their political engagement. There is no doubt that development aid flows can represent a powerful incentive to recipient states to redesign policy in a manner suitable to achieving durable solutions for protracted refugee situations.

The main comparative advantage of development actors in protracted refugee situations is their ability to engage with the underlying conflict at the macro level. In other words, development actors are better positioned than humanitarian ones to address the drivers of conflict which create the political and other impasses to resolving protracted refugee situations. Finding durable solutions for protracted refugee situations is a question not only of designing narrowly targeted programmes, but of working to mitigate social conflict more generally. This is the primary entry point for development actors in responding to protracted refugee situations.

Notes

1. UNHCR, *The State of the World's Refugees 2006: Human Displacement in the New Millennium*, Oxford: Oxford University Press, 2006.
2. *The Role of the World Bank in Conflict and Development: An Evolving Agenda*, Washington, DC: The World Bank, 2004, pp. 4–5.
3. Jeff Crisp, 'Forced Displacement in Africa: Dimensions, Difficulties and Policy Directions', *New Issues in Refugee Research*, Research Paper no. 126, Geneva: UNHCR, July 2006.
4. *Economic and Social Impact of Refugee Populations on Host Developing Countries As Well As Other Countries*, EC/53/SC/CRP.4, UNHCR Executive Committee, 26th meeting, 10 February 2003. For a concise general overview of the issues at stake, see Karen Jacobsen, *The Economic Life of Refugees*, Bloomfield: Kumarian, 2005.
5. Robert Chambers, 'Hidden Losers? The Impact of Rural Refugees and Refugee Programs on Poorer Hosts' in Robert F. Gorman, ed., *Refugee Aid and Development*, Westport, CT: Greenwood Press, 1993.
6. Karen Jacobsen, 'Can Refugees Benefit the State? Refugee Resources and African Statebuilding', *Journal of Modern African Studies* 40, no. 4, 2002, pp. 577–596.
7. Cindy Horst, 'Refugee Livelihoods: Continuity and Transformation', *Refugee Survey Quarterly* 25, no. 2, 2006, pp. 6–22.
8. Oliver Bakewell, 'Repatriation: Angolan Refugees or Migrating Villagers?' in Philomena Essed, Georg Frerks and Joke Schrijvers, eds., *Refugees and the Transformation of Societies: Agency, Policies, Ethics and Politics*, New York: Berghahn, 2004, pp. 31–41. For a more specific critique of the Uganda Self-Reliance Strategy, see Sarah Meyer,

'The "Refugee Aid and Development" Approach in Uganda: Empowerment and Self-Reliance of Refugees in Practice', *New Issues in Refugee Research*, Research Paper no. 131, Geneva: UNHCR, 2006.
9. Jeff Crisp, 'Who Has Counted the Refugees? UNHCR and the Politics of Numbers', *New Issues in Refugee Research*, Working Paper no. 12, Geneva: UNHCR, 1999.
10. Machtelt De Vriese, 'Refugee Livelihoods: A Review of the Evidence', UNHCR, Evaluation and Policy Analysis Unit Report (EPAU/2006/04), February 2006.
11. UNDP/World Bank Joint Guidance on Post-Conflict Needs Assessments, 2004.
12. Georg Frerks, 'Refugees between Relief and Development' in Philomena Essed, Georg Frerks and Joke Schrijvers, eds., *Refugees and the Transformation of Societies: Agency, Policies, Ethics and Politics*, New York: Berghahn, 2004, pp. 167–178.
13. Swarna Rajagopalan, *Within and Beyond Borders: An Independent Review of Post-Conflict Fund Support to Refugees and the Internally Displaced*, Social Development Papers: Conflict Prevention & Reconstruction, Paper no. 17, World Bank, October 2004.
14. The World Bank, *Effective Conflict Analysis Exercises: Overcoming Organizational Challenges*, Washington, DC: The World Bank, Document no. 36446GLB, 21 June 2006.
15. For the World Bank, the seminal conceptual piece in this respect is Paul Collier et al., *Breaking the Conflict Trap: Civil War and Development Policy*, New York: Oxford University Press for the World Bank, 2003.
16. The World Bank, *Toward a Conflict-Sensitive Poverty Reduction Strategy: Lessons from a Retrospective Analysis*, Washington, DC: The World Bank, Document no. 32587GLB, 30 June 2005.
17. IDA allocation mechanisms, http://www.worldbank.org/ida.
18. The World Bank, *Community-Driven Development in the Context of Conflict-Affected Countries: Challenges and Opportunities*, Washington, DC: The World Bank, Document no. 36425GLB, 20 June 2006.
19. Jeff Crisp, 'Forced Displacement in Africa.
20. Jeff Crisp, 'Mind the Gap! UNHCR, Humanitarian Assistance and the Development Process', *New Issues in Refugee Research*, Working Paper no. 43, Geneva: UNHCR, 2001.
21. UNHCR, *Framework for Durable Solutions for Refugees and People of Concern*, Geneva: UNHCR Core Group on Durable Solutions, 2006.
22. Betsy Lippman and Sajjad Malik, 'The 4Rs: The Way Ahead?', *Forced Migration Review* 21, September 2004.
23. UNHCR, *The State of the World's Refugees 2006*, pp. 136–137.
24. Note that for the OECD the concept of SSR includes justice institutions. In this chapter, the two domains are treated separately and SSR refers to military institutions only.
25. *Post-Conflict Security Sector and Public Finance Management: Lessons from Afghanistan*, Social Development Notes: Conflict Prevention & Reconstruction, no. 24, World Bank, July 2006.
26. Kirsti Samuels, *Rule of Law Reform in Post-Conflict Countries: Operational Initiatives and Lessons Learnt*, Social Development Papers: Conflict Prevention & Reconstruction, Paper no. 37, Washington, DC: The World Bank, 2006.
27. Georg Frerks, 'Refugees Between Relief and Development' in Philomena Essed, Georg Frerks and Joke Schrijvers, eds., *Refugees and the Transformation of Societies: Agency, Policies, Ethics and Politics*, New York: Berghahn, 2004, pp. 167–178.
28. The World Bank, *Engaging with Fragile States: An IEG Review of World Bank Support to Low-Income Countries Under Stress*, Washington, DC: The World Bank, 2006.

7
A surrogate state? The role of UNHCR in protracted refugee situations

Amy Slaughter and Jeff Crisp

Established in 1950, UNHCR was charged by the 1951 Convention relating to the Status of Refugees with the protection of their interests: full political and economic rights in the country of asylum, with the hope of eventual voluntary repatriation. As a brutal testament to its contemporary failure, at least 3.5 million of those refugees currently struggle for survival in sprawling camps in Africa and Asia ... If it was originally a guarantor of refugee rights, UNHCR has since mutated into a patron of these prisons of the stateless: a network of huge camps that can never meet any plausible 'humanitarian' standard, and yet somehow justify international funding for the agency.[1]

In a recent article published in the *New Left Review*, quoted in the preceding paragraph, Jacob Stevens provides a scathing critique of UNHCR. According to his analysis, the organization's primary interest lies in its own size and status, and not in the welfare of the refugees it is mandated to protect. By pursuing these interests, the article suggests, UNHCR has been complicit in the perpetuation of refugee situations that might otherwise have been brought to a speedy and satisfactory end.[2] The analysis presented in this chapter, which focuses primarily, but not exclusively, on Africa, where the problem of protracted refugee problems has assumed the most serious dimensions, reaches a different conclusion.

The chapter argues that humanitarian agencies in general, and UNHCR in particular, have been placed in the position of establishing and assuming responsibility for such 'sprawling camps' in order to fill gaps in the international refugee regime that were not envisaged at the time of its establishment after the Second World War.[3] It goes on to

Protracted refugee situations: Political, human rights and security implications,
Loescher, Milner, Newman and Troeller (eds),
United Nations University Press, 2008, ISBN 978-92-808-1158-2

suggest that the United Nations' refugee agency has been limited in its ability to address the problem of protracted refugee situations, mainly because of the intractable nature of contemporary armed conflicts and the policies pursued by other actors, but also because of the other issues which the organization has chosen to prioritize and the limited amount of attention which it devoted to this issue during the 1990s. The chapter concludes by examining the organization's more recent and current efforts to tackle the issue of protracted refugee situations, and identifies some of the key principles on which such efforts might most effectively be based.

Refugee-hosting countries

UNHCR's relationship with host states, and the division of responsibilities it has established with refugee-hosting states, have varied over time and differed significantly from country to country. However, certain patterns of UNHCR engagement have emerged in the four decades since the 1960s, when large-scale refugee movements first began to take place in Africa and other developing regions. According to the predominant model of refugee protection and assistance that has prevailed throughout that period, UNHCR and other humanitarian organizations have assumed a primary role in the delivery and coordination of support to refugees, initially by means of emergency relief operations and subsequently through long-term 'care and maintenance' programmes. Host country involvement has generally been quite limited, focusing primarily on the admission and recognition of refugees on their territory; respect for the principle of non-*refoulement* (which prevents refugees from being returned to a country where their life or liberty would be in danger); and the provision of security to refugees and humanitarian personnel.

Under the terms of this arrangement, the notion of 'state responsibility' (i.e. the principle that governments have primary responsibility for the welfare of refugees on their territory) has become weak in its application, while UNHCR and its humanitarian partners have assumed a progressively wider range of long-term refugee responsibilities, even in countries which are signatories to the 1951 Refugee Convention and which are members of the organization's governing body, the Executive Committee. Such tasks have included those of registering refugees and providing them with personal documentation; ensuring that they have access to shelter, food, water, health care and education; administering and managing the camps where they are usually accommodated; and establishing policing and justice mechanisms that enable refugees to benefit from some approximation to the rule of law. In these respects, it can be argued, UNHCR has been transformed from a humanitarian organization to one that shares certain features of a state.

How did this situation arise? Primarily, this chapter suggests, because the international refugee regime was forged in the specific historical context of the late 1940s and early 1950s, when the international community's primary concern was to address refugee problems in Europe associated with the Second World War and its Cold War aftermath. Despite the devastation caused by conflict with Nazism and fascism, the states most directly concerned with those problems had considerable resources at their disposal. And in their efforts to address the refugee problem, they were assisted by the fact that large numbers of refugees in and from Europe were able to find a solution to their plight elsewhere in the world, by means of resettlement programmes to Australia, Canada, the United States, and to a lesser extent South Africa and South America.

When the focus of the refugee problem shifted from Europe to the developing regions in the 1960s, and when the international refugee regime was extended to those regions by means of the 1967 Protocol to the Refugee Convention, the circumstances were quite different. On the one hand, the states most directly affected by the refugee problem had relatively few resources at their disposal, most of them being former colonial territories with typically dependent and underdeveloped economies. On the other hand, only a small (and privileged) minority of the world's refugees could expect to benefit from the solution of third country resettlement. This was particularly the case in Africa, which between the 1960s and 1980s witnessed a succession of major new refugee emergencies, but which did not benefit from the large-scale resettlement programmes established for refugees from Indo-China.

In the initial phase of the post-colonial period, the people and politicians of Africa demonstrated a significant degree of hospitality towards people who were fleeing from conflict in nearby and neighbouring states. Many of the new arrivals came from countries that were locked in struggles for national liberation and independence – struggles that received strong support from the countries to which they fled, and which played a central role in the emergence of pan-African ideologies and the establishment of the Organisation of African Unity (OAU) in 1963.

Symbolizing this sense of solidarity, in 1969 the OAU established its own Refugee Convention, which broadened the refugee definition included in the 1951 Refugee Convention and made it more relevant to the political circumstances of the African continent. Thus the 1951 Convention limited refugee status to people who had left their own country because of 'a well-founded fear of persecution' for reasons of race, religion, nationality, membership of a particular social group or political opinion. By way of contrast, the OAU Convention stated that 'the term "refugee" shall also apply to every person who, owing to external aggression, occupation, foreign domination or events seriously disturbing public order in either part or the whole of his country of origin or nationality, is compelled to leave his place of habitual residence'.

By the time that the OAU Refugee Convention came into force in 1974, the political and material conditions which had underpinned such expressions of solidarity with the continent's refugees were already being undermined. First, significant changes were taking place in the number of refugees that the continent was obliged to accommodate. While there were only around 1 million refugees in Africa at the beginning of the 1980s, the figure climbed inexorably in the years to come, reaching approximately 6 million by the end of the decade. Throughout this period, the speed and scale of the continent's refugee movements also increased, placing additional strains on the countries and communities where the new arrivals settled.

Second, the capacity of those countries to accommodate an ever growing number of refugees was declining. While their relative prosperity in the early years of independence had allowed them to exercise a degree of generosity to refugees, the newly independent states of Africa now began to suffer from a wide range of interrelated ills: unfavourable movements in the terms of trade for raw materials and oil; high levels of population growth combined with low rates of economic growth; the progressive introduction of structural adjustment programmes that curtailed public services and employment; environmental degradation; the emergence of the HIV/AIDS pandemic; as well as the economic mismanagement and political instability that were both a cause and a consequence of such problems.

Third, the refugee movements witnessed in Africa and other developing regions began to assume a new character. No longer the victims of liberation struggles, a growing proportion of the world's refugees were now forced from their homes by armed conflicts and power struggles taking place within (and to a lesser extent between) independent states. Rather than being considered as victims of external aggression, occupation and foreign domination, refugees were increasingly regarded as a source of political instability and social tension, particularly when, as a result of their nationality, ethnic origins or political allegiance, they were associated with one of the parties to the conflict which had forced them to flee.

Finally, the last two decades of the 20th century witnessed a growing sense amongst the developing countries that they were obliged to bear a disproportionate share of responsibility for the global refugee problem. During the Cold War years, donor countries regarded generous humanitarian assistance programmes as a means of supporting client states and elites, while simultaneously winning the hearts and minds of recipient populations. However, in the unexpectedly tumultuous period that followed the demise of the bipolar world, the refugee policies of donor states were, as the following section explains, driven by other considerations.

The industrialized states

During the 1980s and 1990s, the industrialized states became increasingly preoccupied with the task of reducing the number of people from other parts of the world who were seeking to enter and remain on their territory. Unable to enjoy security or sustainable livelihoods in their own countries, and deprived of any opportunity to move to the industrialized states in a legal and safe manner, growing numbers of citizens in Africa, Asia, the Middle East, Latin America and European countries outside of the European Union attempted to enter the world's more prosperous states, many of them submitting asylum applications once they had reached their destination.

In response to these developments, the countries of Western Europe, North America and the Asia-Pacific region introduced a vast array of measures specifically designed to prevent or dissuade the arrival of these would-be refugees: visa restrictions, carrier sanctions, interdiction and detention, limitations on social welfare and the right to work, as well as restrictive interpretations of the 1951 Refugee Convention. While a limited number of the industrialized states (essentially Australia, Canada and the United States) continued to admit refugees by means of organized resettlement programmes, these countries were the exception that proved the rule. As far as the states of the South were concerned, the countries of the North had turned their back on the notion of 'burden-sharing' (or, as many humanitarian organizations prefer it to be known, 'responsibility-sharing'), a principle which had hitherto underpinned the international refugee protection regime.

Such concerns were reinforced when the industrialized states began to express growing interest in notions such as 'regional solutions', 'protection in regions of origin' and 'extra-territorial processing', all of which could be (and were) interpreted as efforts to ensure that refugees and asylum seekers were confined to the poorer and less stable regions of the world that were already accommodating the vast majority of displaced and exiled people.

In this context, it was no coincidence that developing countries also began to introduce more restrictive refugee policies. Confronted with the circumstances described above, countries of asylum in Africa and other developing regions responded in a number of related ways: by restricting the rights of refugees on their territory; by accommodating them in closed and semi-closed camps rather than open rural settlements; by depriving them of opportunities to become self-reliant and to benefit from the solution of local integration; and, most significantly for the analysis presented in this chapter, by suggesting that they would only admit and refrain from the *refoulement* of refugees if the needs of such populations were fully met by the international community. By the mid-1990s, UNHCR was, as Stevens suggests, left to run 'a network of huge camps', the inhabitants of which had little or no prospect of finding an early solution to their plight, primarily because the armed conflicts which had driven them from their homes went unresolved.

They went unresolved for two principal reasons. First, because they were symptomatic of a new and intractable form of warfare that had emerged in many of the world's failed and fragile states – a form of warfare in which communal identities and the struggle for land and resources played a more important role than ideological differences, and in which militias, warlords and bandit groups replaced conventional armies and military formations. Often described as 'internal armed conflicts', such wars actually involved a mixture of local, national, regional and international protagonists. This trend has been witnessed most graphically in the central portion of sub-Saharan Africa, which for much of the past decade has been afflicted by an interlocking series of conflicts, stretching from Somalia and Sudan in the east to Liberia and Sierra Leone in the west.

A second reason for the failure to resolve such conflicts is to be found in the selective application of the doctrine of 'humanitarian intervention'. Coming to prominence in the years that followed the end of the Cold War, this doctrine suggested that traditional notions of state sovereignty could no longer stand in the way of international action in situations where large numbers of civilians had been placed at risk by human rights violations, armed conflicts and complex political emergencies. In practice, however, the world's most powerful states were generally reluctant to invoke this principle in the deadly conflicts that afflicted Africa. As one of the authors of this chapter has pointed out elsewhere:

> An instructive comparison can be made with Northern Iraq, Bosnia, Kosovo and East Timor – four armed conflicts which produced (eventually) a decisive response from the world's more prosperous states, enabling large-scale and relatively speedy repatriation movements to take place. In each of these situations, the United States and its allies had strategic interests to defend, not least a desire to avert the destabilizing consequences of mass population displacements. In Africa, however, the geopolitical and economic stakes have generally been much lower for the industrialized states, with the result that armed

conflicts – and the refugee situations created by those conflicts – have been allowed to persist for years on end.[4]

The role of UNHCR

Hitherto, this chapter has suggested that the world's protracted refugee situations are to a large extent the outcome of actions taken and not taken by states – both those in developing regions that host the vast majority of the world's refugees, and those in the industrialized world that play a leading role in the United Nations and the international refugee protection regime.

But what role has been played in this scenario by the leading multilateral actor in that regime, namely UNHCR? The allegation made by Stevens – that the 'derelictions of UNHCR' have actively contributed to the problem of protracted refugee situations – is one that deserves to be taken seriously, despite the intemperate language in which it is written. It would be naïve to ignore the fact that the organizational culture of the United Nations can be one that encourages 'safety first' approaches that are acceptable to states, and provides inadequate incentives for the rethinking and reorientation of long-established activities. It is the contention of this chapter, however, that the role assumed by UNHCR in protracted refugee situations is to be found primarily in other factors.

Competing priorities

As indicated by the title of the book published by former High Commissioner for Refugees Sadako Ogata, the 1990s constituted 'the turbulent decade' for UNHCR.[5] During this period, throughout which she directed the organization, UNHCR was confronted with three enormous and simultaneous challenges. The first was to assist with the return and reintegration of the many refugees who had been forced into exile during conflicts that were rooted in Cold War politics, but which had now come to an end, such as Cambodia, El Salvador, Mozambique, Nicaragua and South Africa. The second was to respond to the spate of new crises and refugee emergencies provoked by the unexpectedly violent nature of the post–Cold War world, including those witnessed in the Balkans, the Great Lakes region of Africa and West Africa. The third was to address the rapid growth in the number of people from poorer and less stable parts of the world who were moving to and seeking asylum in the industrialized states, and who were generally unwanted by the receiving states.

The common feature of these challenges was that they all entailed movements of people – movements that were large, rapid and highly visible, and which therefore attracted a great deal of attention from the international community and the global media. With their attention focused on these high-profile and highly politicized situations, UNHCR and other humanitarian actors were able to give less attention to protracted situations in which refugees were moving in no direction, and had effectively become trapped in long-term camps and settlements.

Funding

The relatively low priority given to protracted refugee situations in the years that followed the end of the Cold War was reflected in and reinforced by funding patterns. Reluctant to intervene militarily in many of the world's most serious refugee-producing crises, eager to ensure that refugees and asylum seekers remained within their regions of origin, and under popular pressure to 'do something' about the emergencies that were being played out on television screens across the industrialized world, donor states were now prepared to make unprecedented amounts of funding available to the humanitarian community.

However, relatively little of that funding was earmarked for the more stable and static refugee situations that existed in Africa and other parts of the world, a problem that was in some senses compounded by the fundraising and media relations strategies pursued by the humanitarian community. Images of destitute refugees seeking urgent protection and assistance in countries of asylum proved to be an effective means of attracting international attention and resources, as did images of exiled communities who were going home to begin a peaceful and productive life in their country of origin. By way of contrast, relatively little attention was given to those refugees whose immediate past and indefinite future entailed the monotony of life in a camp.

Time for solutions?

A logical response to this scenario would have been for the international community to recognize the semi-permanence of many refugee situations in the developing world, to assist the populations concerned to attain progressively higher levels of self-reliance during their time in exile, and to promote a process of local development that provided opportunities and brought benefits to refugees and citizens alike. In reality, however, this approach proved very difficult to implement.

With the number of refugees in low-income regions of the world steadily expanding, from the 1970s onwards UNHCR made repeated efforts to promote a developmental and solutions-oriented approach to refugee assistance, incorporating the principles outlined in the preceding paragraph. Perhaps the most prominent example of such efforts was to be found in 'ICARA II' (Second International Conference on Assistance to Refugees in Africa), an initiative co-sponsored by UNHCR and UNDP in 1984, under the evocative slogan 'Time for solutions'.

However, such initiatives met with very limited success. Host governments were generally eager to retain the visibility of the refugee populations they hosted and to discourage those people from settling permanently on their territory. They consequently preferred the exiles to be segregated from the local population, in camps funded by donor states and administered by UNHCR. They were concerned that, if development aid were to be targeted at refugee situations, it would lead to a reduction in the level of international assistance available for their regular development programmes and it would imply their agreement to the long-term or permanent settlement of the refugees concerned. Meanwhile, such states were still struggling to respond to a succession of new humanitarian emergencies, such as that caused by the 1984 famine in the Horn of Africa, which occurred almost immediately after ICARA II. At a time when massive numbers of people were on the move and in urgent need of life-saving assistance, the notion of 'Time for solutions' began to seem very optimistic.

This situation was reinforced by the administrative structures to be found in most donor states, which embody a clear separation between humanitarian assistance on the one hand, and development aid on the other. For these countries, refugee crises such as that witnessed in the Horn of Africa were primarily 'humanitarian' in terms of their nature and required response. As a result, even if those crises persisted for years and transmuted in the process from 'refugee emergencies' to 'protracted refugee situations', they were generally addressed from the limited perspective of emergency relief.

Programme objectives and design

As a result of the considerations outlined above, in the 1990s the objectives and design of the world's long-term refugee programmes received relatively little attention. Indeed, the concept commonly employed to describe these operations, namely 'care and maintenance programmes', was indicative of the rather low level of ambition which the international community brought to the issue of protracted refugee situations during this period.

A defining characteristic of the 'care and maintenance' model was the extent to which it endowed UNHCR with responsibility for the establishment of systems and services for refugees that were parallel to, separate from, and in many cases better resourced than those available to the local population. In doing so, this model created a widespread perception that the organization was a surrogate state, complete with its own territory (refugee camps), citizens (refugees), public services (education, health care, water, sanitation, etc.) and even ideology (community participation, gender equality). Not surprisingly in these circumstances, the notion of state responsibility was weakened further, while UNHCR assumed (and was perceived to assume) an increasingly important and even pre-eminent role.

Some interesting evidence in this respect can be found in the work of two anthropologists who worked amongst Burundian Hutu refugees in Tanzania. Undertaking research in the Kigoma region of the country, Liisa Malkki found that the refugees lionized UNHCR and demonized the Tanzanian authorities and host population, practically equating their hosts with their Tutsi opponents in Burundi's civil war. In their discourse, the Hutus drew parallels between UNHCR and the Belgians in Burundi, perceiving them both as 'benign foreigners' that would shield them from their enemies.[6]

Somewhat similar dynamics were witnessed by Simon Turner, who undertook fieldwork amongst Burundian refugees living in Tanzania's Lukole camp. According to Turner, UNHCR's identity had blended with that of *wazungu* (white people) and the international community at large. Refugee women are quoted as saying that 'UNHCR is a better husband', in the sense that the organization provides for the household what a Hutu man would normally provide for his family. Turner goes on to argue that traditional social structures often break down in this context, with UNHCR assuming the role of the patriarch. According to one refugee man he interviewed, 'there is a change. People are not taking care of their own life. They are just living like babies in UNHCR's arms'.[7]

These circumstances created some serious dilemmas for UNHCR. If the organization was to compensate for the limited capacity of host states by assuming a wide range of responsibilities, it could help to ensure that refugees received the protection and assistance to which they were entitled, but it could also absolve host states of their international obligations. But if UNHCR was to insist upon the principle of state responsibility and to limit its own operational involvement in protracted refugee situations, how could it safeguard the welfare of the people it was mandated to protect? As a senior UNHCR official remarked in a personal communication with the authors, 'many a UNHCR manager has pushed so hard to get reticent and phlegmatic governments involved in refugee administration that in the end they throw their hands up in the air with frustration'. Indeed, much of the refugee legislation adopted by host states in Africa and other developing regions throughout the period under review, as well as the practical arrangements established for protection activities such as refugee registration, documentation and status determination, is the result of UNHCR's 'gap-filling' efforts.

Recent developments

As the preceding section of this chapter has explained, UNHCR became involved in a growing number of protracted refugee situations during the 1990s, many of which involved the confinement of refugees to camps where they enjoyed little freedom of movement and had few opportunities to establish sustainable livelihoods. For the majority of people who found themselves in such situations, the options of voluntary repatriation, local integration and third country resettlement all remained a distant dream.

Regrettably, that continues to be the case for large numbers of refugees around the world. Since the turn of the new millennium, however, three related factors have enabled UNHCR and other members of the international community to become more engaged with the problem of protracted refugee situations and to ask whether it can be approached in alternative ways. First, while a number of new refugee emergencies erupted (most notably those involving Iraq and the Darfur region of Sudan), the scale and frequency of such crises generally diminished from 2000 onwards. This trend, combined with large-scale voluntary repatriation movements to countries such as Afghanistan, Angola, Liberia, Sierra Leone and Somaliland, led to a progressive reduction in the size of the world's refugee population and enabled UNHCR to refocus its attention on issues such as protracted refugee situations which had assumed a lower priority during the previous decade.

Second, UNHCR was confronted with growing evidence with respect to the negative consequences of protracted refugee situations, especially those in which the populations concerned experienced deteriorating conditions of life and could not look forward to a brighter future. Refugees who found themselves in such situations were more likely to engage in onward movements, leaving their camps in order to take up residence in an urban area or to seek asylum in more distant parts of the world. They were more likely to be susceptible to exploitation and to engage in negative survival strategies such as theft and other forms of criminality, the manipulation of assistance programmes, and becoming victims of sexual exploitation. In addition, they were also more likely to become attracted to political and military movements whose activities conflicted with the strictly humanitarian nature of refugee status and of UNHCR's mandate.

Third, the issue of protracted refugee situations became the subject of new research and lobbying efforts, led by UNHCR. Thus, in 1999, the organization's Evaluation and Policy Analysis Unit launched a Protracted Refugee Situations Project which published a wide range of reports and papers on this issue. This led in turn to the establishment of a web-based initiative entitled the 'Refugee Livelihoods Network', which encouraged practitioners and researchers to share ideas and information on the steps that could be taken to promote self-reliance in long-term refugee situations.[8] Similar themes were subsequently taken up by other organizations, including the US Committee for Refugees and Immigrants, which launched a vigorous 'anti-warehousing campaign', and by a number of academic groups which established research projects on similar themes.

Prompted by these developments, from 2000 onwards, UNHCR began to adopt a more assertive and proactive role in relation to protracted refugee situations than had been possible during the previous decade. A new High Commissioner, former Dutch Prime Minister Ruud Lubbers, launched a series of initiatives ('Convention Plus', 'Development Assistance to Refugees' and 'Development through Local Integration'), all of which were indicative of a new institutional focus on the durable solutions dimension of the organization's mandate. At the same time, UNHCR brought the issue to the attention of the agency's governing body, the Executive Committee, organized a special meeting of African states to consider how the problem might be more effectively addressed, and began for the first time to collect and publish statistics on protracted refugee situations.

These initiatives had a number of important operational outcomes. Working in cooperation with the governments concerned, UNHCR established a Self-Reliance Strategy for refugees in Uganda and launched the development-oriented Zambia Initiative for refugees living in that country. The organization sought to reinforce the rights and improve the material circumstances of long-term refugees in countries such as Kenya, Tanzania and Thailand by means of a new Strengthening Protection Capacity Project.

Under the leadership of another new High Commissioner, António Guterres, a former Prime Minister of Portugal, UNHCR also began to explore the opportunities for local integration for refugees in areas such as West Africa, a solution that had been largely ignored in the preceding decades. While these different initiatives have not been an immediate or unqualified success, and have indeed attracted some criticism, they nevertheless provide some tangible evidence of a new commitment on

UNHCR's part to addressing the problem of protracted refugee situations.[9]

Elements of a humanitarian strategy

Now that the plight of the world's long-term exiles has assumed a more central place on the international humanitarian agenda, what can be done to formulate a more effective and equitable response to the issue of protracted refugee situations? The final section of this chapter offers some suggestions with respect to the approaches that might be pursued if this question is to be answered in a positive manner.

Promoting interaction between refugees and local populations

First and foremost, there is a continued need to revisit established approaches to refugee protection and assistance, especially the care and maintenance model, which tends to maximize the role of UNHCR and other humanitarian organizations, but minimizes that of host states and other actors. Ideally, exiled populations should not be obliged to live an isolated existence in internationally administered enclaves, but should be able to engage in positive interactions with people and communities living in the same area. Of course, the establishment of safe and demilitarized areas where refugees can benefit from life-saving forms of protection and assistance may be required in the early days of an emergency, but the negative aspects of separation often begin to outweigh the advantages as time goes on.

The adoption of alternative approaches to the administration of protracted refugee situations will not be easy. As earlier sections of this chapter have suggested, large and long-term refugee camps have become the norm in many parts of the world because of the interacting priorities of host governments, donor states and humanitarian organizations. Recent advocacy efforts intended to challenge the practice of 'warehousing' have also tended to gloss over the fact that refugees themselves are sometimes averse to leaving their camps or to forging closer connections with the local population. Refugee camps, even if the services they offer are minimal, provide an important safety net for many refugees, especially the more vulnerable members of the population. Remaining in a camp may also have perceived benefits for refugees who hope to participate in an organized resettlement or voluntary repatriation programme, as well as for political activists who wish to mobilize the refugee population in support of their cause or to give their cause greater international visibility.

Despite these constraints, a number of steps could be taken to approach the issue of protracted refugee situations in a more constructive manner. The delivery of services to refugees and local people could be structured in a way that avoids the establishment of separate and parallel systems, thereby improving the interaction that takes place between the two groups. Refugees could be offered better access to local markets for both the sale and purchase of goods, an approach that would boost the local economy and demonstrate the positive impact of the refugees' presence. As was recognized as long ago as ICARA II, refugee-populated areas as a whole should be properly incorporated into national and local development plans, so as to avert the establishment of camps that are disconnected from the surrounding state and society.

Humanitarian and human rights agencies could organize bridge-building seminars between refugee and local populations and, if necessary, conduct conflict resolution sessions between the two groups. There is a common assumption that such initiatives are not needed and that refugees in developing regions invariably share the language and culture of their local hosts, but this is not always the case. Moreover, refugees and local populations may actually have complex histories and strained social or political relations as a result of their proximity. In such situations, a process of mutual adaptation will be required, supported by efforts to ensure that the local community is receptive to the refugees' presence and that the refugees themselves feel secure in their country of asylum.

Such efforts need not entail a great deal of expense, but they do require some initiative on the part of the humanitarian community and some political will on the part of the host country authorities. Refugees in Ghana, for example, have been issued with photo identity cards, bearing the seal of both the authorities and UNHCR. As a result, they state that they feel more secure in the country and more confident in their interactions with the host community and local officials. They also experience less harassment when they encounter the police, which has facilitated their freedom of movement outside their camp and boosted their potential for self-reliance.

Supporting the role of the state

As the preceding example demonstrates, UNHCR and other humanitarian actors should be instrumental in supporting the role of the authorities in relation to protracted refugee situations. Of course, such a role must be based on a strict respect for the principles of refugee protection, and must therefore be supported by practical initiatives that encourage and enable host states to uphold their obligations under international and regional refugee law, as well as human rights and customary law. As noted earlier, UNHCR has a particularly important role to play in the establishment of national refugee legislation that is in accordance with the 1951 Refugee Convention, and in supporting capacity-building efforts that enable the agents of the state, including the police, military, judiciary and local government officials, to adhere to such legislation.

More generally, UNHCR should lose no opportunity to underline the twin principles of state responsibility and international solidarity, pointing out that the latter is a necessary condition of the former in low-income countries with significant refugee populations. As Betts and Kaiser explain in their respective chapters of this volume, external support has played an important role in prompting and enabling host countries to pursue relatively progressive refugee policies, especially when conditionalities are attached to such assistance.

Humanitarian actors could play a more active role in ensuring that relevant stakeholders understand the responsibilities and authority of the state which has admitted them to its territory. The ubiquity of UNHCR's personnel, offices, vehicles and logo in many long-term refugee camps often leads to confusion on this matter, a situation exacerbated by the fact that many government assets also carry the prominent inscription, 'donated by UNHCR'. When coupled with the physical separation of refugee camps, it is hardly surprising that refugees, local people and government officials should perceive such locations as extra-territorial entities, administered by an international organization with greater visibility and resources – and even legitimacy – than the state.

Recognizing the difficulties and dangers associated with this situation, in 2003, Kenya's newly appointed Home Affairs Minister, Moody Awori, referred to the 'hands-off refugee policy' pursued by the previous administration, observing that this approach 'caused more harm to our hospitable people'. 'It should', he said, 'be the responsibility of the Government to undertake refugee issues seriously'.[10] In many countries, the failure of governments to 'undertake refugee issues seriously' has been based on an assumption that exiled populations do not strive to meet their own needs and invariably have damaging consequences for the local economy, environment and security, and that to avert such outcomes refugees should be induced to return to their country of origin, even if it is not safe for them to do so.

Rather than reinforcing such assumptions by references to the 'dependency syndrome' and the 'negative impact' of refugee movements, UNHCR and its humanitarian partners should challenge and change them by means of public and private advocacy efforts. In this respect, the collection and analysis of empirical data are essential. UNHCR could, for example, devote more effort to supporting research on the efforts that refugees make to establish their own livelihoods, on the difficulties that they encounter in this process, and on the opportunities that are opened up when host government policies provide refugees with greater freedom of movement, better access to land and increased opportunities to engage in the local economy.

Communicating UNHCR's capacities and limitations

Efforts to reorient UNHCR's role in protracted refugee situations must also, as one of the organization's staff members has suggested, be based on 'a clear statement of the limits of humanitarian action'. Such an approach, he goes on to suggest, 'may help governments understand (and even assume) their political responsibilities'.[11] If it is to pursue such an approach and is to engage in the careful management of the expectations placed on it, UNHCR must recognize the dangers of overstating its own capabilities. In the competition for 'brand recognition' and 'market share', UNHCR has emphasized the extent to which the world's refugees rely on the services which it provides. Given the realities of humanitarian funding, UNHCR will have to tell both sides of the story. The organization should underline its strengths and successes, while simultaneously acknowledging its limitations and emphasizing the need for other actors to play their part in addressing the problem of protracted refugee situations.

Such efforts should be directed not only at host governments, donor states and the international media, but also at refugees themselves. In many long-term refugee situations, there is an information vacuum which breeds misinformation and inflated expectations. It should become a high priority for UNHCR to communicate systematically and clearly to refugees the terms of their rights, entitlements, obligations and future options, as well as the extent to which the organization can realistically support them in these respects.

Working with other actors

In order to address the outsized role of UNHCR in protracted refugee situations, there must be a broader recognition that the organization is not the only member of the humanitarian community or the UN system that has a substantive role to play in this area. When people flee from their own country, cross an international border and acquire the status of refugee, they naturally become of direct and immediate concern to UNHCR. But, in becoming refugees, they do not cease to be of concern to other actors within and outside the United Nations – actors whose mandate and activities lie in areas other than humanitarian relief, such as socioeconomic and community development, education and training, agriculture and micro-finance. The search for effective responses to protracted refugee situations should be regarded not as the sole fiefdom of UNHCR, but as a responsibility to be shared with – and amongst – these other actors.

Hitherto, UNHCR's ability to engage with these other actors has been limited.[12] As explained in an earlier section of this chapter, this is partly because of the artificial way in which the international aid machinery is structured. But it also derives from UNHCR's mandate-driven preference to retain the leading role in refugee situations. Thus, when the United Nations' Emergency Relief Coordinator established an ambitious process of humanitarian reform in 2005, designed to establish a better coordinated response and a more effective division of labour amongst the organizations concerned, UNHCR successfully insisted that refugee situations be excluded from the exercise.

UNHCR has an obligation to uphold its protection mandate, and thus has a legitimate concern to avoid any coordination arrangements that might compromise that mandate. At the same time, the organization cannot act in isolation from the rest of the UN system and humanitarian community. The humanitarian reform initiative has already led to a new inter-agency coordination model in non-refugee emergencies, whereby designated organizations within and outside the United Nations assume responsibility for specific sectors or 'clusters'. UNHCR has agreed to lead three of those clusters (protection, camp management and camp coordination, and emergency shelter) in situations involving internally displaced persons (IDPs). If the cluster approach really does enable the international community to pool and deploy its resources more effectively in IDP situations, then perhaps a similar arrangement could be established in relation to refugees, thereby enabling a wider range of actors to be involved in the search for solutions to their plight?

The dynamics of the UN system would appear to be pointing in that direction. In addition to the introduction of the cluster approach, there is growing international support for the 'One UN' concept, which requires the different United Nations agencies to function in a more integrated manner at the country level, with a common programme and budgetary framework.[13] At the same time, the United Nations has become increasingly committed to the establishment of 'integrated missions' in war-affected and post-conflict situations, bringing together the humanitarian, human rights, development, peacekeeping and political functions of the world body under the overall authority of the Secretary-General. These developments have an evident relevance to the task of resolving the problem of protracted refugee situations, both in supporting countries of asylum that have large numbers of refugees on their territory, and in supporting countries of origin from which those people have fled, and to which many will eventually return.

Notes

1. Jacob Stevens, 'Prisons of the Stateless: The Derelictions of UNHCR', *New Left Review*, no. 42, November–December 2006.
2. An incisive rejoinder to Stevens can be found in Nicholas Morris, 'Prisons of the Stateless: A Response to New Left Review', *New Issues in Refugee Research*, no. 141, Geneva: UNHCR, April 2007.
3. In addressing these issues, the chapter draws upon the analysis presented in Jeff Crisp, 'No Solutions in Sight: The Problem of Protracted Refugee Situations in Africa', *New Issues in Refugee Research*, Working Paper no. 75, Geneva: UNHCR, January 2003; and Jeff Crisp, 'Refugees and the Global Politics of Asylum', in Sarah Spencer, ed., *The Politics of Migration*, Oxford: Blackwell, 2003.
4. Jeff Crisp, 'No Solutions in Sight', p. 3.
5. Sadako Ogata, *The Turbulent Decade: Confronting the Refugee Crises of the 1990s*, New York: Norton, 2005.
6. Liisa Malkki, *Purity and Exile: Violence, Memory, and the National Cosmology among Hutu Refugees in Tanzania*, Chicago: University of Chicago Press, 1995.
7. Simon Turner, 'Angry Young Men in Camps: Gender, Age and Class Relations Among Burundian Refugees in Tanzania', *New Issues in Refugee Research*, Working Paper no. 9, Geneva: UNHCR, June 1999, p. 6.
8. For details of both initiatives, see http://www.unhcr.org/research/3b850c744.html.
9. See, for example, the critique of Development Assistance to Refugees and the Uganda Self-Reliance Strategy in 'Giving Out Their Daughters for Their Survival: Refugee Self-Reliance, Vulnerability and the Paradox of Early Marriage', Working Paper no. 20, Kampala, Uganda: Refugee Law Project, April 2007 (no author cited).
10. Kurgat Marindany, 'Govt: Refugees Starving Due to Iraq War', *Eritrean News Wire*, 16 April 2003.
11. Nicholas Morris, 'Protection Dilemmas and UNHCR's Response: A Personal View from Within UNHCR', *International Journal of Refugee Law* 9, no. 3, 1997, p. 497.
12. For a more detailed examination, see Jeff Crisp, 'Mind the Gap! UNHCR, Humanitarian Assistance and the Development Process', *International Migration Review* 35, no. 1, 2001.
13. See, for example, *Delivering as One*, Report of the Secretary-General's High-Level Panel on UN System-Wide Coherence, New York: United Nations, November 2006.

8

A realistic, segmented and reinvigorated UNHCR approach to resolving protracted refugee situations

Arafat Jamal

States create refugees by failing to protect citizens, while asylum countries, donors and UNHCR perpetuate protracted refugee situations by failing to offer adequate responses. The suspension of millions in these liminal states subverts the intentions of the 1951 Refugee Convention, and UNHCR, as the guardian of this convention, has an urgent responsibility to mitigate and resolve such situations. The objective of this chapter is to outline a realistic approach to be pursued by UNHCR, the pre-eminent international refugee agency, in responding to and resolving protracted refugee situations. The problem of protracted refugee situations is by now well recognized, both within UNHCR and amongst others involved in refugee issues. All observers agree that, since the causes of protracted refugees are inherently political, durable solutions must ultimately be sought in the political sphere. However, this analysis is as complicated as it is correct, and making progress through this track has proved frustratingly difficult.

While political actors are to blame for the persistence of long-standing refugee problems, UNHCR has as its primary purpose 'to safeguard the rights and well-being of refugees'.[1] In addition to this statutory function, it also assumes, or is bequeathed, an 'outsized role' in long-term refugee situations[2] – often crowding out the space for others to act. Thus it follows that UNHCR bears a unique responsibility towards this group of often forgotten people, and as such this chapter focuses primarily on the role of UNHCR in alleviating the plight of refugees in these uncomfortable states. This role should be both bold – it must accept the obligations

---

Protracted refugee situations: Political, human rights and security implications,
Loescher, Milner, Newman and Troeller (eds),
United Nations University Press, 2008, ISBN 978-92-808-1158-2

imposed by its perceived centrality in such situations – and modest – it should attempt to devolve functions responsibly to the host state and other actors.

Without neglecting the wider political picture, but to avoid miring refugees in care and maintenance programmes while awaiting elusive political solutions, UNHCR should focus on what can realistically be achieved under current circumstances. UNHCR must take the lead in ensuring that refugees are able to enjoy secure conditions of asylum, and that they are able to enjoy their due rights and freedoms and develop their human capabilities, no matter what the long-term prognosis is. It could do so by segmenting the population and focusing on specific responses to receptive sub-groups, and by elaborating longer-term visions that are both principled and specific to each given refugee situation. UNHCR must both let go, by giving more leeway to others, and take a firmer grip, by being more responsible for seeing through dynamic plans. This approach should reward flexibility and imagination, and should be constantly revised and calibrated in light of regular evaluations and measurements of progress.

This chapter is written in a personal capacity and does not purport to represent the views or opinions of the UN High Commissioner for Refugees. However, the author is a UNHCR staff member who has been and continues to be involved in policy and operations concerning protracted refugee situations, and the arguments presented are grounded in work on, and experience with, the issue and current organizational thinking. In particular, this chapter draws upon the author's evaluations and other studies on Kakuma, Kenya, Afghanistan and Guatemala; policy speeches and papers on the issue for UNHCR's Executive Office and its Executive Committee; and his current work in promoting the strategic use of resettlement in North Africa and the Middle East. This piece is written with a view to being reinserted into continuing UNHCR debates and policy formulation on the topic.

Background: Recognizing the phenomenon and placing it on the agenda

After a 'turbulent decade' of expansion and prominence gained largely as a result of decisive and visible involvement in a series of emergencies,[3] UNHCR at the beginning of the millennium confronted an atomized displacement landscape, marked not by burning crises but by smaller, more complex and seemingly intractable problems. Although in the early days High Commissioners had dealt with long-lasting refugee problems, the issue had retreated during the Ogata years, and it was not until 2000 that protracted refugee situations once again featured on its agenda. In that year, UNHCR's Evaluation and Policy Analysis Unit (EPAU) embarked on a series of studies of such situations,[4] its Africa Bureau convened a major symposium on the subject,[5] and High Commissioner Ruud Lubbers' Convention Plus Unit provided tools to bring together in a strategic manner a mix of durable solutions.[6] The state of UNHCR's thinking on the subject was summarized comprehensively in a 2004 Executive Committee policy paper,[7] which defined the problem, described its dimensions and consequences and suggested some ways forward.

Today, UNHCR recognizes protracted situations as a mainstream policy priority. UNHCR's current Global Strategic Objective 3 is about 'redoubling the search for durable solutions, with priority given to developing and implementing comprehensive strategies to resolve protracted situations'. High Commissioner António Guterres has pledged to undertake a yearly review of protracted refugee situations with a view to identifying opportunities for resolving part or all of long-running crises.[8] UNHCR's aim is to put in place:

> multi-year, comprehensive durable solutions strategies to resolve protracted refugee situations, developed in collaboration with relevant actors, which contemplate the strategic use of resettlement and local integration ... At the same time, UNHCR will promote refugee livelihoods wherever possible, and make self-reliance and empowerment a policy priority in situations where solutions are not available.[9]

Concurrently, the issue also gained ground in other circles, notably in the works of academics such as Gil Loescher and James Milner,[10] in the US Committee for Refugees' 2004 anti-warehousing campaign[11] and even in the United Nations' selection in 2006 of protracted refugee situations as one of the '10 stories the world should hear more about'.[12]

Political solutions, meagre results

All observers have recognized the political factors that contribute to the protraction of refugee situations, and the consequent need to search for lasting solutions in the political sphere. As former UNHCR Assistant High Commissioner Kamel Morjane observed in a 2002 policy speech:

> For every protracted situation, there is a political origin. Camps and idle populations do not simply appear as a natural consequence of forced displacement – they are established in response to political realities and constraints.... Solutions, then, must ultimately be sought in the political arena – be it in the country of origin, of asylum or in third countries.[13]

This recognition regarding the ultimate, political causes of protracted refugee situations is also clearly stated in other major UNHCR policy pronouncements on the topic, and by non-UNHCR experts on the subject. But if the analysis has been correct, the prognosis has been difficult to follow, perhaps precisely because of the political focus. Three recent examples – Somalia, Afghanistan and North Africa[14] – illustrate some of these difficulties.

The 2004 Somalia Comprehensive Plan of Action (CPA), part of the Convention Plus Initiative, was an ambitious attempt at formulating an overarching response to a problem that had persisted for over 16 years and that saw half of the Somali population displaced. The overall objectives of the CPA were firstly to promote voluntary repatriation, ensure sustainability of return, support local integration and seek to enhance resettlement opportunities. The motor for the CPA was to have been an international conference on Somali refugees.[15] However, in spite of the use of solid staff members, concerted efforts and high-level meetings, it failed to deliver. Alexander Betts – in this volume – finds this unsurprising, noting that the Somalia CPA followed what he terms a programmatic rather than a political engagement model, and failed to adequately link humanitarian factors with underlying political and economic issues.[16] It was carried out on a largely technical and apolitical basis. As one otherwise upbeat description of the plan put it, 'What is currently lacking, but sorely needed, is the political will in the international community to develop an integrated approach to Somalia, spanning security, economic development and humanitarian assistance.'[17]

Approaches towards the long-enduring Afghanistan refugee situation understand and account for political undercurrents, but also acknowledge that many of the preconditions for a successful comprehensive plan are not yet in place. Ewen Macleod's chapter illustrates the magnitude and complexity of the situation, recognizing that, ultimately, bilateral political agreements underpinned by international commitments offer the best hope of progress, but current conditions in Iran, Pakistan and Afghanistan militate against this. While the time may not yet be propitious, at least Afghanistan – unlike Somalia – has some of the ingredients in place that could enable a comprehensive approach to be implemented, including international engagement with the country and a central – albeit weak – government. Macleod's contribution to this volume recommends that, even before it is possible to launch the more ambitious strategies, incremental work on new approaches that go beyond standard refugee and humanitarian paradigms be tried out.

In 2006, UNHCR developed a '10-Point Plan of Action' for refugee protection and mixed migration, with an initial focus on North Africa and Southern Europe.[18] North Africa presents a very different type of protracted refugee situation. A traditional transit space for migration to Europe, it also experiences relatively smaller but still significant refugee flows. When these refugees fail to enter Europe, they find themselves stranded in North Africa, where they sink into a desperate protection and economic state. The 10-point Plan attempts to address the issue of both new and protracted refugees through a series of measures, including different processes and procedures to identify and process those in need of protection, and a comprehensive mix of solutions including the three classic ones (voluntary repatriation, local integration and resettlement), plus such options as legal migration. It is too early to say whether the programme will yield results. There is a certain level of political awareness in the plan, combined with a willingness to consider the angle of receiving governments (for example, security and economic concerns), and there is a plan to launch it with a series of visits by the High Commissioner to concerned countries. That said, UNHCR is venturing into an arena heavily dominated by security departments and bilateral agreements, in which its message will be difficult to convey. Meanwhile, the refugees in North Africa remain in an uncomfortable and largely unknown situation, with no effective system yet in place to identify them and respond to their needs.

Comprehensive plans of action for refugees, involving a mix of the three durable solutions, have a clear conceptual logic. However, they require, as Ogata evoked in relation to refugee protection in general, 'a convergence of interests covering humanitarian, political, and security action by major international and regional powers'.[19] Without these, UNHCR – even if politically aware – will be unable to push forward a comprehensive solution on its own and by appealing to humanitarian interests alone. While such convergences take shape (or fail to), the millions of refugees in protracted limbo stagnate, and deserve a more immediate, if less encompassing, response.

Black spots, camps and deprived capabilities

One High Commissioner referred to protracted refugee situations as 'black spots' that should 'burn holes in the consciences of all those privileged to live in better conditions',[20] while another stated that when refugees are accommodated in remote, economically marginalized and insecure areas, and are confronted with unacceptably low levels of assistance, 'this, to me, is a violation of human rights'.[21] The main UNHCR document on the subject bluntly lists the debilitating consequences of having so many human beings in a static state: wasted lives, squandered resources and future problems.[22]

External critiques of protracted refugee situations, and UNHCR's role in not resolving them, have been astringent, with Barbara Harrell-Bond, for example, questioning the underpinnings of the care and maintenance model,[23] and the US Committee for Refugees noting that the phenomenon, which it terms 'warehousing', violates both the letter and spirit of the 1951 Convention and reduces refugees to enforced idleness, dependency and despair.[24]

Camps – the most visible manifestation of refugee problems, even though the majority of refugees do not stay in them – are a powerful line of defence for refugees in the heat of an emergency. Obliged to lift normal alien entry requirements in the face of humanitarian emergencies, first asylum states nonetheless try to control security, stability and economic concerns by making entry conditional upon encampment.[25] For refugees, camps can provide visibility and keep their plight in the public eye.[26] No matter how clearly one might recognize the dangers and slippery compromises involved in camp creation, UNHCR staff time and again resort to camps because they see them as the most effective and initially uncontroversial means of responding to mass influxes.[27]

The tragedy is that the camp that once ensured the life of a refugee becomes, over time, the prime vehicle for denying that same refugee the rights to liberty, security of person and other rights enshrined both in the Universal Declaration of Human Rights and in the refugee instruments.[28] The price of extending this short term measure year after year is paid in terms of rights frustrated, capabilities deprived and expectations unmet. That these same camps have come to embody the refugee experience, to represent the content of international protection for refugees, is grimly ironic, and demonstrates how desperately new approaches to responding to refugee situations are needed.

Capability deprivation in the 'Fourth World'

Most humanitarian workers dealing with protracted refugee populations feel that, whatever the drawbacks of the care and maintenance approach, at least the refugees are protected. By this, they mean protected from reconduction: from being forced back to the countries in which they may have been hounded, tormented, tortured and raped. So tenuous can the right of non-*refoulement* seem that UNHCR accepts a degraded state of affairs – far from anything envisaged as a minimum standard of treatment in the refugee instruments – simply to avoid the realization of any lurking, implied or explicit threat to deport refugees to their home countries.

Amartya Sen, the influential Nobel-winning economist, equates development with freedom. The expansion of freedom is both the primary end and the principal means of development. Development, in this view, 'requires the removal of major sources of unfreedom: poverty as well as tyranny, poor economic opportunities as well as systematic social deprivation, neglect of public facilities as well as intolerance or overactivity of repressive states'.[29] 'Unfreedom', as Sen terms it, restricts human capabilities – the substantive freedom or ability people have to lead the lives they value and to enhance their real choices. Therefore, Sen emphasizes the agency of individuals to shape their own destiny, rather than simply being seen as the 'passive recipients of cunning development programs'.[30]

Encamped refugees have limited rights and endure many restrictions and as such – using a Sen perspective – suffer unfreedom and capability deprivation. The individual agency that they should normatively be able to experience, and which is instrumental in enabling them to achieve freedom, is severely limited. Moreover, camps suffer from an acute form of unfreedom – poverty – which can make one 'helpless prey in the violation of other kinds of freedom'.[31]

Caroline Moorehead, in her book on refugees, spoke of the effect of this particular capability deprivation thus:

> The poverty of camp refugees is about more than just not having things; it is about having no way in which to get them, and no means of altering or controlling one's own life. Their poverty curbs and crushes all hope and expectation. [Camp refugees] are destitute in possibilities.[32]

This destitution of possibilities can drive a refugee mad. Cindy Horst, who has conducted extensive research amongst Somali refugees in camps and in Europe and the United States, has described one poignant symptom of stasis and lack of horizons: *buufis*. This is the term used to describe encamped Somali refugees who are so bereft of options on the ground that they place all their hope, longing and desire into the dream of resettlement. *Buufis* describes this longing, and also 'the madness that at times occurs when the dream to go overseas is shattered'.[33]

Former High Commissioner Sadruddin Aga Khan described refugees as members of the 'Fourth World', 'a world without representation in [the General Assembly] or any other assembly, yet peopled by millions: refugees, the displaced and often stateless, and others in similar circumstances'. He recognized that there was an organic link between the functions of his Office and the Human Rights Declaration, and observed that 'to the extent that the international community makes progress and harmony [in human rights fields], the problem of refugees will be eclipsed, as it should'.[34]

The denizens of this Fourth World have a guardian appointed for them by the international community: UNHCR. UNHCR's function is not limited to protecting refugees by ensuring non-*refoulement*; it is an integral part of the United Nations and an upholder of its principles, including encouraging respect for human rights and fundamental freedoms – in particular, what a recent high-level UN report described as the cause of larger freedom, including freedom from fear and the freedom to live in dignity.[35] When refugees, by virtue of their being refugees and in order for them to continue enjoying that status, are constrained in their rights and capabilities, UNHCR has an obligation to act.

UNHCR's ideal role: responsible, bold and realistic

A gloomy convergence of interests – amongst host states which prefer refugees in camps in order to avoid security and other problems; donors who will pay to ensure survival but nothing more; refugees who fear invisibility and inattention more than confinement; and a UNHCR concerned about its own standing and eminence in an increasingly crowded field – perpetuates the deprivation that is the hallmark of a long-standing refugee situation. But because of its 'outsized' role, the responsibility for this state of affairs tends to be thrust on UNHCR. As Harrell-Bond put it, 'It is assumed that states have the primary responsibility for upholding the rights of refugees, but when greater power over their welfare is in the hands of [the] UN and ... NGOs, the direct obligation falls on *both*'.[36]

Unfair as this may be, it means that UNHCR has the opportunity, and the responsibility, to move beyond protraction. Rather than accepting enduring situations and declining interest, or – even with justification – claiming that responsibility for such situations lies elsewhere, UNHCR should assume the mantle thrust upon it, recognize its vital role and work in ways that are bold enough to conceptualize comprehensive responses, yet realistic enough to work on immediately applicable solutions.

UNHCR's recent role in resolving protracted refugee situations has not yielded too many results. To generalize rather broadly, it has been:
- facilitative and catalytic, as in its involvement with successful comprehensive plans of action such as that for Indo-Chinese refugees (1989) and the International Conference on Central American Refugees (1989); certain resettlement operations (such as that for Uganda Asians in 1972); and repatriations, such as those to Hungary (1950s) and Mozambique (1990s);
- persistent and partially successful, as in its efforts to push for various durable solutions (mostly voluntary repatriation) in, for example, Afghanistan and South Sudan today;

- ineffective, as when its best efforts fail to make a dent on the problem (for example, the Somalia CPA, and attempts to solve the problem of Bhutanese refugees in Nepal); or
- irrelevant, when populations unmix or manage to resolve issues on their own (for example, the Lebanese refugees who repatriated *en masse* to Lebanon as soon as a ceasefire was declared in summer 2006, or the Turkmen Afghans who sought sanctuary in Turkmenistan and received UNHCR assistance, but who returned home without any international guarantees as soon as their leader had struck a deal).[37]

Amy Slaughter and Jeff Crisp, in this volume, argue that established modes of humanitarian response aggravate protracted refugee situations, and that, therefore, agencies such as UNHCR should alter these modes in order to engage host states more effectively and to ensure that their actions do not further contribute to or entrench these situations. They recommend that UNHCR 'right-size' expectations of itself, and supplement rather than supplant the efforts of host states to respond to the needs of refugees on their territory.

While their analysis is correct, it is also clear that hope and responsibility will continue to be pinned on UNHCR for the foreseeable future. Successive High Commissioners have continually expanded UNHCR's role and space,[38] and it is not easy to see how much UNHCR could or should retreat, or to whom it could reasonably expect to hand over its functions.

Given the above, UNHCR should use its central position to advocate for reinvigorated, but realistic, responses. Rather than permit others to establish expectations that it will be unable to meet, UNHCR must be at the forefront in advocating different strategies (as outlined below) that recognize the long-term political solutions but are grounded in making an immediate difference on the ground, whatever the prevailing realities.

Owing to its particular organizational ethos, its desire and ability to react swiftly in refugee emergencies, and its conditional funding structure, UNHCR finds itself – time and again – striking implicit deals that ensure immediate and basic protection without accounting for longer-term effects.[39] These deals, and their underlying premises, must be revisited constantly. Numerous studies have pointed to the need for UNHCR to begin planning for the medium term right at the start of refugee emergencies;[40] this is one way to begin breaking away from such compacts.

Strategy: Focus on the present, provide security and freedom in exile

This chapter argues that, while the ultimate causes of and solutions to protracted refugee situations are political, 'caring and maintaining' refugees until political conditions are ripe for a solution condemns millions to a state of unfreedom. Given the untenable nature of such a situation, and given UNHCR's central role in finding solutions, what sort of strategy should be adopted to move away from this condition and ensure that refugees are protected, able to enjoy essential freedoms even in exile, and presented with a cogent vision for their future?

Any strategy must take as its starting point respect for refugees and their potential. As a refugee, one is entitled to protection from *refoulement* and to a durable solution. As an individual – particularly one who has come to the direct attention, and possibly care, of the United Nations – one must be able to enjoy the rights and freedoms enshrined in the Universal Declaration on Human Rights, and must be allowed to develop one's capabilities to the maximum extent.

The 2003 Executive Committee document on protracted refugee situations[41] suggested a policy along these lines, which consisted of providing refugees with physical, legal and economic security; removing barriers to self-reliance; and creating opportunities. The rationale behind this approach was to propose a policy that was fully in line with UNHCR's protection mandate, while at the same time able to offer a principled approach centred on the individual that could be applied even in the absence of a durable solution. The first element – security – involves core UNHCR concerns. The second element – removal of barriers – takes an Amartya Sen approach (grounded in the 1951 Refugee Convention and the human rights instruments) in its focus on removing unfreedoms and thereby enabling the greater realization of human capabilities. Opportunity creation, the third element, involves building on refugee capacities through provision of loans, income-generation projects and the like.

Besides being rights and capability based, this approach helps to build a feeling of security in asylum. Contrary to some arguments put forward by host states, such security is a key component for the realization of truly durable solutions, whenever these present themselves. A refugee who feels secure in asylum, and who has been able to develop his or her capabilities during exile, is a confident and dynamic candidate for voluntary repatriation (or local integration or resettlement). Such refugees tend to exercise their agency to take decisions based on evidence and from a secure position. In Guatemala, for example, an evaluative meeting on the repatriation and reintegration exercise concluded that 'security of asylum, and a choice amongst durable solutions, greatly enhances the voluntariness of repatriation. It enables refugees to make meaningful choices and enter into negotiations with the government from a relatively firm position'.[42] Similarly, in Iran, a UNHCR evaluation suggested that Afghans in Iran 'with considerable resources, and who consequently enjoy the security required to take a calculated risk', were one of the groups likely to repatriate most readily.[43] This type of secure and planned return is often an essential component to building peace in war-torn societies, as High Commissioner António Guterres recognized when he observed to the UN Security Council that 'refugees return with schooling and new skills ... Over and over, we see that their participation is necessary for the consolidation of both peace and post-conflict economic recovery'.[44]

'Any measures calculated to improve the situation of refugees': Applying the strategy through segmenting and targeting

If it is clear that the prolongation of refugee situations in camp or other restrictive situations results in crippling rights and capability deprivations, and if UNHCR is the entity most directly involved and implicated in such situations, then it must approach them with the same urgency it applies to other, more photogenic emergencies. Since it is equally clear that comprehensive solutions are often difficult and remote, it follows that – without neglecting the bigger picture – UNHCR must focus on the immediate condition of the individual refugees. It must, in the words of its Statute, promote 'the execution of any measures calculated to improve the situation of refugees and to reduce the number requiring protection'.

What does this mean in practice? UNHCR should take a detailed look at the composition of the different populations in protracted exile and decide what the best strategies for them are, sub-group by sub-group. Dealing with an entire population can be frustrating because, unless there is real political change in the country of origin, progress is unlikely. Segmenting that population is more fruitful, for it entails separating it into sub-groups with different profiles and for whom different strategies might lead to improvements. This approach requires six elements to be in place: a solid understanding of the constellation of interests and opportunities that determines the scope of intervention; a focus on attainable goals in asylum; the provision of safety nets; the implementation of durable solutions for sub-groups able to benefit from them; more funds, not less; and the incorporation of all elements within a larger and forward-looking strategy.

Determining the scope for achievement

Even in the most dispiriting and seemingly intractable situations, there are opportunities and interests; the challenge is to identify and pursue them. Most staff working in refugee camps recognize some of these possibilities, and in some places they are able to act upon them. But in many other situations part of the problem is that a combination of an established approach (care and maintenance), coupled with diminishing funds, stymies the impulse to discover and act upon such opportunities.

There are a number of means for UNHCR to map out a situation and derive modest but workable solutions for sub-groups. Of particular value in this respect are the participatory planning exercises undertaken together with all main partners in certain selected situations. These provide a good platform from which to plan, but have not been especially successful because their straightforward attempts at deriving needs and not resource-based budgets have encountered funding constraints, disillusioning participants at these events. UNHCR needs to experience a considerable change in its outlook, and in its approach to funding, for these to be fully effective exercises. Nevertheless, it is a good place to begin to identify solutions in exile and durable solutions.

Results depend upon traction and interests. What do host states really want when they strike the infamous 'deal' with UNHCR? Are there means of convincing them to loosen restrictions? And is UNHCR always speaking with the right state organ on these issues? To take the Kakuma example, Turkana province, where the camp is located, is remote, difficult to access and without a significant, local entrepreneurial class. While Nairobi was insisting on refugee sequestration, the local District Commissioner told a UNHCR evaluator that he would love to see some of the skilled refugees running a *matatu* (taxi) service within the district, which would not only generate refugee employment but provide a sorely needed service to the local population.[45]

Jeff Crisp has noted four areas where refugee and local interests could converge sufficiently to enable local integration in African contexts: when refugees are in an area inhabited by people of the same ethnic origin; when refugees are in areas with surplus land and economic opportunities; when refugees are *de facto* self-reliant, but do not have legal status or residence rights; or when a residual caseload has established strong links to the country of asylum.[46]

Focus on attainable goals in asylum

All too often, asylum is viewed in a static manner, a place for refugees to remain in a state of animated suspension – cared for and maintained until something better comes along. This is wrong, particularly when viewed through the Amartya Sen prism of capability deprivation and unfreedom. The 2003 Executive Committee paper recognized the need to focus on attainable goals in asylum, and suggested the above-mentioned three-pronged approach of providing security, removing barriers and creating opportunities.

Self-reliance for refugees is the cornerstone of any such approach, a fact well understood at UNHCR. Several Executive Committee conclusions have recognized the importance of self-reliance for refugees,[47] and it underpins the three main approaches in UNHCR's Framework for Durable Solutions (Development Assistance for Refugees, Development through Local Integration, and sustainable Repatriation, Reintegration, Reconciliation and Reconstruction). UNHCR's *Handbook for Self-Reliance* calls it 'a key component in any strategy aimed at avoiding or addressing protracted refugee situations, enabling agencies and refugees to find durable solutions that are truly sustainable'.[48] The handbook goes on to furnish a number of cases where self-reliance has worked, and provides ideas on how to help self-reliance to occur. It situates self-reliance firmly within the rights-based approach to tackling protracted refugee situations, and the first two arguments in its favour are about ensuring that refugees are treated in accordance with human rights principles and about addressing human development amongst refugees. Enabling self-reliance provides refugees with options and permits their human capabilities to develop.

Other strategies can also be applied to ease dire conditions and build capabilities. Education is an obvious example, for learning is the ultimate transportable and intrinsic asset. Education is not only a right in and of itself, but also a portal to the realization of other rights, including the right to return. This goes against those who feel that education makes refugees too comfortable in the countries of asylum, but is undeniable. Many senior African government officials have received university scholarships from UNHCR, and their presence today in their countries of origin, in consequential positions, is testimony to the effectiveness of this option.[49] Scholarships have been a prime victim of funding cuts at UNHCR but, without a truly effective replacement from other sources, this option has receded, much to the detriment of university age refugees.

There are also more modest approaches, such as the organization of athletic and artistic activities for refugee youth.[50] Also on the front-line when budget cuts are made, these activities go a long way towards lifting gloom in refugee settings, enabling young refugees to develop in healthy ways that will benefit them in the future, and equip them to take firmer control of their destinies when the opportunity arises.

To a certain extent, these approaches entail *de facto* local integration. True local integration is difficult to realize, and its mere mention is often enough to derail discussions with a host government. That said, it is still possible to make considerable progress in improving a refugee's plight and productivity, even in the absence of local integration. UNHCR recognizes this, which is one reason it tries to use other concepts. A recent discussion amongst UNHCR and NGOs observed that 'the notion of local integration has to be deconstructed. In some cases, full integration, involving naturalization and citizenship, might be possible, but in many situations, approaches such as local integration and self-reliance are more appropriate and more workable'.[51] The terminology is important, because it pre-empts the thorny legal questions and puts the spotlight on making progress in the current restricted, but not entirely sterile, context.

The provision of safety nets

A UNHCR document states that: 'Governments normally guarantee the basic human rights and physical security of their citizens. But when civilians become refugees this safety net disappears'.[52] Refugee status implies a particular vulnerability – that of having neither state nor regular family and social networks to fall back upon. Assistance programmes, and the camps in which they are often administered, serve as powerful nets. Their utility, and the lack of alternatives, explains why so many refugees cling to camps even when offered other opportunities – the risks in case of failure are too great. Thus many refugees use camps as their safety mechanism, with heads of family often keeping the family in the camps while themselves looking for more viable economic opportunities.

UNHCR and other international agencies need to be able to offer security without tying it to encampment. The World Food Programme has tried such an approach in northern Uganda, in which it provided support to refugees to become self-reliant and cut off regular food distributions, while at the same time maintaining the capacity to effect short-term distributions during lean months.[53]

The targeted and incremental execution of durable solutions

Even in the least promising conditions, it is always possible to help some sub-groups of refugees benefit from one of the three traditional durable solutions. UNHCR should pursue these limited but significant options because it is mandated to do so, but also because progress towards solutions, even if modest, creates a different dynamic and atmosphere in protracted situations, and places the emphasis on realizable change.

Local integration

The previous section has described some of the issues surrounding *de facto* versus *de jure* local integration. The former is easier to implement, and will remain the main strategy for most. However, *de jure* local integration is sometimes a definite legal possibility for some groups, and when this is the case UNHCR should push hard to obtain it. Certain groups that might qualify include spouses, descendants of citizens and children.

Voluntary repatriation

The preconditions for voluntary repatriation are well known, and UNHCR cannot promote repatriation to countries where the original causes of flight persist. But there are situations where parts of a country may be peaceful, or where political circumstances have changed but mass repatriation is not occurring (South Sudan is a good example of the latter). UNHCR has to have a detailed sense of the population in question and their motivations. Using this information, it can begin to design programmes that target those who would return, but for the lack of one or two elements (that UNHCR can provide). This approach, used from the start and involving more concentrated efforts and greater resources, is likely to be more fruitful than an immediate application of a uniform – and usually relatively low – repatriation assistance policy. Assuming that such return proceeds well, such movements can build up a positive momentum for more general voluntary repatriation. This approach does not always work, particularly if political conditions deteriorate or if the economic absorption capacity does not exist; as High Commissioner Guterres put it regarding South Sudan: 'Despite the resilience of the people, it is naïve to expect that pots, pans and hope are enough to begin life over'.[54]

Resettlement

After some years of relative neglect, resettlement is once again a major UNHCR priority, with UNHCR Global Strategic Objective 3.4 being to 'enhance the use of resettlement as a strategic protection tool, durable solution, and burden and responsibility sharing mechanism'. Resettlement is a particularly potent tool in situations of protracted displacement because – unlike the other two durable solutions – UNHCR has much more control over it and it is easy to measure. The centrepiece of UNHCR's current thinking on resettlement is to use it strategically in order to maximize the benefits to those other than the refugee being resettled – other refugees, hosts, other states and the international protection regime in general. This is done primarily through leveraging protection for other refugees through the demonstration of tangible international burden-sharing which, it is hoped, will convince the host state to relax asylum restrictions.[55]

Resettlement has the potential both to 'clear out' certain situations, but also to create unmanageable expectations. In some camps, it has been possible to segment the population and extract certain well-defined groups through resettlement. Some groups identified through this approach include the Sudanese Lost Boys and the Somali Bantus.

In the Middle East and North Africa region, UNHCR has urged resettlement countries to help it resolve intractable, or 'sticky', situations through targeted resettlement. These situations are usually protracted, involving difficult but finite populations who share common claims and are often 'residual', and who are located in countries with very tight asylum climates where resettlement is one of the few possible options. Populations under this category include ex-Iraq Palestinians in Ruwaished, Jordan; Iraqis in Rafha camp, Saudi Arabia; Darfurians in Iraq; and sub-Saharan Africans in Algeria. The numbers are relatively small and the groups are finite, so UNHCR is pushing hard for a resolution to these problems, which otherwise weigh down operations and relations in the region. However, for larger populations, such as Sudanese in Egypt or Iraqis in Syria and Jordan, resettlement has to be used with considerable precision – focusing on specific vulnerability criteria – in order to avoid the problems associated with raised expectations.

Other solutions

Finally, UNHCR must also pursue other solutions. The three classic solutions are well defined, and UNHCR knows when to push for them and what the legal and protection implications are of going with one or other of them. However, there are other opportunities, such as legal and labour migration. The former is being contemplated in, for example, the 10-point Plan. The latter is somewhat less appealing to UNHCR because there are no guarantees about what happens to a refugee issued with a worker visa once that visa expires. Return to the country of origin must be avoided at all costs, but readmission to the first country of asylum is usually not possible. Nonetheless, given that labour migration is a hallmark of today's world, and that many of the refugees come from countries where their compatriots do travel for work purposes, it makes sense for UNHCR to devote more energy and attention to trying to come up with a protection-compliant strategy for channelling certain categories of refugees into a labour migration stream.

More funds, not less

UNHCR's refugee programmes are funded through voluntary contributions, and these contributions almost inevitably decrease the longer a refugee situation endures. This is probably the single most important factor inhibiting UNHCR's ability to do something dynamic and effective for refugees marooned in protracted situations. The Office does not lack for ideas or handbooks, but the availability of funding forces programmes into specific, constricting contours.

To emerge from protracted situations, more funds are required in order to accommodate the labour intensiveness of taking a segmented, targeted approach, and to put in place new systems that, while economical in the long term, will cost more in the short term. Funding cuts completely undermine this and force UNHCR into what has been called 'plastic-sheeting mode'. This term describes a situation where UNHCR keeps buying plastic sheeting for refugees, year after year, rather than making a one-time investment in a more durable shelter simply because that durable shelter would cost more in a given year (although the cost would be amortized over the medium term).

Although long-term refugees are in acutely constrained situations in which they are unable to enjoy basic rights or develop capabilities, their plight seems to elicit frustration and non-interest. UNHCR is caught in a dilemma for, when it decides to take dramatic and imaginative action in such situations, it finds that it cannot because it barely has sufficient funds to keep humdrum but vital care and maintenance programmes running. Donors see this as a lack of initiative on UNHCR's part, and observe only a static situation with no solution in sight, and consequently keep cutting funds.

What can be done about this? UNHCR's funding structure is not about to be changed. An equivalent of the CERF (the OCHA-administered Central Emergency Revolving Fund) for protracted situations would be ideal, as it would provide the Office with seed money with which to begin the type of targeted and initially costly interventions that might be successful and thus attract additional funds. Furthermore, taking a leaf from its early days, when it got a kick-start with funding from the Ford Foundation,[56] UNHCR needs to continue to deepen its relationship with private sector funders, such as the Gates and Nike foundations.

A vision and an approach

A segmented approach is not a piecemeal one. Rather, it is deeply embedded within a long-term strategic vision for each particular refugee situation and driven by a UNHCR that is far-sighted in its thinking, firm in its leadership, limber in its recognition of the role of other actors, customized in its actions, and flexible and responsive in its evolution.

This chapter has focused on ad hoc, modest and segmented approaches, rather than comprehensive ones. This is not due to a disagreement with the latter's underpinnings, but rather because, while the principles may be sound, these are not always applicable, immediately, in protracted refugee situations. If the grand, political approaches are unattainable, then – this chapter suggests – UNHCR must persevere in a lower gear in order to be part of the solution to refugee situations and not to perpetuate problems. It should always keep in mind what actions are needed to ensure that refugees are given a fighting chance to enjoy their rights and freedoms, and which of these actions can be realized – in the short, medium and long term. It needs to be able to articulate its customized visions effectively and imaginatively if it is to succeed in funding them.

The politically grounded approaches, the CPAs, remain firmly on the table, as elaborated in various UNHCR documents and also in Loescher and Milner's solutions-oriented approach outlined in chapter 16. Such conceptual frameworks should always inform segmented approaches, but there should be clear-sightedness regarding the elements that could trigger their successful application and give them traction, and about the time-frames needed to achieve them. Moreover, when it seems that a political solution could be found, UNHCR should make more use of the United Nations system, including the moral authority of the Secretary-General or such bodies as the Peacebuilding Commission, to move refugee issues onto the agenda and push for firm political, integrated support to resolve them.

Finally, once UNHCR has gone ahead and elaborated visions and plans for all protracted refugee situations, as per its Global Strategic Objective, it must take the initiative to constantly evaluate progress and to modify plans in light of actions deemed to have worked or not, or assumptions validated or invalidated.

Conclusion

'The awkward truth about human deprivation', Prince Sadruddin observed near the end of his term as High Commissioner, 'is that it demeans those who permit or ignore it, more than it does those who are deprived'.[57] If refugees are but a sliver in the wider spectrum of human deprivation, they are nonetheless a group recognized by the international community as being in need of particular attention. When their plight slips from protected to protracted, and UNHCR finds itself 'administering misery', then it needs to act with responsibility, accountability, imagination and 'constructive impatience'[58] to bring about immediate changes in their condition.

Notes

1. UNHCR Mission Statement.
2. The Crisp and Slaughter chapter in this volume.
3. Described thus by former UN High Commissioner for Refugees Sadako Ogata. Reflecting on the crises facing the Office in the 1990s, and her own approach to dealing with them, Ogata focuses on emergencies, politics, relations with the military, repatriations and peacebuilding. Camps are viewed primarily through a protection and security prism, and there is little discussion of protracted refugee situations as such. Ogata, *The Turbulent Decade: Confronting the Refugee Crises of the 1990s*, New York: Norton, 2005.
4. EPAU conducted evaluations of the situation in Kakuma, Kenya; Ukwimi, Zambia; Guinea; Côte d'Ivoire; southern Uganda; Kigoma, Tanzania; Ghana; Dadaab, Kenya; Western Province, Zambia; and Mexico. A summary of some of the main findings is contained in J. Crisp, 'No Solutions in Sight: The Problem of Protracted Refugee Situations in Africa', *New Issues in Refugee Research*, Working Paper no. 75, Geneva: UNHCR, January 2003.
5. 'Addressing Protracted Refugee Situations in Africa', paper prepared for the 'Informal Consultations on New Approaches and Partnerships for Protection and Solutions in Africa', Geneva, 14 December 2001.
6. This complemented a number of existing tools, in particular the May 2003 Framework for Durable Solutions, which brought together three initiatives: Development Assistance for Refugees, Development through Local Integration and the 4Rs (Repatriation, Reintegration, Rehabilitation and Reconstruction).
7. UNHCR, 'Protracted Refugee Situations', paper presented to UNHCR's Executive Committee (EC/54/SC/CRP.14), Geneva, 10 June 2004.
8. Closing Statement by Mr. António Guterres, United Nations High Commissioner for Refugees, at the 56th Session of the Executive Committee of the High Commissioner's Programme, Geneva, 7 October 2005.
9. UNHCR, *Global Appeal 2007: Strategies and Programmes*, Geneva: UNHCR, 2007.
10. Notably 'The Long Road Home: Protracted Refugee Situations in Africa', *Survival* 47, no. 2, 2005; and G. Loescher and J. Milner, 'Protracted Refugee Situations: Domestic and International Security Implications', Adelphi Paper 375, Abingdon: Routledge for the International Institute for Strategic Studies, July 2005; and chapter 5, 'Protracted Refugee Situations: The Search for Practical Solutions', in UNHCR, *The State of the World's Refugees: Human Displacement in the New Millennium*, Oxford: Oxford University Press, 2006.
11. US Committee for Refugees, *World Refugee Survey 2004: Warehousing Issue*, Washington, DC: USCR, 2004.
12. http://www.un.org/events/tenstories_2006.
13. 'Protracted Refugee Situations: Impact and Challenges', Statement by Kamel Morjane, UN Assistant High Commissioner for Refugees, Copenhagen, 23 October 2002.
14. Also of note is the Mexico Plan of Action, which aims to address the protection needs of refugees and internally displaced persons in Latin America, including through regional resettlement.
15. UNHCR, 'Information Note: Preparatory Project for the Elaboration of a Comprehensive Plan of Action for Somali Refugees', High Commissioner's Forum (FORUM/2004/8), 24 September 2004.
16. *The State of the World's Refugees*, 2006, p. 124.
17. Ibid., p. 59.

18. UNHCR, 'Refugee Protection and Mixed Migration: A 10-Point Plan of Action' (revised version), Geneva: UNHCR, 1 January 2007.
19. Ogata, *The Turbulent Decade*, p. 317.
20. High Commissioner Gerrit van Heuven Goedhart, quoted in Loescher, *UNHCR and World Politics: A Perilous Path*, Oxford: Oxford University Press, 2001, p. 50.
21. Address by Ruud Lubbers, United Nations High Commissioner for Refugees, to the 58th Session of the United Nations Commission on Human Rights, Geneva, 20 March 2002.
22. UNHCR, 'Protracted Refugee Situations'.
23. See, for example, B. Harrell-Bond, 'Towards the Social "Integration" of Refugee Populations in Host Countries in Africa', presented at the Stanley Foundation Conference 'Refugee Protection in Africa: How to Ensure Security and Development for Refugees and Hosts', Entebbe, Uganda, 10–14 November 2002.
24. US Committee for Refugees, 'Statement Calling for Solutions to End the Warehousing of Refugees', 2004.
25. For one perspective on the gains and compromises involved in camps, see A. Jamal, 'Camps and Freedoms: Long-term Refugee Situations in Africa', *Forced Migration Review* 16, 2003.
26. This is the case with some of the Iraqi refugees in the Middle East. Estimated at some 2 million, this non-camp population has been largely invisible, recently prompting one well-educated Iraqi refugee in Syria to appeal to High Commissioner Guterres to establish camps to protect them from deportation and the hardship of paying high rents for squalid and insecure lodgings (Damascus, 9 February 2007).
27. This author, for example, was directly involved in decisions to establish camps in at least two situations: Herat, Afghanistan, and Takhta Bazar, Turkmenistan.
28. A. Jamal, 'Minimum Standards and Essential Needs in a Protracted Refugee Situation: A Review of the UNHCR Programme in Kakuma, Kenya', Geneva: Evaluation and Policy Analysis Unit, UNHCR, 2000.
29. A. Sen, *Development as Freedom*, Oxford: Oxford University Press, 1999.
30. Ibid., p. 11.
31. Ibid., p. 8.
32. C. Moorehead, *Human Cargo: A Journey Among Refugees*, New York: Henry Holt and Company, 2005, p. 156.
33. C. Horst, 'Connected Lives: Somalis in Minneapolis, Family Responsibilities and the Migration Dreams of Relatives', *New Issues in Refugee Research*, Research Paper no. 124, Geneva: UNHCR, July 2006.
34. Statement by Prince Sadruddin Aga Khan, United Nations High Commissioner for Refugees, to the Third Committee of the General Assembly, 26 November 1973.
35. United Nations, *In Larger Freedom: Towards Development, Security and Human Rights For All*, New York: United Nations Publications, 2005.
36. B. Harrell-Bond, 'Towards the Social "Integration" of Refugee Populations in Host Countries in Africa', p. 2.
37. See A. Jamal, 'Negotiating Protection in a Central Asian Emergency', Working Paper no. 17, *New Issues in Refugee Research*, Geneva: UNHCR, February 2000.
38. See G. Loescher, *UNHCR and World Politics: A Perilous Path*.
39. See, for example, A. Jamal, 'Camps and Freedoms and Slaughter and Crisp in this volume for descriptions of the 'deal'.
40. See, for example, Tom Kuhlman, 'Responding to Protracted Refugee Situations: A Case Study of Liberian Refugees in Côte d'Ivoire', UNHCR, Geneva, EPAU/2002/07, July 2002.
41. UNHCR, 'Protracted Refugee Situations'.

42. A. Jamal, 'Refugee Repatriation and Reintegration in Guatemala: Lessons Learned from UNHCR's Experience', Geneva: UNHCR, EPAU/2000/03, p. 2.
43. The other group was those with scarcely any resources, such as day labourers, who had little to lose by returning. A. Jamal and E. Stigter, 'Real-time Evaluation of UNHCR's Response to the Afghanistan Emergency: Bulletin no. 3', Geneva: UNHCR, 31 May 2002.
44. Statement by António Guterres, United Nations High Commissioner for Refugees, to the United Nations Security Council, New York, 24 January 2006.
45. A. Jamal, 'Minimum Standards', p. 29.
46. J. Crisp, 'No Solutions in Sight', p. 25.
47. In particular, Conclusions no. 50 (XXXIX)–1988, no. 88 (L)–1999, no. 93 (LIII)–2002, no. 95 (LIV)–2003, no. 100 (LV)–2004, and no. 101 (LV)–2004.
48. UNHCR, *Handbook for Self-Reliance*, Geneva: UNHCR, August 2005. p. xi.
49. Author's encounters and discussions with colleagues.
50. A fascinating account of the positive role of football amongst a group of resettled refugees is contained in the *New York Times* article 'Refugees Find Hostility and Hope on Soccer Field', by W. St. John, 21 January 2007.
51. UNHCR, 57th Session of the Executive Committee: Report on the Consultations with Non-Governmental Organizations, 27–29 September 2006, Geneva, p. 19.
52. UNHCR, *Protecting Refugees & the Role of UNHCR: 2007–08*. See http://www.unhcr.org/basics/BASIC/4034b6a34.pdf.
53. For a nuanced view of this approach see B. Broughton, et al., 'Evaluation of PRRO Uganda 6176', Rome: WFP, 31 August 2001.
54. High Commissioner's speech at the 2006 Executive Committee.
55. UNHCR, 'The Strategic Use of Resettlement', Standing Committee, 3 June 2003 (EC/53/SC/CRP.10/Add.1).
56. Loescher, *UNHCR and World Politics: A Perilous Path*.
57. Statement by Prince Sadruddin Aga Khan, United Nations High Commissioner for Refugees, to the Third Committee of the General Assembly, New York, 14 November 1977.
58. A. Sen, *Development as Freedom*, p. 11.

9

Historical lessons for overcoming protracted refugee situations[1]

Alexander Betts

Neither protracted refugee situations nor attempts by the international community to address them are new. Throughout the 1950s and until the mid-1960s, UNHCR's work focused on overcoming the protracted situation of those displaced in Europe by the Second World War. By appealing to governments to provide funding or resettlement, the work of UNHCR contributed to ensuring access to durable solutions for Europe's long-term displaced. With the geographical expansion of the scope of the 1951 Convention beyond Europe in 1967, UNHCR's work increasingly focused on addressing the consequences of protracted refugee situations in the global South. Throughout the Cold War, long-term exile was a common experience of Africa, Latin America and Asia, and UNHCR and the wider UN system attempted to develop multilateral approaches to find solutions to protracted situations that had similarities to many contemporary examples of long-standing refugee situations. Understanding and drawing upon the insights offered by these historical precedents has a great deal to offer current and future attempts to address long-standing exile in the global South.

The notion of comprehensive plans of action (CPAs) has recently been revived within the context of UNHCR's Convention Plus initiative as a suggested means of overcoming protracted refugee situations. Broadly speaking, CPAs represent multilateral approaches to ensure access to durable solutions for refugees within a given regional context. Such approaches can be regarded as 'CPAs' insofar as they are *comprehensive* in terms of drawing on a range of durable solutions simultaneously;

Protracted refugee situations: Political, human rights and security implications,
Loescher, Milner, Newman and Troeller (eds),
United Nations University Press, 2008, ISBN 978-92-808-1158-2

cooperative in terms of involving additional burden- or responsibility-sharing between countries of origin and asylum, and third countries acting as donors or resettlement countries; and *collaborative* in terms of working across UN agencies and with NGOs. Drawing upon the legacy of the two most prominent historical examples of CPAs – the International Conference on Central American Refugees (CIREFCA) and the Indo-Chinese Comprehensive Plan of Action, both of 1989 – UNHCR explicitly attempted to use the CPA approach as a means to address protracted refugee situations. Most notably, the revived CPA concept was re-launched by UNHCR in 2004 in the pilot project the 'Comprehensive Plan of Action for Somali Refugees', in order to address the protracted refugee situation of Somali refugees within East Africa and the Horn of Africa. By 2006 it was clear that this pilot project had failed to mobilize significant political or economic support for enhancing access to durable solutions in the region. However, the limitations of this pilot were not due to the irrelevance of the CPA concept for overcoming protracted refugee situations. Rather, the initiative's shortcomings owed a great deal to the failure of UNHCR and the wider UN system to learn from past multilateral attempts to address protracted refugee situations.

In order to assess the political and institutional preconditions for a successful CPA, this chapter examines the most significant examples of UNHCR-led attempts to develop multilateral approaches to overcome regional protracted refugee situations in the global South during the last 25 years: the International Conferences on Assistance to Refugees in Africa (ICARA I and II) of 1981 and 1984, CIREFCA and the Indo-Chinese CPA. The chapter argues that, assessed in terms of their ability to promote international cooperation to improve access to durable solutions and end long-term encampment, ICARA I and II constitute a relative failure and CIREFCA and the Indo-Chinese CPA represent relative successes. Furthermore, the chapter suggests that the two 1989 conferences represent an archetypal approach to addressing regional refugee situations that may be referred to as the *political engagement model*, while the ICARA I and II experience represents a contrasting model that may be referred to as the *programmatic model*. These two archetypes represent contrasting approaches to attempting to address protracted refugee situations through UNHCR-led multilateral approaches. Although both models can legitimately be claimed to be CPAs, and the models are not entirely mutually exclusive, they have contrasting implications. The former is based on sustained UN-facilitated political dialogue, culminating in political agreement between a range of governmental and non-governmental stakeholders, and including, but not being confined to, addressing the refugee issue. Meanwhile, the latter can be characterized as a technical process of identifying projects that address the situation of the displaced, compiling them into programmatic areas and seeking funding for them through a pledging conference.

This chapter argues that the limitations of UNHCR's attempt to revive the CPA approach in the CPA for Somali Refugees are largely due to its failure to learn from the important historical insights of these past precedents. In particular, it claims that although UNHCR used the examples of CIREFCA and Indo-China to legitimize its revival of the CPA concept, the model it adopted in the CPA for Somali Refugees was the failed programmatic model applied in ICARA I and II. The chapter suggests that the limitations of the Somali pilot project should therefore not discredit future attempts to develop CPAs; however, these should be based on the political engagement model drawn from the insights offered by CIREFCA and the Indo-Chinese CPA. Drawing upon material from UNHCR's archives and interviews with stakeholders in the various initiatives, the chapter therefore teases out what those lessons are and how they might be applied to address contemporary protracted refugee situations. In order to make these arguments, the chapter is divided into three sections. Firstly, it outlines the context, content and outcome of the three historical case studies. Secondly, it will extrapolate from the cases studies two diverging ideal-type models through which the UN system can develop multilateral approaches to respond to protracted refugee situations in the global South. Thirdly, it will draw upon the examples of CIREFCA and the Indo-Chinese CPA to examine the institutional and political conditions for developing future responses to protracted refugee situations that are based on the political engagement model.

The three case studies

ICARA I and II (1981 and 1984)

Although the ICARA process is not commonly referred to as a CPA, it has many of the characteristics of one, and is an example of a multilateral attempt to respond to a regional protracted refugee situation.[2] The initiative represented an African-led response to the growing protractedness of what had been previously perceived to be a temporary refugee population. Until the late 1970s, the majority of African refugees had fled the continent's independence wars. Given the assumption that these wars would eventually result in victory for the independence movements, voluntary repatriation was assumed to be a viable durable solution. Yet this began to change in the late 1970s. By 1979, the majority of Africa's 3–4 million spontaneously settled rural refugees emanated from conflicts in

Chad, Ethiopia, Angola, Uganda and Zaire, and there appeared little immediate prospect of repatriation. The majority of these refugees were supported by the state's own resources and infrastructure, with little support from the international community.

The 1979 Arusha Conference, convened by the OAU, attempted to address the issue of how African states could equitably share the burden amongst themselves and this, in turn, led to a call from the African states to convene an international conference to explore how the international community could compensate the host states for the developmental costs of long-term hosting. This led to two Geneva-based conferences in which African states sought assistance to help reception states cope with the impact on their economic and social infrastructures of hosting (and having hosted) large rural refugee populations. These conferences – ICARA I and II – were conceived as one-off pledging conferences. In each case, UNHCR worked with UNDP and the African host states to develop a series of project proposals and programmes that could be submitted to the donor community for consideration. The projects were intended to have a focus on integrated development that would benefit both refugees and host country citizens by building infrastructure and social services. It was envisaged in both conferences that such an approach – based on the concept of Refugee Aid and Development (RAD) – would contribute to durable solutions, through facilitating both local integration and self-sufficiency pending repatriation. The ICARA conferences merit analysis, both because they failed to meet their aims and because their approach contrasted so greatly with that of CIREFCA and the Indo-Chinese CPA.

ICARA I's key objective was to 'aid countries of asylum in bearing the burden imposed upon them by the large number of refugees'.[3] Reflecting the African states' emphasis on the need for greater 'burden-sharing', it was primarily a pledging conference, setting out few ideas, principles or guidelines. The conference, held 9–10 April 1981, had three stated objectives: 1) to 'focus attention on the plight of refugees in Africa'; 2) to 'mobilize additional resources to assist both refugees and returnees'; and 3) to 'aid countries of asylum in bearing the burden imposed upon them by the large number of refugees'.[4] African states were invited to submit development and emergency relief projects, compiled in collaboration with UNHCR, to the conference for funding.

ICARA I failed to meet the host states' expectations for additional resources. Equally, however, it failed to satisfy Northern donor states because the financial commitments did not translate into durable solutions for refugees but were largely spent on supporting basic needs.[5] Although US$560 million was pledged at the conference, it was only later that the extent to which these pledges had been earmarked by states became increasingly apparent, leaving UNHCR with an estimated US$40 million available for the high-priority projects that did not fall into its regular or specific programmes.⁶ When the UN General Assembly reflected on the achievements of ICARA I, it regretted 'that, in spite of efforts made, the assistance provided to an increasing number of African refugees is still very inadequate'.⁷ In many ways, this is unsurprising. ICARA I had established no process for political dialogue but had simply made the assumption, firstly, that donor states would altruistically write a blank cheque for funding and a list of projects compiled by African states and only nominally verified by UNHCR and, secondly, that these projects would contribute to durable solutions for refugees, even in the absence of clear follow-up mechanisms or funding conditions.

ICARA II was a response to the failure of ICARA I. In the words of the Austrian Ambassador, it was conceived more as a 'think tank' than a 'pledging conference'.⁸ It purported to focus more on the conceptual areas of finding durable solutions through developing the principle of refugee-related development assistance. The central theme was 'Time for Solutions'.⁹ It had a strong focus on projects designed to promote the self-sufficiency and local integration of refugees. However, in practice, ICARA II was not dissimilar to the first conference in that it amounted to a one-off pledging conference that submitted the projects compiled by the African states, in collaboration with UNHCR and UNDP, to the donor community. Once again, it worked on the assumption that, if a list of projects were submitted to a conference, the donor community would altruistically and equitably allocate funding, even outside of a process of sustained political dialogue.

Ultimately, only US\$81 million of the US\$392 million sought was pledged at the conference. Once again, the conference was therefore a failure. A significant reason for the failure was that, while the African states wished to focus on burden-sharing, the donor states wished to focus on the durable solutions focus reflected by the conference theme. Although donors did not reject the notion of expanded burden-sharing *per se*, for them an increased economic commitment needed to be directly linked to expanded access to local integration. They wanted enhanced access to durable solutions for refugees rather than 'an open-ended claim on their resources'.¹⁰ However, in the absence of a sustained political dialogue there was no adequate means to ensure that both of these concerns were met in a way that could be mutually beneficial to African and donor states. No clear principles were established to link African states' interest in burden-sharing to Northern states' interests in durable solutions, and hence the only basis on which Northern states provided funding was, once again, their own perceived strategic interests in the context of the region's proxy conflicts. Consequently, ICARA II again led to a relatively small amount of money going into basic needs programmes in areas in which Northern states selectively earmarked their contributions based on their Cold War interests.

CIREFCA (1987–1994)

CIREFCA represented a response to the protracted refugee situation of Central American refugees displaced by the region's civil conflicts of the previous decade.[11] It dealt primarily with the 150,000 refugees in the region but also had links to initiatives that addressed the region's 900,000 undocumented 'externally displaced' and 900,000 IDPs. CIREFCA was convened by the governments of Belize, Costa Rica, El Salvador, Guatemala, Honduras, Mexico and Nicaragua. The main Conference took place in Guatemala City, 29–31 May 1989. It adopted a Declaration and a 'Concerted Plan of Action in Favour of Central American Refugees, Returnees and Displaced Persons' (CPA). This three-year plan, which was eventually extended by two further years, was adopted by 58 countries and represented a flexible strategy for the development of each of the seven convening states' own programmes.[12] The CPA provided an initial portfolio of 36 projects, requiring US$375 million over a three-year period, which was later added to. The initial project submissions were compiled by states with the support of a five-week UNHCR Mission to the region in mid-1988. The CPA also provided a set of 'Principles and Criteria for Protection and Assistance'. Implicitly, the adoption of policies, standards and legal norms was posited by UNHCR as a condition for states receiving financial support through CIREFCA. Like ICARA, CIREFCA was based on the idea of Refugee Aid and Development, adopting an integrated development approach based on UNDP partnership in order to promote local integration, self-sufficiency and sustainable repatriation by targeting refugees and returnees, on the one hand, and local host populations on the other. However, unlike ICARA, CIREFCA proved to be a great success in terms of providing durable solutions, enhancing protection in the region and contributing to peace and security. Indeed, CIREFCA represents the single most successful historical example of a multilateral response to a protracted refugee situation. This success was in no small part due to its differences in comparison to the ICARA process.

CIREFCA contrasted markedly with ICARA in the institutional and political approach adopted by UNHCR. Firstly, unlike the ICARA conferences, CIREFCA was based on a sustained political process. Rather than being based on a one-off conference, CIREFCA built up momentum over time based on a series of informal consultations with host states in the region and donors. Significantly, the Guatemala City Conference, as the focal point of CIREFCA, was explicitly *not* conceived as a pledging conference. Instead, its primary aim was to establish a political consensus upon which UNHCR could build a multi-year process. The initial stages of the strategy explicitly shunned a financial emphasis in favour of fostering political support. It noted of the 'lead-up phase':

> The top priority must be promotion of policy/political/diplomatic support for the Conference as such and for the strategies it represents. In this perspective, fund-raising of any active or specific kind is dangerous. Too much pressure on the fund-raising issue now could even affect the yet-to-be determined level and quality of political/policy support for the Conference.[13]

Instead of encouraging pledging, support for the process, high-level participation at the conference and 'mention[ing] discretely that 'it is ... the hope of UNHCR that policy support would be translated at a later date into a financial contribution/commitment' were highlighted as pre-conference priorities, and CIREFCA itself was seen primarily as a political event, with the Declaration and Concerted Plan of Action being an inter-state consensus rather than a programmatic list intended to attract money.[14] Contributions were only explicitly solicited at a series of international follow-up conferences. The CPA itself represented a set of political commitments, establishing clearly the responsibilities of all the major stakeholders and the principles underlying those responsibilities.

Secondly, unlike ICARA, CIREFCA institutionalized a range of wider partnerships and linkages to initiatives in other areas of the UN system. In contrast to ICARA, UNHCR's collaboration with UNDP was institutionalized within the San Jose-based Joint Support Unit (JSU). The JSU allowed the two agencies to develop a working relationship on coordination and implementation where, during ICARA II, the division of labour between UNHCR's responsibility for durable solutions and UNDP's responsibility for development had led to little direct coordination of the crucial relationship between the two elements of RAD. Both organizations provided the seven regional states with technical support in developing their own 'priority projects', both for initial submission to CIREFCA and for submission to the international follow-up conferences. Alongside CIREFCA, the Italian government's simultaneous PRODERE ('The Development Programme for Displaced Persons, Refugees and Returnees in Central America') project also provided significant funding to address other categories of the displaced, such as IDPs, that fell outside UNHCR's immediate mandate, and helped to provide another institutional link between UNHCR and UNDP.

CIREFCA also enjoyed the support of the wider UN system. Notably, the clear institutional relationship between CIREFCA and the wider peace process for the region proved crucial to CIREFCA's success. CIREFCA was institutionally linked to both the Esquipulas II Accords and the wider UNDP-led development plan for the region (the 'Partnership for Economic Cooperation', or PEC), in a way that created an opportunity for UNHCR to make CIREFCA a part of these wider initiatives and so channel the interests of states in these other areas into CIREFCA.[15] These relationships were explicitly referred to in the Concerted Plan of Action and the reports of the Secretary-General on the two related initiatives.[16] The institutional link to the peace agreement and wider development initiative also brought an immediate commitment from the Secretary-General to the issue of displacement. The decision for the Concerted Plan of Action to stand in for chapter 1 of the PEC, in particular, created an immediate institutional link between CIREFCA and the Office of the Secretary-General, immediately giving the refugee issue high status within the wider UN context.[17]

These wider linkages were the very reason why many states were willing to make commitments to the search for durable solutions, identifying the 'refugee issue' as institutionally and practically related to areas in which they held clear perceived interests. Indeed, both the Central American states and the donor states, which contributed the majority of the total project funding for CIREFCA, were largely committed to the peace process and the region's security and development rather than the refugee issue *per se*. Such linked interests were missing in ICARA, in which the 'refugee issue' was addressed largely in isolation from other issue areas. Indeed, the European commitment to funding CIREFCA stemmed from a broader interest to support the conditions which would facilitate Central America emerging as a viable European trade partner.[18] Italy's commitment to provide US$115 million to PRODERE was largely based on its own concerns to maintain solidarity with left-wing, Christian democratic governments in the region. Meanwhile, the main European bilateral donors – Sweden, Norway and Finland – were mainly concerned with the promotion of democracy, human rights and development in line with the Nordic states' broadly cosmopolitan approach to overseas assistance. In contrast, the contribution of the US government to CIREFCA was limited, at least until the fall of the Sandanista regime in Nicaragua, when, under the Reagan Administration, the US began to provide highly selective support for CIREFCA, which was largely directed towards supporting the new regime in Nicaragua. Hence, for all of the main state contributors, CIREFCA was important not because of an altruistic concern for the welfare of refugees *per se*, but because of a perceived relationship between the 'refugee issue', on the one hand, and security and development on the other.

The combination of states' perceived linked interests, the creation of institutional links to the wider peace process and regional development initiative, and a sustained process of political facilitation based on high-level UNHCR leadership therefore contrasted markedly with the ICARA experience and underpinned the initiative's success. Although there has been little formal monitoring of the projects implemented under CIREFCA, the extent to which it raised a significant proportion of its required funding clearly distinguishes it from the limited legacy of ICARA I and II. A General Assembly Resolution on CIREFCA passed at the 85th Session in late 1993 expressed 'its conviction that the work carried out through the integrated conference process could serve as a valuable lesson to be applied to other regions of the world'.[19] The most obvious contribution of CIREFCA was the projects which it developed, implemented and financed. Although the total amount of additional funding attracted by CIREFCA is difficult to estimate accurately because of difficulties in 'tracking' bilaterally funded NGO projects implemented 'in the framework of CIREFCA', a total of US$422.3 million was recorded by the CIREFCA Joint Support Unit by 1994, which amounts to an estimated 86% of the total project requirements. By 1993 this funding had provided full or partial financing for 72 'priority projects' in the seven countries.[20] The projects focused on a range of areas including immediate assistance, rehabilitation, economic development and institution-building.

In terms of durable solutions, CIREFCA contributed to voluntary repatriation through the protection principles it elaborated in the Plan of Action, through the resources it allocated to support reintegration and notably through political dialogue in relation to the Tripartite Agreements. This work allowed the repatriation of some 27,000 Salvadoreans and 62,000 Nicaraguans and the return of 45,000 Guatemalans from Mexico.[21] These returns were supported by what might be considered to be the precursor of UNHCR's 4Rs framework.[22] Within the framework of CIREFCA, UNHCR and UNDP also developed the notion of Quick Impact Projects (QIPs), supporting basic needs and short-term productive infrastructure for 70,000 returnees in Nicaragua.[23] The projects were also notable for the extent to which they facilitated self-sufficiency and local integration. The most obvious case study for successful self-sufficiency was in Mexico, in Campeche and Quintana Roo in the Yucatan Peninsula, where consolidation of the local agricultural settlements and the development of integrated service provision benefited both the 18,800 refugees and the host communities. In Chiapas, self-sufficiency was also encouraged, but a shortage of land was an obstacle to allowing refugees to become equally engaged in agricultural activities. In Campeche and Quintana Roo, local integration and repatriation were promoted simultaneously from 1996, while in Chiapas local integration *followed* repatriation from 1998 onwards. The self-sufficiency and local integration projects ultimately provided education, health services, access to markets and sustainable livelihoods. For the Mexican government the projects were seen as an attractive means to develop the poorest areas of the country, particularly in the Yucatan Peninsula.[24] CIREFCA also provided local integration for Salvadorean refugees in Belize, particularly through the Valley of Peace project. Although the project had begun in 1983 and had been widely criticized for relocating refugees to a jungle area with poor roads and poor-quality land, CIREFCA helped to resurrect the Valley of Peace project.[25] By 2003, some 300 families remained and were integrated alongside the Belizeans of predominantly Maya Quechi ethnicity. Initially supported with food aid, a fund to build housing, tools and seeds, many of the Salvadoreans now work in the tourism industry or in local employment, receiving social services alongside the Belizean community.[26] There was also a degree of local integration in Costa Rica. This took place on a smaller scale and was mainly for Salvadorean refugees in urban areas, who were few in number and were perceived to be 'hard working'. This contrasted with the Costa Rican approach to the Nicaraguan refugees, who, although they were given a degree of self-sufficiency in agricultural production, had been largely confined to camps and were not given the same level of opportunities to integrate.[27]

Indo-Chinese CPA (1988–1996)

Although the Indo-Chinese CPA responded primarily to a long-standing mass influx situation, rather than an initiative designed to address long-term encampment, it is nevertheless relevant for considering responses to protracted refugee situations.[28] This is because, firstly, it was about enhancing access to durable solutions for refugees, many of whom were in an intractable state of limbo, and, secondly, it represents one of the most successful examples of UNHCR-facilitated multilateral cooperation.

The CPA represented a follow-up to the first international conference on Indo-Chinese refugees in 1979 which had agreed that the US, along with other Northern states, would commit to resettle all the Indo-Chinese refugees offered asylum in the region. However, by the late 1980s, this agreement had largely broken down, the US commitment to resettlement was dwindling, and the countries of first asylum were beginning to revert to 'pushing back' the arriving boats. Despite the resettlement of large numbers of refugees since 1979, roughly 150,000 remained in camps in Southeast Asia at the end of 1988.[29]

In contrast to the previous decade, a new dimension emerged in the process, in which, for the first time, the Socialist Republic of Vietnam

(SRV), as part of its wider attempts to repair its ties with ASEAN, declared itself willing to engage in the process and to repatriate without punishment or persecution those who voluntarily agreed to return.[30] The end of the Cold War and the general thaw in US–Vietnamese relations therefore meant that the CPA introduced two significant new elements: the screening of asylum seekers in countries of first asylum, and the possibility of return to Vietnam for non-refugees. In the words of Sergio Vieira de Mello, a former UN High Commissioner for Human Rights, there was therefore a need for 'a new solutions-oriented consensus involving the cooperation of countries of origin, first asylum and resettlement'.[31] The CPA's combination of consensus between host countries of first asylum, the country of origin and third countries beyond the region makes it an important case study.

The CPA adopted in Geneva relied upon a three-way commitment by countries of first asylum in the region, counties of resettlement beyond the region, and the main country of origin. For the CPA to be successful, each group of stakeholders had to perceive that their own contribution directly underpinned the overall aim of finding a comprehensive solution to the 'problem' of the Indo-Chinese 'boat people'. The resettlement states, led by the US, agreed to resettle all those already in the asylum countries up to a 'cut-off' date and all those determined to be 'refugees' by individual refugee status determination after the cut-off. The cut-off dates varied from state to state but began from as early as 14 March 1989. In return, the ASEAN states and Hong Kong agreed to maintain the principle of first asylum and cease engaging in 'push backs'. Meanwhile, Vietnam committed to accept the voluntary return of non-refugees, to work to limit clandestine departures, and to continue with the Orderly Departure Procedure (ODP), to allow people to emigrate from Vietnam via an alternative migratory channel.

As with CIREFCA, the main conference was intended to be a political focal point at which the CPA, as a clearly elaborated political consensus, could be launched. The actual CPA was compiled by a small Drafting Group of the major stakeholders, based on these informal consultations. The document was extremely concise, highlighting the main obligations of the different groups of states. It elaborated the principal commitments of the host states, resettlement states and country of origin, while leaving the details to be agreed after the main 1989 conference had ended.

Following the CPA, a Coordinating Committee comprising a 'core group' of states was assembled. The Committee provided a focal point to which the three Sub-Committees on, firstly, 'Reception and Status Determination', secondly, 'Departures and Repatriation', and, thirdly, 'Resettlement' could report their work.[32] This work established the substantive details for how the CPA, as a basic political agreement, would be implemented in practice following its adoption.

The CPA was possible because all of the main stakeholders had perceived interests in linked issue areas. The ASEAN states were largely concerned with migration control. Malaysia and Indonesia, in particular, were keen to limit their ethnic Chinese populations so as not to alter the demographic balance, which they perceived in terms of national security in the context of their wider relationship with China. Meanwhile, other states, even those largely unaffected by the 'boat people' such the Philippines, were keen to uphold solidarity within ASEAN by backing the affected states. For the US, the legacy of the Vietnam War, the subsequent commitment to resettlement in light of the withdrawal from Saigon in 1975, the growing Vietnamese diaspora, and a concern with regional security gave it an ongoing stake in finally ending the mass exodus. For the SRV, the decline of the USSR and the end of the Cold War created a need to seek development assistance, trade and political rehabilitation from the international community. These underlying interests, present at the end of the Cold War, created the preconditions for inter-state agreement. However, by themselves they were not a sufficient condition for international cooperation.

Achieving the initial consensus of the CPA and then agreeing the means to implement the initial political agreement relied upon UNHCR providing significant and high-level organizational leadership in order to facilitate inter-state agreement. The Office explicitly identified itself as playing a 'catalytic role' in the preparatory process.[33] Much of this role is attributed to the contribution of Sergio Vieira de Mello, whose role many members of UNHCR staff identify as the single most significant reason underpinning inter-state agreement in the CPA.[34] Vieira de Mello's charismatic approach to conflict resolution fostered dialogue and his deft diplomatic skills helped to facilitate agreement. The extent to which UNHCR facilitation was important is particularly evident from the way in which UNHCR saved the CPA from near-collapse in 1990. One of the most significant unresolved issues in the CPA had been the US's unwillingness to countenance involuntary deportation to the SRV because it was still perceived as a Communist state. This, combined with the way the SRV initially limited return, meant that the ASEAN states quickly became disillusioned and began to threaten to continue with 'push backs' of arriving boats or to cease offering first asylum altogether. These concerns reached their most divisive level in 1990. The most serious impasse concerned the issue of return for those not recognized as refugees, with the US and Vietnam continuing to insist that return be voluntary. The unwillingness of Vietnam to allow returns at a satisfactory rate, and to reduce clandestine arrivals, led to crisis talks at the Steering Committee Meeting in Manila in mid-1990. Opening the meeting, Vieira de Mello suggested that 'Seldom ... have we been so close to a breakdown of this otherwise exemplary process'.[35] Indeed, complaining about the lack of cooperation from Vietnam, a joint statement from the countries of asylum threatened abandoning the principle of non-*refoulement*: 'In the event of failure to agree even an intermediate solution to the VBP [Vietnamese boat people] problem, countries of temporary refuge must reserve the right to take such unilateral action as they deem necessary to safeguard their national interest, including the abandonment of temporary refuge'.[36] The ASEAN states placed the blame squarely with the US:

> The United States, which opposes involuntary repatriation for its own reasons, has not been helpful either. In fact the United States' position provides comfort and protection to the Vietnamese intransigence ... It is the United States' insistence on treating the Vietnamese economic migrants differently that is putting the very principle of first asylum in peril.[37]

However, on the basis of the meeting in Manila, and thanks largely to the conflict resolution skills of Sergio Vieira de Mello, a 'Near Consensus Note' emerged. Significantly, this provided the basis for compromise on the issue of the return of non-refugees, which put the CPA back on track. In the words of Dennis McNamara, 'the consensus [on return] was not to call it forced and not to call it voluntary; just to say that those who were found to be refugees could not be sent back'.[38] The agreed compromise was that non-refugees should be actively encouraged to return on the basis of three months' counselling, would not be coerced, and would be monitored by UNHCR upon their return to the SRV.[39] It further noted that, while 'conditions of safety and dignity' should be upheld, 'the modalities of return ... would be a matter for first asylum countries to resolve with the country of origin, with the guidance and involvement of UNHCR and other appropriate agencies'.[40] In other words, UNHCR passed responsibility for return over to the countries of first asylum based on the understanding that it would be 'return respecting human rights' but tacitly acknowledging that strict voluntarism might need to be compromised for the CPA to be viable.[41] Having restored consensus, the CPA was duly reaffirmed by the Fourth Steering Committee in April 1991.[42] Although the details for implementation needed ongoing refinement and the Vietnamese refugees remained in protracted detention in Hong Kong throughout much of the 1990s, the reaffirmation that followed the Manila meeting represented the achievement of a lasting consensus which ultimately led to the resolution of the 'boat people' issue.

From 1991 the rate of voluntary returns increased rapidly and the number of new arrivals began to decline. As UNHCR increased the level of the reintegration grants for returnees and began implementing QIPs, the SRV was gradually persuaded that its interests lay in supporting return and cooperating to reduce clandestine departures. This strategy, the SRV realized, would attract the greatest bilateral and multilateral support for development assistance, trade and political engagement. Although UNHCR attempted to uphold the CPA's commitment to 'voluntary' return for non-refugees, in practice, from around 1992, the countries in the region increasingly engaged in coerced return, an approach which UNHCR tacitly acknowledged as the process drew to a close in 1996.[43] The CPA was widely criticized from a human rights perspective and there are doubts about the extent to which the screening process adequately respected the principle of non-*refoulement*. However, by 1996 UNHCR's process of high-level, inter-state political facilitation had contributed to ending the mass exodus and clearing the camps and detention centres of the host states of first asylum.

Two archetypal models

The three historical case studies highlight two contrasting approaches to UN-led multilateral responses to protracted refugee situations: a programmatic model and a political engagement model. These represent ideal-type models extrapolated from the cases and are by no means mutually exclusive. However, they help to highlight some of the contrasts in a way that illustrates why UNHCR's approach to ICARA had so little legacy and why CIREFCA and the Indo-Chinese CPA contributed to overcoming collective action failure. Identifying the characteristics of each model is also particularly useful because it highlights the way in which, in reviving the CPA concept, UNHCR has reverted to the failed programmatic model rather than the political engagement model. These characteristics are summarized in Table 9.1 and explained below.

In the first instance, the political engagement model implicit in the relatively successful CIREFCA and Indo-Chinese CPA has a number of characteristics. Firstly, it is based on political facilitation by UNHCR. In both cases, UNHCR engaged with and recognized states' interests within and beyond the refugee regime. It chaired and organized a series of formal and informal meetings in order to promote interest convergence and consensus amongst the main stakeholders. Rather than being based on one-off pledging conferences, CIREFCA and the Indo-Chinese CPA were based on a sustained political process carried out over time.

Table 9.1 Main characteristics of the two archetypal models for CPAs

	Political engagement model (e.g. CIREFCA/Indo-Chinese CPA)	Programmatic model (e.g. ICARA and the CPA for Somali Refugees)
Main UN actors	UN system (incl. Secretary-General)	UNHCR (and UNDP)
Time period	Sustained	One-off conference
Initial focus	Political	Financial
UNHCR role	Facilitative	Technical
Linkages	Broad	Narrow

UNHCR employed high-level staff with diplomatic, political and analytical skills to work with states and the wider UN system. In both cases, the CPAs and the main conferences held in Guatemala City and Geneva were not so much about attracting immediate donor commitments but rather centred on securing initial political commitment on general principles and establishing a political basis on which momentum and credibility could be built. Secondly, the approach was based on seeing the refugee regime within its broader context. In both cases, UNHCR recognized that states do not look at refugee issues in isolation and that durable solutions can best be found alongside addressing states' concerns in other issue areas, such as security, migration, development and peacebuilding. As such, CIREFCA in particular developed strong institutional links to the relevant peace process and regional development plan. Linking the 'refugee issue' to these wider concerns had a number of benefits. For example, it appealed to the linked interests of host and donor states in these wider areas and helped channel them into a focus on the 'refugee issue'. Furthermore, it gave the wider UN system, and particularly the Office of the Secretary-General, a stake in overcoming Central America's protracted refugee situations. The success of both CIREFCA and the Indo-Chinese CPA was therefore underpinned by strong and wide-reaching political facilitation by UNHCR and the wider UN system that recognized and appealed to states' wider interests in linked areas such as security, peacebuilding, migration and development.

In the second instance, the programmatic model applied in the case of the ICARA conferences contrasted markedly with the approach of CIREFCA and the Indo-Chinese CPA. Firstly, UNHCR's role was largely technical. In both conferences, it focused mainly on working with UNDP and the African host states to develop project submissions for the main conference. The advice and support they provided in the short-term missions and in-country consultations were based mainly on trying to ensure that the projects and programmes submitted to the conference reached certain technical standards. Beyond this, UNHCR's role was in serving as the secretariat for the international conferences. At the conferences, the projects and programme areas were compiled into country submissions that were put to donor states. The assumption was that, by providing a 'laundry list' of areas requiring funding, donors would provide the necessary funding. Secondly, the approach was largely apolitical. Reflecting the 'non-political character' of UNHCR, ICARA did not establish any framework for sustained dialogue between the African states and donor states. Instead, inter-state interaction was confined to the two conferences. In contrast to CIREFCA and the Indo-Chinese CPA, there was no possibility for a range of formal and informal meetings within and beyond the region to contribute to the emergence of trust, legitimacy and political consensus over time. Furthermore, the 'refugee issue' was largely addressed in isolation. Although UNDP was reluctantly involved in ICARA II, this was to provide technical support within the refugee context. Addressing the refugee issue in isolation greatly reduced the scope for issue linkage that had underpinned political agreement in the other two case studies.

Given the recognition that the ICARA process constituted a failure and CIREFCA and the Indo-Chinese CPA are regarded to have been relatively successful, it is surprising that the approach adopted by UNHCR in reviving the CPA concept largely replicated its approach to ICARA. The CPA for Somali Refugees developed by UNHCR in the context of the Convention Plus initiative had many of the characteristics of the second model described above.[44] The logic of the initiative was to develop a series of programmatic areas and projects to enhance access to durable solutions and protection capacity within the four main hosting states in the region: Kenya, Ethiopia, Djibouti and Yemen. These were compiled through a series of 'Gaps Analyses', establishing where protection gaps existed within each state, and 'National Consultations', discussing with those states the findings of the analyses and the type of projects and programmes that could respond to those needs. The intention was that these programmatic areas, with a range of representative projects, would then be submitted to a one-off Geneva-based donor conference in order to attract donations. The entire process was, however, conducted on a largely technical and apolitical basis, with extremely limited inter-state dialogue and with little facilitation or leadership being offered by high-level UNHCR staff. In contrast to CIREFCA and the Indo-Chinese CPA, the Somali CPA was managed by only one middle-ranking, full-time member of staff. Meanwhile, because the host states were approached by UNHCR on an individual basis, potential donor and host states were not at any stage brought face-to-face by UNHCR in order to discuss the general principles underpinning the CPA. This, in turn, limited the opportunity for UNHCR to offer a coherent, conceptual vision of the CPA to the main stakeholders. With little political momentum generated amongst potential donors, the CPA conference was indefinitely postponed and the CPA appears to have been largely abandoned. The failure of the CPA for Somali Refugees was in part attributable to the poor choice of region for the pilot and also the deteriorating political situation within Somalia. However, the main reason relates to the approach used by UNHCR, which ignored the lessons of history and implicitly replicated the problems of ICARA. The problem underpinning the programmatic model is that it assumes that donor states will altruistically commit resources to enhance access to durable solutions or to improve protection. In reality, states are rarely concerned with the 'refugee issue' for its own sake unless they are persuaded, as part of a political process, that contributing is likely to meet their linked interests in other related areas. Furthermore, the absence of a sustained political process does little to allow a growing sense that an initiative will be credible in terms of its viability, or the likelihood that it will achieve its stated aims.

Preconditions for a CPA based on the political engagement model

The CIREFCA and Indo-Chinese CPA case studies, and their contrast to the ICARA experience, point to a number of political and practical preconditions for a successful comprehensive approach to resolving protracted refugee situations. On a political level, they require interests, linkages and leadership. On a practical level, it appears important that a number of further conditions are met by UNHCR and the wider UN system: country of origin involvement, ownership, inter-agency collaboration and a strong UN regional presence.

Interests

Firstly, both CIREFCA and the Indo-Chinese CPA relied upon the existence of state interests in resolving the long-standing refugee situation. In neither case did the main stakeholders have significant interests in altruistically resolving the 'refugee issue' for its own sake. They did, however, have perceived interests in issue areas that related to the wider context within which the refugee situation was more broadly embedded. In particular, both the states within the region and the donor and resettlement states outside the region had perceived interests in related issues such as security, migration, peacebuilding and development. In the case of

CIREFCA, the interests of the main European donors lay in the wider concern to promote security, democracy and human rights in Central America, partly for ideological reasons and partly as a means to promote inter-regional trade. In the case of Indo-China, the interests of the US lay mainly in supporting regional security and promoting political change within the SRV, the interests of the ASEAN states mainly related to migration control, and the interests of the SRV related to attracting development assistance and political recognition.

Linkages

Secondly, though, the existence and recognition of these underlying interests were not by themselves a sufficient condition for inter-state agreement. Rather, the CPAs relied upon UNHCR facilitating the creation of a perceived linkage between the 'refugee issue' on the one hand and these wider interests on the other. This was particularly effectively achieved in the case of CIREFCA, in which UNHCR contributed to institutionalizing a number of these 'linkages'. A direct relationship was formed between CIREFCA and UNDP's development initiatives, such as PEC and PRODERE, which created an association between development assistance and durable solutions. Similarly, a clear link was formed with the Esquipulas II Peace Accords as a result of their direct reference to displacement. Developing such linkages brought not only state commitment but also wider support from across the UN system, notably from the UN Secretary-General.

Leadership

In both cases, UNHCR committed high-level staff to work on the CPAs. In particular, the Directors of the relevant regional Bureaux – namely Sergio Vieira de Mello and Leonardo Franco – were highly committed and led the processes both internally and externally. They were also able to draw on significant support from the High Commissioner and high-level staff throughout headquarters. Although it is rarely argued in grand theories of international politics, the role of individual personalities was crucial, as it has been throughout the history of the refugee regime.[45] However, what is also crucial to recognize is that UNHCR created an enabling environment which allowed such leadership to emerge. Political momentum was particularly important in both cases and this relied upon having a clear vision that could be conveyed to states and a message that the end goal was achievable and in the interests of all stakeholders. Perhaps most notably, the two main conferences in Geneva and Guatemala City and the CPA documents did not address every detail of implementation. Rather, they were used as political focal points upon which the wider process could build. They were consciously conceived as politically focused commitments, rather than pledging conferences, and were used primarily to build momentum and credibility for the process.

Country of origin involvement

In both CIREFCA and the CPA, the countries of origin were active partners within the negotiations, making the promotion of voluntary repatriation or return a viable component of each comprehensive approach. There is no practical reason why a CPA need necessarily have to include return. However, in practice, given that states tend to regard repatriation as 'the preferred durable solution', and approaches to the asylum–migration nexus are only likely to be meaningful if non-refugees are returned, the viability of the country of origin, as a negotiating partner and a recipient of returnees, would appear to be an important precondition for a successful CPA. In the case of Indo-China, the SRV's role was what made resolution of the impasse on the 1979 agreement possible. In the case of CIREFCA, the positive impact of the change of government in Nicaragua and the evolving role of the Guatemalan government, for example, show the importance of the countries of origin as partners in the process.

Ownership

Rather than being passive recipients of external support, the countries in the region were active participants throughout the two processes. The active involvement of not only the countries of origin but also the countries of asylum ensured that there was 'buy-in' on the part of all the relevant actors. In both cases the states had an identifiable stake in the success of the process and were vocal in promoting the initiatives and engaging the donor and resettlement countries. In CIREFCA the projects were compiled by the Central American states themselves, with the technical support of UNHCR, ensuring that they had clear interests in implementation. Similarly, the availability of additional development assistance created an incentive for them to drive the process. In the case of Indo-China, the 'countries of temporary refuge' were directly involved in identifying their own cut-off date and developing their own reception and status determination procedures in consultation with UNHCR. This, and their collective bargaining position through ASEAN, gave them a central role throughout the process. The notion of 'ownership' was therefore significant inasmuch as it meant that UNHCR did not have to provide all of the political momentum for the initiative in isolation.

Inter-agency collaboration

In both initiatives the scope of the comprehensive approach necessarily went beyond the bounds of UNHCR's mandate. In order to address these concerns, inter-agency collaboration was required. During CIREFCA, UNDP's role allowed an integrated development approach that could simultaneously provide for the needs of groups who fell outside UNHCR's mandate; notably IDPs, the 'externally displaced', and local populations. The Indo-Chinese CPA was one of the first examples of UNHCR–IOM (International Organization for Migration) partnership. IOM's role was important in relation to the logistical aspects of resettlement and providing alternative migratory channels for non-refugees. Although no clear division of labour was established, the debates within UNHCR at the time reveal that a role for IOM was considered to be important so that UNHCR would not be directly implicated in the deportation of non-refugees. Although IOM ultimately refused to play a role in return, meaning that UNHCR largely had to renounce the role to states, the organization assumed a significant role, particularly with respect to the Orderly Departure Procedure, providing an alternative migration channel for non-refugees wishing to leave Vietnam. Together, the Indo-Chinese and Central American cases therefore show the importance of UNHCR partnership with development actors when a CPA focuses on integrated development, and the importance of IOM partnership in cases related to addressing an asylum–migration nexus. The CIREFCA experience in particular also highlights how important it is not to see UN approaches to resolving protracted refugee situations in purely UNHCR-centric terms. Rather, by seeing responses to protracted refugee situations as part of a broader context, the Office of the Secretary-General and the New York-based UN system can play a significant leadership role and facilitate inter-agency collaboration.

Strong regional presence

An important element of both initiatives was the strong UNHCR presence within the region, supported by frequent and high-level visits to the region by Headquarters staff. Part of CIREFCA's success has been attributed to the presence of much of the process in Central America. The JSU was present in San José, the Representatives were particularly strong, and Spanish provided a common working language. During the Indo-Chinese CPA, the majority of the intergovernmental meetings were held in the region. As with CIREFCA, this allowed high-level participation by, for example, the region's foreign ministers.

Conclusion

Historical analysis has a significant number of insights to offer for developing UN-led responses to current protracted refugee situations. Unfortunately, in reviving the CPA concept within the context of Convention Plus, UNHCR did not draw adequately upon these insights. Indeed, the recent history of the refugee regime suggests that there are broadly two ideal-type models for developing multilateral responses to protracted refugee situations: the programmatic model and the political engagement model. The former is technical and apolitical and addresses the refugee issue in isolation from other areas of global governance. It is based on identifying projects and programme areas for a given protracted refugee situation and submitting them to a one-off donor conference. This approach proved a resounding failure in the case of ICARA I and II. In contrast, the latter is based on sustained UNHCR-led political facilitation and addresses the 'refugee issue' within the broader context of states' wider concerns in areas such as migration, security, development and peacebuilding. It is based upon generating political agreement through informal and formal high-level dialogue, culminating in an agreement of general principles. It provided a highly successful means to promote inter-state cooperation in both CIREFCA and the Indo-Chinese CPA. Yet, in re-launching the CPA concept to address the situation of Somali refugees, UNHCR's Africa Bureau developed an approach based on the programmatic model. Unsurprisingly, it led to failure because it was based on the same flawed assumptions about state behaviour as ICARA I and II.

Nevertheless, the failure of the CPA for Somali Refugees should not discredit the idea of using a CPA-like approach to address contemporary protracted refugee situations. The CPA approach offers the greatest potential means to overcome protracted refugee situations. However, crucially, it must be based on the political engagement model and draw its insights from cases such as CIREFCA and the Indo-Chinese CPA. Critical analysis of these initiatives reveals a number of political and institutional preconditions for developing such approaches. In particular, the 'refugee issue' should not be seen in isolation. CPAs rely upon states having perceived interests in contributing economically and politically to improving access to durable solutions. However, states' perceived interests are rarely in improving refugees' welfare for its own sake. Rather, they are generally based on interests in linked issue areas, such as peacebuilding, security, development and migration. Channelling these wider interests into a focus on durable solutions is more likely to be successful than hoping that states will altruistically commit resources to a 'laundry list' of programme areas. That said, it is crucial that this approach relies upon, firstly, a sustained political process in which momentum, trust and credibility are built over time. Such a process cannot be based on a one-off conference but needs to be a much longer process that builds both political consensus between states and a sound basis for collaboration between UN agencies. Secondly, it relies upon approaching protracted refugee situations from a broader institutional perspective that recognizes the wider context in which they are perceived by states. Given that such situations are politically and practically related to other issue areas of global governance, leadership must come not only from UNHCR and states but also from the wider UN system and the Office of the Secretary-General, just as it did in CIREFCA in particular.

Notes

1. This chapter draws upon the empirical work of two working papers published in UNHCR's *New Issues in Refugee Research* series: A. Betts, 'International Cooperation and Targeting Development Assistance for Refugee Solutions: Lessons from the 1980s', *New Issues in Refugee Research*, Working Paper no. 107, Geneva: UNHCR, 2004; and A. Betts, 'Comprehensive Plans of Action: Insights from CIREFCA and the Indo-Chinese CPA', *New Issues in Refugee Research*, Working Paper no. 120, Geneva: UNHCR, 2006.
2. For an overview of ICARA I and II see, for example, B. Stein, 'ICARA II: Burden-Sharing and Durable Solutions' in R. Rogge, ed., *Refugees: A Third World Dilemma*, New York: Rowman and Littlefield, 1987; R. Gorman, 'Beyond ICARA II: Implementing Refugee-Related Development Assistance', *International Migration Review* 8, no. 3, 1986; R. Gorman, *Coping with Africa's Refugee Burden: A Time for Solutions*, The Hague: Martinus Nijhoff, 1987.
3. UN General Assembly Resolution 35/42 of 25 November 1980.
4. Ibid.
5. R. Gorman, 'Linking Refugee Aid and Development in Africa', in R. Gorman, ed., *Refugee Aid and Development: Theory and Practice*, London: Greenwood, 1983, p. 63.
6. 3rd Draft of Steering Committee of Post-ICARA Coordination Meeting, held 15 September 1981, New York, HCR/NY/572, Fonds UNHCR 11, 391.62/460.
7. UN General Assembly Resolution 36/124 of 14 December 1981, cited in J. Milner, 'Golden Age? What Golden Age? A Critical History of African Asylum Policy', paper presented at the Centre for Refugee Studies, York University, 28 January 2004.
8. Note for the File: Summary of Statements Relating to ICARA II Made at Informal Meetings of ExCom Representatives, 27 May 1983, Fonds UNHCR 11, 391.78/215.
9. High Commissioner's Opening Remarks at the 3rd Steering Committee Meeting on ICARA II, 14 November 1983, Jessen-Petersen's summary of the debate, Fonds UNHCR 11, 391.78/398A.
10. Stein, 'ICARA II'.
11. For an overview of CIREFCA see, for example, UNHCR, *State of the World's Refugees*, Oxford: Oxford University Press, 2000, chapter 6; UNHCR, 'Review of the CIREFCA Process', EPAU Working Paper, 1994, see http://www.unhcr.org; A. Betts, 'Comprehensive Plans of Action: Insights from CIREFCA and the Indo-Chinese CPA', *New Issues in Refugee Research*, Working Paper no. 120, Geneva: UNHCR, 2006.

12. UNHCR, 'Review of the CIREFCA Process'; UNHCR, 'Questions and Answers About CIREFCA', prepared for Seminar on the Implementation of a Human Development Approach for Areas Affected by Conflict in Central America and Related Strategies for the Post-CIREFCA Process, June 1993, UNHCR, Fonds 11, Series 3, 391.85.5.
13. Memo, Kevin Lyonette to Leonardo Franco, 'Tactical Proposal for Promotion of Funding of CIREFCA Projects', 19 April 1989, UNHCR Fonds 11, Series 3, 391.86.3.
14. Ibid.
15. Article 8 of the Esquipulas II Accords referred explicitly to displacement, meanwhile CIREFCA was conceived as chapter 1 of PEC. Section X of the UNHCR report to the main Conference makes clear the relationship of the initiative to other linked issue areas, claiming CIREFCA to be a 'converging point' for these issues. The relationship is outlined in, for example, 'Procedures for the Preparatory Activities of the Conference Itself and the Establishment of Follow-Up Mechanisms – Proposal Submitted to the Organizing Committee Meeting, Guatemala, 24 January 1989', UNHCR Fonds 11, Series 3, 391.86.3 HCR/NYC/0102.
16. 'From Conflict to Peace and Development: Note on Implementation of the Concerted Plan of Action of CIREFCA', Pablo Mateu JSU to K. Asomani RBLAC, 17 March 1992, UNHCR Fonds 11, Series 3, 361.86.5.
17. Letter from Jean-Pierre Hocké to Javier Perez de Cuellar, 20 December 1988, UNHCR Fonds 11, Series 3, 391.86.3 HCR/NYC/1553.
18. B. Garoz and M. Macdonald, 'La Politica de Cooperacion de la Union Europa Hacia Guatemala', Estudio elaborado para Asamblea de la Sociedad Civil (ASC), Guatemala, 1996, on file with the author, p. 5.
19. 'International Conference on Central American Refugees', GA Resolution A/RES/48/117, 85th Plenary Session, New York, 20 December 1993, UNHCR Fonds 11, Series 3, 391.86.5.
20. UNHCR, 'Questions and Answers About CIREFCA'.
21. For an evaluation of UNHCR's repatriation reintegration programmes in Guatemala, see UNHCR, 'Lessons Learnt from UNHCR's Involvement in the Guatemalan Refugee Repatriation and Reintegration Programme (1987–1999)', EPAU Evaluation, 1999, http://www.unhcr.org.
22. UNHCR, *Framework for Durable Solutions for Refugees and Persons of Concern*, Geneva: UNHCR, 2003.
23. For an evaluation, see UNHCR, 'Quick Impact Project: A Review of UNHCR's Returnee Reintegration Programme in Nicaragua', EPAU Evaluation Report by Jeff Crisp and Lowell Martin, Geneva: UNHCR, 1992.
24. Interview with Ana Low, intern, researching self-sufficiency and local integration in southern Mexico, UNHCR, 25 October 2005.
25. Interview with Pablo Mateu, former Programme Officer in the JSU, UNHCR, 18 October 2005.
26. *El Diario de Hoy*, 'From Conflict to the Valley of Peace', 18 October 2005.
27. Interview with Pablo Mateu, former Programme Officer in the JSU, UNHCR, 18 October 2005.
28. For an overview of the CPA, see, for example, C. Robinson, *Terms of Refuge: The Indochinese Exodus and the International Response*, London: Zed, 1998, pp. 187–230; C. Robinson, 'The Comprehensive Plan of Action for Indochinese Refugees, 1989–1997: Sharing the Burden and Passing the Buck', *Journal of Refugee Studies* 17, no. 3, 2004, pp. 319–333; S. Bronee, 'The History of the Comprehensive Plan of Action', *International Journal of Refugee Law* 4, no. 4, 1992, pp. 534–559; P.-M. Fontaine, 'The Comprehensive Plan of Action (CPA) on Indochinese Refugees, Prospects for the post-CPA and Implications for a Regional Approach to Refugee Problems', *Pacifica Review*

(formerly *Interdisciplinary Peace Research*) 7, no. 2, 1995, pp. 39–60; P. Jambor, *Indochinese Refugees in Southeast Asia: Mass Exodus and the Politics of Aid*, Bangkok: Ford Foundation, 1992.
29. *New York Times*, 'Vietnam and Laos Finally Join Talks on Refugees', 30 October 1988.
30. Ibid.
31. Memo, Sergio Vieira de Mello to Refeeudin Ahmed, Secretary-General's Office, 'Recommended Opening Speech for Kuala Lumpur Meeting, 7–9 March', 22 February 1989, UNHCR Fonds 11, Series 3, 391.89 HCR/NYC/0248.
32. 'Report on the Meeting of the Coordinating Committee for the International Conference on Indochinese Refugees', 19–20 April 1989, UNHCR Fonds 11, Series 3, 391.89 HCR/THA/0516.
33. 'UNHCR Informal Consultations on Indochinese Asylum Seekers in South-East Asia, Bangkok, 27–28 October 1988', 10 November 1988, UNHCR Fonds 11, Series 3, 391.89 100.Ich.gen.
34. Interview with Anne Dawson-Shepherd, based in Hong Kong and Singapore during the CPA, UNHCR, 19 October 2005.
35. 'Introductory Remarks by Sergio Vieira de Mello at Informal Steering Committee Meeting', Manila, 17 May 1990, UNHCR Fonds 11, Series 3, 391.89.
36. 'Joint Statement by Countries of Temporary Refuge', Manila, 16 May 1990, UNHCR Fonds 11, Series 3, 391.89.
37. Statement by Malaysian Minister of Foreign Affairs, 23rd ASEAN Ministerial Meeting, Jakarta, 24–25 July 1990, UNHCR Fonds 11, Series 3, 391.89.
38. Interview with Dennis McNamara, Deputy Director of the Department of Refugee Law and Doctrine at the time of the CPA, Geneva, 28 November 2005.
39. 'Draft Consensus Note', 18 July 1990, UNHCR Fonds 11, Series 3, 391.89.
40. 'Revised Version of the "Near Consensus Note"', 12 July 1990, UNHCR Fonds 11, Series 3, 391.89.
41. Interview with Dennis McNamara, Deputy Director of the Department of Refugee Law and Doctrine at the time of the CPA, Geneva, 28 November 2005.
42. 'Report of the Fourth Steering Committee', Geneva, 30 April–1 May, UNHCR Fonds 11, Series 3, 391.89.
43. Interview with Brian Lander, UNHCR, based in Hong Kong and Indonesia during the CPA, 1 November 2005.
44. For an overview of the initiative see UNHCR, *The State of the World's Refugees 2006: Human Displacement in the New Millennium*, Oxford: Oxford University Press, 2006, p. 59.
45. C. Skran, *Refugees in Inter-war Europe: The Emergence of a Regime*, Clarendon: Oxford University Press, 1995, pp. 287–292.

Part II
Case studies

10

Palestinian refugees

Michael Dumper

Palestinian refugees constitute one of the longest-standing and numerically largest refugee situations in the world. Some estimates conclude that there are up to 7 million Palestinian refugees and displaced persons comprising approximately 75% of the Palestinian population and 30% of the world's refugees.[1] With a growth rate of 3.1%, the *registered* refugee population of 4 million is increasing at approximately 124,000 per annum. It is therefore an issue that will not fade away over time. Indeed, delay will only increase the magnitude of the problems to be solved.

In writing this chapter along the themes suggested by the organizers, one is struck by the extent to which the Palestinian case appears to some extent unique, or at least very different from many of the other refugee cases. One can posit at least five ways in which such differences should be noted.[2] First, the most striking difference or uniqueness of the Palestinian refugee situation, as has already been indicated, is its sheer longevity. Created as a result of the establishment of the state of Israel in 1948 – 60 years ago – the Palestinian case is thus a multi-generational one, with a fourth generation of descendants of the original displaced Palestinians currently being born.[3] This longevity produces specific dynamics of exile. On the one hand, there are greater opportunities of integration and economic and social ties being established with the host community. On the other, there can be a greater forging of nationalist consciousness as communal solidarities are maintained in a foreign environment. While a degree of political and economic integration has been permitted, especially in Syria and Jordan (but not in Lebanon), it is clear

---

Protracted refugee situations: Political, human rights and security implications,
Loescher, Milner, Newman and Troeller (eds),
United Nations University Press, 2008, ISBN 978-92-808-1158-2

that in the Palestinian case there has been a strong growth in nationalist feeling and Palestinian self-identity.

The second point of difference concerns demography. The exact number of Palestinians displaced by the 1948 war is disputed. Estimates range from 600,000 to 957,000 but the long duration has meant that the numbers have multiplied. As already mentioned, at the end of 2002, it is estimated that there were more than 7 million Palestinian refugees and displaced persons. This makes the Palestinian refugee and displaced persons population the largest refugee and displaced persons population in the world. It is more than the combined total for all refugees in Asia under the responsibility of UNHCR.[4] What is important to remember is that the proportion of refugees to the total Palestinian population is significantly higher than in most other refugee situations. In total, the Palestinian refugee and displaced population comprises nearly three-quarters of the entire Palestinian population worldwide of approximately 9.3 million.[5]

A third point of difference is that the legal framework for Palestinians of refugee status and protection is quite exceptional. Most Palestinian refugees are registered with UNRWA and not UNHCR. This was partly due to historical and political reasons which will be dealt with below. UNRWA was made responsible for Palestinian refugees only, and for those living in the countries bordering the new state of Israel: Lebanon, Syria, Jordan, Egyptian territory in the Sinai Peninsula, the West Bank and the Gaza Strip (see Figure 10.1). This has meant that the provision of services and institutional development has been outside the UNHCR framework for over 50 years, leading to a highly separate culture and ethos and a close association between UNRWA and the sense of 'refugee-ness' and identity felt by Palestinians.

The fourth point is the nature of the Palestinian displacement. Israel was established as a state of the Jewish people and the return of the indigenous Palestinian population would undermine its *raison d'être*. To put it simply, if Israel is to remain a Zionist and a Jewish state it cannot accept the repatriation of a large number of refugees. Thus the transition from refugee to citizen in the Palestinian case is more complex and politically charged than in many other refugee cases, implying as it does the dismantling of the Jewish nature of the state.[6] The Palestinian refugee case turns the principle of non-*refoulement* on its head. The issue is not whether the conditions are safe for repatriation, as in many other refugee cases, but whether they will ever be allowed to return.

Closely connected to this is the fifth point, which concerns the lack of Palestinian sovereignty over its historical territory. Because of the establishment of Israel on 72% of the land of historical Palestine, the existing Palestinian leadership is in an ambiguous position. Its main constituency is the refugee population, virtually all of whom have claims to return to an area that is within the borders of Israel and which is not under the actual and projected jurisdiction of a new Palestinian state. Thus a repatriation programme will be to a new state of Palestine which is not where the majority of refugees have come from. In this sense, the term 'repatriation' is a misnomer. Much of the political discussion and policy planning regarding repatriation, as a durable solution, relates, therefore, not to the place or the country of origin but to a different part of historical Palestine (the West Bank and Gaza Strip) and a new state that was not in existence prior to 1948.

The main objective of this chapter will be to argue that the protracted nature of the Palestinian case is derived from two main causes. First, it is derived from a lack of agreement over whether the conflict is about the displacement that took place in 1948 (the Palestinian view), or is concerned with the Israeli occupation of the West Bank and Gaza Strip in 1967 (the Israeli view). Second, it is derived from the transformation of the country of origin (Palestine) into a state based upon ethnicity (Israel). Unless there is an agreement on the first cause there will be a continuing mismatch between the remedies being suggested and the causes being addressed. Furthermore, unless there is a Palestinian acceptance of this second cause, or an Israeli modification of its concept of citizenship to include Palestinians, it is unlikely there will be a solution to the conflict and to the refugee situation.

It should be stated at the outset that the problems in achieving a resolution to the refugee issue in the Palestinian–Israeli conflict are not centred on the desire of every Palestinian refugee to return to their homes or those of their grandparents. It is clear that the vast majority would prefer to remain within the cultural and socioeconomic networks that they have built up in exile. However, for reasons of self-identity, justice and respect they are refusing to accept their exile without some form of recognition of their rights to the land they have left and some form of recompense. The challenge to policymakers is how to square the circle of recognizing the ethnic exclusivity or Jewishness of Israel while at the same time satisfying the legitimate demands of the Palestinian refugees for justice and recompense.

Finally, despite these specific differentiating features, it is still nevertheless important for the Palestinian case to be analysed comparatively. Comparative study will help to contextualize the Palestinian case and draw out the key elements that need to be addressed. UNHCR statistics indicate that no more than 25% of refugees have returned to their countries of origin. In a comparative study we would see, for example, that global patterns of actual repatriation suggest that refugees, while desirous of achieving their political rights, are often wary of returning to their place of origin after many years in exile. Similarly, it would be apparent that the repatriation of refugees rarely entails a mass flow of refugees back to their homes, but is often a carefully managed process involving local institutional capacity building, training and human resource development, prior investment and a series of consultation mechanisms before the first refugees leave their exile. In this way, comparative studies can be used to unpack some of the myths and fears associated with refugees, exile and repatriation, and lead to negotiations based on actual realities.[7]

Origins of the Palestinian refugees

Palestinian refugees are, in the main, refugees from those parts of Israel that were formerly part of Palestine. Prior to the arrival of the first Zionist Jewish settlers, Palestine had a mixed population with some Jewish communities but was predominantly inhabited by Palestinian Arabs who were subjects of the Ottoman Empire (1517–1918). At the time of the establishment of the British Mandate for Palestine (1921–1948), there were close to 60,000 Jews living in Palestine, mostly Zionist settlers, and 500,000 Palestinians, or 90% of the population. On the eve of the 1948 War, the Jewish population in Palestine amounted to around 600,000, with Palestinians comprising approximately 1,230,000, representing 33% and 67% of the total population respectively.[8]

Through the issuing of a letter to the Zionist movement, known as the Balfour Declaration, in 1917, the British government supported the development of a Jewish 'national' homeland in Palestine. There was, however, a proviso that the civil and religious rights of the existing non-Jewish communities – that is, mainly Palestinians – would be respected. Nevertheless, the Zionist settler movement interpreted the Declaration as a pledge to establish a Jewish state, regardless of the obligations to maintain the political and demographic status of the Palestinians.

The incompatibility of British promises to both communities led to increased tensions and open conflict. Unable to devote the resources to confront Zionist aspirations, the British government invited the UN to resolve the problem and, in 1947, the United Nations General Assembly approved the partition of Palestine into a Jewish state and an Arab state, with an international enclave around Jerusalem[9] (see Figure 10.1). This resolution led to the British withdrawal in May 1948 and the formation of an Israeli government by the Zionist settlers, which in turn became the catalyst for intercommunal fighting in Palestine and the intervention of Arab states opposed to the new state. A series of armistice agreements between Israeli forces and the Arab states was agreed in 1949 which established the ceasefire lines and the *de facto* borders of the Israeli state. The remaining pieces of Palestine – the West Bank and Gaza Strip – were annexed by Jordan and administered by Egypt respectively.

As a result of the fighting prior to the armistice agreements, there was a major exodus of refugees from Palestinian cities and villages. The refusal of the Israeli government to allow them to return to their homes became known by the Palestinians as '*Al Nakba*' – 'the Catastrophe'. International concern for their safety and conditions led to the adoption by the General Assembly (GA) of UN Resolution 194, which stated:

> Refugees wishing to return to their homes and live at peace with their neighbours should be permitted to do so at the earliest practicable date, and that compensation should be paid for the property of those choosing not to return and for the loss or damage to property.[10]

The Palestinians have always interpreted Resolution 194 as their 'right to return' and as a resolution which has foundations in international law. Israel, on the other hand, disagreed, claiming that as a GA resolution it is non-binding.

By the time of the signing of the armistices agreements, 750,000 Palestinians had fled to neighbouring countries to be temporarily housed in refugee camps or with relatives. Initially, Palestinian refugees received assistance from NGOs, such as the American Friends Service Committee (AFSC) and the International Committee of the Red Cross (ICRC), and under a temporary ad hoc UN agency, the United Nations Relief for Palestine Refugees. However, when it became obvious that the newly founded state of Israel would not agree to the return of all the refugees created by the conflict, the UN decided that a dedicated agency was required to provide emergency relief. Between 1948 and 1949, the United Nations General Assembly established two separate agencies to provide protection, assistance and durable solutions to the Palestinian refugees. The first was the United Nations Relief and Works Agency for Palestine Refugees in the Near East (UNRWA) and the second, the United Nations Conciliation Committee for Palestine (UNCCP). UNRWA was established as a temporary agency and its mandate was renewed at regular intervals.[11] Most Palestinian refugees were registered with UNRWA, whose role was to supply services to the refugee camps, to provide education, and to support the communities in five locations: Jordan, Syria, Lebanon, the West Bank and Gaza. It is important to note that in contrast to the work of the main UN refugee agency set up some months later, the United Nations High Commissioner for Refugees (UNHCR), UNRWA's work did not involve permanent solutions or the explicit provision of international protection to Palestinian refugees.

At the same time, the new Israeli state took advantage of the flight of the refugees and transferred Palestinian refugee property and land to new Israeli state bodies.[12] These actions allowed Israel to accommodate new Jewish immigrants in refugee property and thus demographically consolidate the new state. Henceforth, Israel was willing to contemplate compensation for refugees and after 1948 it introduced a limited programme of 'family reunification' for refugees, but it ruled out a return programme with any restitution.[13] Approximately 120,000 Palestinians remained in the area of Palestine that became Israel, and, while having Israeli citizenship imposed upon them, many of these became internally displaced if they had fled or been evicted from their homes.

Resolution 194 also set up the UNCCP in 1949. Among other provisions its aim was to facilitate the return or resettlement, restitution and compensation of the refugees based on their individual choices. The role of the UNCCP was to act as mediator between Israel, the Arab states and the Palestinians, and to provide protection and facilitate durable solutions for persons displaced as a result of the 1947–1948 conflict in Palestine. This includes internally displaced Palestinians inside Israel. In 1950, the Assembly – Resolution 394(V) – specifically requested the UNCCP to protect the rights, properties and interests of the refugees. Its work encompassed gathering basic information and seeking solutions to the political, legal and economic aspects of the question. The UNCCP developed compensation plans based on both global and individual evaluations of property, culminating in studies which identified and valued every parcel of Palestinian- and Arab-owned land in Israel.[14] However, as a result of lack of support for its work, primarily by the US, by 1966 the agency ceased to operate effectively and existed only on paper. The cessation of UNCCP operations was significant in that it, and not UNRWA, had been given the mandate to protect Palestinian refugees. Thus, when it ceased to function, that protection activity also fell into abeyance, leaving Palestinian refugees without an effective protector of their rights. UNRWA has tried to fill this gap incrementally.

A further spate of refugees was created in the 1967 War. This resulted in the rapid destruction by Israel of the combined armies of Egypt, Jordan and Syria, and the capture of the West Bank and Gaza Strip, in addition to the Egyptian Sinai Peninsula and the Syrian Golan Heights. A new wave of Palestinian refugees fled from the West Bank and Gaza Strip to Jordan and Egypt. Palestinians, particularly those on the floor of the Jordan valley, fled or were expelled from their villages or refugee camps. Approximately 335,000 crossed the river into the East Bank of Jordan, two-thirds of whom were refugees for a second time in their lives.[15] There have been subsequent displacements and expulsions of Palestinians from the West Bank and Gaza Strip, which have come to be known as the Occupied Palestinian Territories (OPTs), as a result of resistance to the Israeli occupation.

Factors contributing to prolongation of the situation

The two main factors contributing to the prolonging of the refugee situation have been Israeli policies to defend the Jewishness of their state and Palestinian refusal to accept the loss of their land and homeland. Before examining these points in more detail we should first clarify some of the terms used with regard to Palestinian refugees.

One of the most contentious issues in the study and politics of Palestinian refugees is the issue of numbers. Determining who exactly is a refugee is fraught with problems and has important political consequences. Seriously complicating discussion of this issue is the fact that there is no single authoritative source for the global Palestinian refugee and IDP population, and figures have been disputed by both Palestinians and Israelis. Figures on Palestinian refugees come mainly from UNRWA and the Palestinian Central Bureau of Statistics (http://www.pcbs.gov.ps/DesktopDefault.aspx?tabID=1&lang=en). There are also other databases to be found in limited formats, including the Norwegian Institute for Applied Social Science (FAFO) (http://www.fafo.no), but these depend mostly on the two sources mentioned above.

Much of the difficulty lies in the fact that many refugees are not registered as such. Whether one is registered or not has financial and social implications for camp-dwelling refugees as, unlike non-registered refugees, registered refugees are entitled to social and financial services. As of March 2006 there were 4,375,000 refugees on UNRWA's books.[16] However, the UNRWA definition of a 'Palestinian refugee' was meant to work as a definition for the purposes of establishing assistance procedures, and not for determining the legal status of refugees or their rights. It was elaborated for operational purposes to determine who was eligible for its services. Thus, Registered Refugees (RR) are not the total refugee population. For example, the RR figure does not include IDPs in Israel – that is, refugees displaced in 1948 but who did not cross any international borders and are now living in a political entity (Israel) which did not exist in 1947. Neither does it include those displaced in the 1967 War, who are referred to as the 1967 Displaced Persons and were the subject of specific provisions in the Oslo peace process. And finally, nor does it include an estimated 1.5 million refugees and their descendants who did not register with UNRWA as refugees.[17] In addition to this lack of clarity, Israeli negotiators have disputed the whole notion of the descendants of refugees being regarded as refugees and therefore being part of the totals being discussed. Their argument is that there is little precedent in international law for such a position. Turning to the UNHCR definitions does not clarify the situation. The 1951 Convention relating to the Status of Refugees refers to a 'well-founded fear of persecution' as a basis for being a refugee but does not refer to descendants.[18] This is very likely due to the fact that the Convention did not anticipate refugee situations lasting for as long as the Palestinian one has. However, the Convention does include the children of refugees, which some have deemed as providing support for the Palestinian position.

The ethnic basis of Israel

Through colonization, successive Israeli governments sought to counter the prospects of Palestinian return to the areas acquired in 1948 and demographic growth in those areas and the OPTs. Following the establishment of the new state, the Israeli government encouraged a rapid programme of mass immigration to absorb as many Jewish refugees from the Second World War and from the hostile Arab states as possible. During 1949 and the 1950s, 47,000 Jews were flown to Israel from Yemen, with a further 120,000 Jews flown from Iraq. Between May 1948 and 1951, the Jewish population of Israel doubled, with an average of 172,000 new immigrants arriving per year.[19] In the OPTs a similar policy was employed after 1967. Geographically, Israeli settlements cut off Palestinian population centres from each other through a road network system and by land acquisition for Jewish-only residency. West Bank and Gaza Strip water resources of the territory were integrated into the Israeli national grid system.[20] Israeli control over Palestinians in the OPTs was extended to virtually every aspect of their lives. In terms of the refugee issue, not only has the creation of a vibrant and strong Israeli state hampered the prospect of Palestinian refugees returning to their homes of 1948, but the ongoing colonization of newly acquired territory in 1967 both threatens to create more dispossession, and also pushes a resolution of the 1948 refugee issue even further away. From an Israeli perspective, Israelis had little alternative but to view Palestinian refugees as an integral part of the enemy forces opposed to the creation of the Israeli state, and their return would jeopardize the ethnic basis or Jewishness of the state.

The persistence of Palestinian nationalism

Palestinian national leadership collapsed after the events of 1948 and Arab states either attempted to co-opt the refugee cause or competed with each other to represent the interests of the Palestinians in formal talks. In 1964, the Arab League established the Palestinian Liberation

Organization but this was seen by refugees as elitist and unrepresentative. Popular and autonomous Palestinian action was boosted after an Israeli incursion into Jordan was repulsed by Palestinian guerrilla forces at the refugee camp of al-Karama in 1968, and the dominant guerrilla group, known as al-Fatah, took over the Palestine Liberation Organization (PLO). Jordan was increasingly used by the PLO to carry out operations against Israel, which resulted in Jordan expelling the organization. The main centre of PLO operations moved to southern Lebanon and the weakness of the state apparatus there allowed the PLO to establish quasi-state functions in and around the refugee camps. The increasing military build-up of the PLO prompted Israel to invade southern Lebanon in 1978 and 1982 and it succeeded in expelling the organization's leadership and fighters. The high civilian costs of this operation and the massacre of camp refugees by Israeli proxies in the refugee camps of Sabra and Shatilla alienated the indigenous Arab population, whose resistance to Israel eventually forced it to withdraw its forces in 2000. (The Israeli attack on Lebanon in 2006 stemmed largely from the animosity and legacy of the 20-year Israeli occupation of south Lebanon.)

The eviction of the PLO as an armed force from Lebanon in 1982 led to a change in its strategy. Partly recognizing the limits of armed struggle and partly through the absence of a base in an adjoining territory from which it could carry out military operations against Israel, the PLO concentrated on building up political support for its goals. In 1974, it obtained observer status at the UN and also recognition as the sole legitimate representative of the Palestinian people from the Arab League, ending Syrian and Jordanian hopes of representing Palestinian interests. It also amended its programme, changing from the maximalist positions of a secular democratic state in all of Palestine to a Palestinian state in any part of liberated Palestine, which was understood to mean in the OPTs. In 1988 it recognized the state of Israel and sought negotiations on the basis of UN Security Council Resolution 242. As Resolution 242 did not mention refugee rights, such a step appeared to soften the PLO's adherence to the earlier Resolution 194, which called for the return of refugees.

Recent contributions to the prolongation

Continued Israeli settlement and consolidation policies in the OPTs led to mounting Palestinian political resistance to the occupation. The first Palestinian *intifada*, or uprising (1987–1991), lasted for four years, forcing the Israeli government to recognize the limits of its military superiority over the Palestinians and that a political agreement was the best way to safeguard its position. The realignment in the region, as a result of the collapse of the Soviet Bloc and the 1st Gulf War (1990), resulted in the first peace negotiations between Israelis and the Arab states since 1948. Held in Madrid in 1991, Israel still refused to meet with PLO representatives and the talks failed to progress substantially. This position was superseded by the Oslo Accords in 1993 after an intensive round of secret negotiations between the Israeli government and the PLO. At their core, the Accords comprised an Israeli recognition of the PLO as their negotiating partner and the introduction of a staged withdrawal of Israeli forces from the OPTs. However, they did not specify the creation of a Palestinian state, nor did they specify a solution to the refugee issue, deferring these issues to 'Permanent Status' talks to be held after a five-year interim period.

Disagreements over the continuing Israeli colonization activity and land expropriations in the OPTs, and over continued Palestinian attacks on Israel by militant Palestinian groups, led to long delays in the timetable of implementation and a failure to come to agreement on 'Permanent Status' issues at the Camp David summit in 2000 between the US, Israeli and Palestinian leaderships. A new wave of violence and political conflict began in September 2000, known as the al-Aqsa *intifada*. It resulted in violent clashes, and confrontations continued in most of the major towns of the occupied territories. It differed from the first *intifada* inasmuch as the resistance was armed and quickly monopolized by militant factions rather than popular protest. It was also characterized by the increased incarceration of Palestinians, severe restriction of movement, escalating violence, suicide bombings and the re-occupation by Israel of those parts of the West Bank it had withdrawn from under the Oslo Accords. The downward spiral of violence was compounded by the al-Qaeda attacks on the World Trade Center towers and the Pentagon of 11 September 2001. Media coverage of the Middle East conflict, especially in the United States, obscured the underlying dynamics of the conflict, with the Palestinian–Israeli conflict being placed in the larger context of the US war on terrorism.

Impact of prolongation on refugees and states in the region

The impact on refugees

The Palestinian experience of exile has been both varied and collective. Palestinian refugees in Syria and Jordan have had greater political freedoms and privileges than those in Lebanon. Refugees in the OPTs, and IDPs in Israel, have been subject to a range of restrictions, from land confiscation to the proscription of political activity. Some refugees have advanced economically, particularly those who have left the region to work in the Gulf or in North and South America and Western Europe. Others have had to contend with extremes of poverty, violence and repression.

Most Palestinian refugees live in permanent housing, including those in refugee camps. Here the infrastructure has been the responsibility of the host governments, but UNRWA has financed and introduced services. Nearly all camps have electricity, water and sewerage, but the stability of the electricity and drinking water supply in the camps is often considerably worse than in surrounding areas. Crowding is higher in the camps than elsewhere, and around 30% of the households have three or more persons per room. The camps in Jordan and the Gaza Strip are particularly under-resourced, with 40% of the households having three persons or more per room.[21] Infant mortality is usually an indicator of the general health of a society and its conditions. Amongst Palestinian refugees infant mortality ranges from 20 to 30 deaths per 1,000 live births. Camps in Syria show particularly low rates, while the Lebanese rates are the highest. Maternal mortality rates are also highest in Lebanon (240 maternal deaths per 100,000 live births) and lowest in Syria.[22] However, due to the special hardship programmes of UNRWA, there is little acute malnutrition among children. There is, on the other hand, more reported psychological distress, as well as somatic illness, among adults in camps than elsewhere in the region, and most of this occurs in Lebanon.[23]

Lebanon

Possibly the most difficult situation has prevailed in Lebanon. Lebanon hosts approximately 400,000 registered Palestinian refugees, half of whom reside in 12 official refugee camps. Comprising 12% of the entire population of Lebanon, Palestinians were forbidden, until 2005, to enter the formal Lebanese workforce and were obliged to take low-paid jobs, mostly in agriculture or construction work. Such marginalization forced many to seek work overseas, which further fragmented the community. The Palestinian experience is unique, partly due to these restrictions, but also because of the civil war in Lebanon and the repeated Israeli incursions.

Syria

In Syria, the main influx of Palestinian refugees took place between 1948 and 1949. A second, smaller wave also took place after the 1967 War. Palestinian refugees in Syria, who number approximately 420,000, are largely accepted and helped by the government. In the main, they share the same duties and responsibilities as Syrian citizens, including joining the Syrian military. They do not require work permits and have the right to own businesses and to receive state education at secondary and university level. However, they have not been completely absorbed, in that while they can vote in local elections they cannot vote in parliamentary or presidential elections or run as a candidate for political office.

Jordan

There are currently 1.8 million Palestinian refugees in Jordan, fewer than 20% of whom live in the 10 official camps. Their arrival in 1948 virtually doubled the population of Jordan and the government pursued an ambivalent policy toward them. On the one hand, it provided humanitarian assistance but did not want to take over the responsibility of UNRWA, whose contribution became a very important part of the government budget.[24] Despite a degree of wariness and hostility from the indigenous population of Transjordanian Arabs, the Jordanian government offered full citizenship to all refugees and their descendants. Formally, they enjoy the same rights and responsibilities as Jordanians. They are allowed access to most employment, except sensitive public service and military positions, and have formed an important part of the private sector in the Jordanian economy. An exception to this is the status accorded to Palestinians from the Gaza Strip, who were displaced in 1967 and are not considered Jordanian citizens.

OPTs and Israel

In the OPTs, there has been a very different set of experiences for the refugees. Gaza was characterized foremost by its dense population and its continuous confrontations with the Israel Defense Forces (IDF). Traditionally, the residents of the West Bank have been more economically and educationally advanced, resulting in a tendency for West Bank Palestinians to dominate the political arena and have greater access to external funds. Refugees in the OPTs total nearly 1.6 million, or 15% of the total Palestinian population, and aspire either to return to their homes in Israel or to be compensated. While treated by the Israeli occupying authorities in exactly the same way as other non-refugee Palestinians in the OPTs, there are some differences in their experiences. In addition to camp residency, unemployment in the refugee camps is 4% higher than the rest of the OPTs, and 32.8% of Palestinians living in refugee camps are classified as poor. In 1998, despite only accounting for 15% of the population, they comprised 25% of the poor in the OPTs.[25]

The position of Palestinians in Israel who were internally displaced as a result of the 1948 War is often forgotten in the literature on Palestinian refugees. Around 30,000–40,000 Palestinians currently living in Israel are considered displaced.[26] Israel does not recognize IDPs or their rights and such displacement and dispossession have had a very visible effect on the socioeconomic status of Palestinian citizens of Israel.[27] Discrimination is rife and IDPs are not recognized as a separate section of society, as refugees or as IDPs. There is no registration system for them, which makes these internally displaced people liable to be ignored in any permanent status negotiations.

Finally, there are those who have undergone displacement in the OPTs but are not recognized as refugees. Their displacement has largely been the result of the ongoing low-level conflict between Palestinians and Israelis. Israel has ordered the demolition of thousands of homes in the OPTs and has confiscated land, such as in East Jerusalem, for security and developmental reasons and also to deter militants from attacks on Israel.

The central role of UNRWA

The second major impact has been the evolution of UNRWA as a quasi-state system, both providing essential humanitarian assistance but also freeing the international community from getting to grips with the political issues which require resolution.

UNRWA is the largest agency of the UN system. Currently, it administers 59 refugee camps and employs 24,000 people, the majority of whom are Palestinian refugees. The actual administration and policing of camps are the responsibility of the host authorities, with UNRWA providing the services. It provides basic health, education and social services for 4 million Palestinian refugees, or about three-quarters of the entire Palestinian refugee population, residing in the five areas of its operation: the Gaza Strip, West Bank, Jordan, Lebanon, and Syria. In addition to its emergency relief, the agency programmes have also, through work programmes, focused on human resource development and improvement of the social infrastructure. Camp residents run their own activities, and camp committees in each camp are regarded as an official body representing the camp population.

Due to the longevity of its existence, UNRWA has had to adapt and face challenges from different state policies and changing donor funding priorities, and to respond to the changing demands of the peace process. These all made new demands on UNRWA in terms of increased expenditure and new programmes, as well as subjecting its personnel to danger or harassment. Such challenges have all taken place within the wider context of an increasing Palestinian refugee population. Thus, as an agency it has been constantly in a condition of transition, change and crisis over the period under review.

Some of UNRWA's original plans in the 1950s were viewed with suspicion by refugees, as they were seen as aimed at absorbing the refugees into the regional economies at the expense of their right to return. From 1957, UNRWA's regular programme of activities focused on education, health, relief and social services. More recently, it has introduced microfinance and micro-enterprise opportunities to encourage greater economic independence.

UNRWA has played a significant role in the field of primary and preparatory education as well as vocational and technical training. Education is by far the agency's largest activity, with 500,000 pupils in 658 schools in UNRWA areas of operation. Despite some criticisms and concern that its focus on humanitarian activities distracts the international community from the fundamental political questions at the core of the conflict, in the main, refugee groups have been supportive of the work of UNRWA.

Its central role in supporting Palestinian refugees has led both Israel and some members of the United States Congress to see the existence of UNRWA as a major contribution in the growth of Palestinian consciousness and identity and, consequently, they have argued that it should be closed down at the earliest opportunity. This was the position taken by the Israelis at the Taba talks and has been accepted by some Palestinian officials eager to extend the role of the Palestinian National Authority (PNA) and the PLO. Clearly, without UNRWA, the onus of providing for the refugees would fall on the host governments in the Lebanon, Jordan and Syria and, since 1993, the PNA. If they were unable or unwilling to do so it would pose a major challenge to international donors and the international NGOs. Since the election in 2006 of an Islamicist Palestinian interim government in the OPTs, which has been subject to economic sanctions, a new consensus has emerged that UNRWA provides an alternative route for funding support to ensure that minimum safeguards in health and other services can be met for refugees and the wider population, particularly in Gaza.

Impact on the region

Regionally, the impact of the Palestinian case has been profound. The refugee cause was adopted by the Arab states in their refusal to allow Israel to be absorbed into the region. It has also has been used to further their own political goals, and differing means of support to the Palestinians have been the grounds for disagreement between them. Furthermore, the ongoing conflict with Israel has been both the cause and excuse for high expenditures on security and the military, the suppression of moves towards democratization and growth in civil society, and the low investment in human resources, particularly education. The regional impact was mostly obvious during the Cold War period, when the Soviet Bloc and the Arab states acted together to put pressure on Israel to meet Palestinian demands and to withdraw from the OPTs. This culminated in the 1973 War and the defeat of the Arab armies.

A major impact of the ongoing war between Israel and its neighbours, therefore, has been the militarization of both Israeli and Arab society. Conscription, the role of the military in public life, state encouragement for reproduction and population growth and the presence of intrusive surveillance over citizens are features of all states in the East Mediterranean littoral. Israel has been able to progress towards a limited democracy, but at the expense of political freedoms for Palestinians inside Israel and even greater restrictions for those in the OPTs. The lack of democracy in its Arab neighbours is even more pronounced. Although we should not attribute such militarization and lack of democratization in the region entirely to the Arab–Israeli conflict and its root cause, the Palestinian refugees, there is no doubt that a resolution of the refugee issue would strengthen democratic and civil society forces in the region.

Another example of the impact on the region has been the period of the 1990s which saw breakthroughs in the stalemate and the beginnings of an international framework to deal with the issue. As already mentioned, the collapse of the Soviet Bloc and the 1st Gulf War (1990–91) brought important changes to the region and the promotion of a US vision of a New World Order. Partly as an incentive in obtaining Arab support for the coalition against Iraq, the US secured the agreement of Israel to participate in an international peace conference on the Palestinian issue. Although Israel benefited from the neutralization of Iraq as a military force, it was economically weakened and dependent upon US economic support for a new wave of immigration of Russian Jews. In addition, the war had left the PLO diplomatically isolated and almost bankrupt. Both parties, therefore, were under pressure to resolve their differences, which resulted in the Madrid peace conference in October 1991. Despite relocating to Washington and continuing for nine rounds of negotiations, very little was actually achieved.

One of the main problems was Israel's refusal to recognize the PLO as a negotiating partner, which led to various diplomatic contortions, such as including a Palestinian delegation in the Jordanian one, and, finally, to an impasse in the negotiations. At the same time, for Israel, the Palestinians were bringing what they thought were unrealistic demands with regard to options for the refugees, which included a return to their homes in Israel. As we saw in the previous section, the Madrid framework was soon superseded by the Oslo Accords.

Resolution to the protracted refugee situation

Peacemaking

One legacy of the Madrid conference was the establishment of a body known as the Refugee Working Group (RWG), which established a network of contacts that proved to be invaluable in the post-Oslo period. At its first meeting in Ottawa in May 1992, the RWG, chaired by Canada, decided to organize its work on a thematic basis and countries were allocated as 'shepherds' for each theme. The themes primarily addressed humanitarian questions, such as family reunification, training and job creation, public health, child welfare, and social and economic infrastructure. Nevertheless, its work soon became mired in controversy as the Palestinians and Israel failed to agree either on what could be placed on the agenda or on definitions of 'refugees' and 'displaced persons'.[28]

Despite the failure of further talks in 2000, known as the Camp David Summit, to deal substantively with the refugee issue, some of the groundwork prepared by the RWG fed into a new set of talks during 2001 at the Egyptian seaside resort of Taba. The Taba talks made significant progress on the refugee issue but nevertheless it remained the sticking point. It also built on what is known as the 'Clinton parameters', which was an attempt by US President Bill Clinton to outline the fundamentals of a compromise between the Israeli and Palestinian positions and which have become a reference point for discussions on how to re-start the peace negotiations. With respect to the refugees, Clinton proposed that refugees wishing to leave their country of asylum would be directed to the new Palestinian state in the OPTs, including areas ceded to it in any land exchanges with Israel, with some being permitted into Israel at the discretion of Israel. Refugees would also be compensated, with the US offering to make a major contribution. However, the election in the same year of a new government in Israel, led by the hardliner Ariel Sharon, and ongoing violence in the OPTs meant that the talks finished without agreement.

In an attempt to bring a new regional framework to the issue, the Arab League's Beirut Declaration of March 2002 offered a trade-off between normal relations with the Arab world in return for a withdrawal from the occupied Arab territories, the creation of an independent Palestinian state with East Jerusalem as its capital, and the 'return of refugees'. The section on refugees called for an 'achievement of a just solution to the

Palestinian refugee problem to be agreed upon in accordance with UN General Assembly Resolution 194'.

The continuing deterioration of the situation on the ground in the OPTs threatened to destabilize the region. As a result, the European Union, the United Nations and Russia attempted to breathe new life into the negotiation process by pressing the US to take some joint action. Known as the 'Quartet', it issued a statement regarding a 'road map' for peace in September 2002. In October of the same year, US President George W. Bush publicly expressed his support for an independent Palestinian state based on the territory of the OPTs and living in peace beside Israel as the ultimate objective of the road map. This was an important step in establishing the 'two-state' solution as the preferred option of the international community, with the implication that the refugees would 'return' to the new state in the OPTs. Despite this international initiative, actions by both the Palestinians and Israelis on the ground prevented confidence-building measures necessary for the road map from being implemented. The Israeli government continued its colonization policies in the OPTs, which in return fuelled the ongoing violent resistance of the Palestinians.

It is important to recognize the role of civil society in the attempts to maintain the momentum towards a negotiated solution. During this post-Oslo period, a number of civil society initiatives or Track II ventures were also launched. The most significant in terms of the backing it received from the political elites was the Geneva Accord. Sponsored by the Swiss government, the Geneva Accord was conducted by leading but unofficial Palestinian and Israeli negotiators and it amounted to an unofficial blueprint for a permanent status agreement. In essence, the Accord expanded further on the concessions Israel offered at Taba and Camp David, such as recognition of a Palestinian state in the OPTs and withdrawal from territories occupied in 1967, including parts of East Jerusalem. Israel would also cooperate in facilitating the relocation of refugees to the new state and contribute to the organization and financing of both a 'return' package and compensation. In exchange, the Palestinians would not implement their right of return.[29]

However, the issue of refugee consultation began to re-surface as an important issue. As a result of all these negotiations – the Oslo process, Camp David, Taba, Beirut, Geneva and other official and unofficial negotiations – many refugees began to feel that their rights were being sacrificed in order to obtain the basic territorial components of a Palestinian state in the OPTs. A number of refugee groups were formed in the OPTs, the Arab host countries and the wider diaspora to uphold the rights of Palestinians to return, compensation and restitution of property, and a 'Coalition of the Right of Return Groups' was formed.[30]

These groups objected to compensation being offered as an alternative to a return, which should be dealt with separately from compensation and property questions. While official PLO positions include both concepts, Palestinian NGOs have been more forceful in insisting upon restitution instead of compensation.

Changes needed

Previous comparative studies on refugee situations and their relevance to the Palestinian case have identified four main elements which could be described as prerequisites in the search for durable solutions to the Palestinian case. They are: international involvement, refugee participation, the consolidation of local and regional structures, and measures to promote justice and reconciliation.[31]

Any peace agreement will require high-level international involvement. The positions of the parties are too entrenched for them to trust an agreement that does not have international guarantors. Indeed, there has already been a long period of quite intensive international involvement on the official level which can be utilized. The Madrid peace conference in 1991 was sponsored by the two superpowers, and the 'multilaterals' that flowed from it, such as the Refugee Working Group, included a wide range of states. The Euro-Mediterranean Partnership, set up by the EU in 1995, provides an important regional forum for coordinating development plans. In addition, the role played by the EU's Special Envoy to the Middle East Peace Process, Ambassador Miguel Moratinos, in the Taba peace talks in 2001, and the role played by the Swiss government in drawing up the unofficial Geneva Accord in 2003 are measures of close European interest in and support for the peace process. Finally, the 'Quartet', comprising the UN, the US, Russia and the EU, is an attempt to follow on from the failed Oslo process with a three-stage 'road map', thus encouraging negotiations over the refugees issue. In this context, it is significant that in both the Taba talks and the Geneva Accord there was no reference to international practice and the experience of UNHCR, or to a regional framework that includes the neighbouring and host countries.[32]

The Beirut Declaration of 2004 seems to provide a basis for the participation of host countries in these discussions, but the lack of clarity regarding what durable solutions were envisaged has stymied efforts in that direction. However, where these discussions in the Middle East peace process do coincide with international practice is in the degree of consensus concerning the need for international funding for both compensation and repatriation. Both the target locations for repatriation – Israel or the new Palestinian state – will require assistance to absorb refugees. All parties recognize that a Palestinian state in the remaining territories of the West Bank and Gaza Strip will not be able to generate sufficient internal revenues to simultaneously construct a new state, provide services for the current population rate of growth and fund a large-scale repatriation programme. Whether in the form of individual or collective payments, external financial support for a repatriation programme will be essential.

The second prerequisite identified was the importance of refugee choice and participation in planning repatriation activities. International experience has shown that a return to the *status quo ante* is rarely possible, and is frequently not the preferred option for the refugees themselves. The critical issue is that the options available are transparent. They may be an overall package which in totality can be deemed a just one and therefore acceptable. The package can include actual but limited return, symbolic acts of return and restitution, or a mix of return compensation and 'return' to the new Palestinian state. What is unlikely to be acceptable is the continuing absence of any formal recognition of the refugees' right of return.

Another issue that can be derived from international experience is creating a sense of ownership of the decision-making process. In the Palestinian case this has been difficult to achieve, due to the dispersal of Palestinian refugee communities, factional divisions and the monitoring of political activity by the host countries and Israel. Nevertheless, the attempts to sideline the right of return have produced, as we saw earlier, a popular reaction. It is clear that, in the Palestinian case, new channels for refugee participation need to be created. The representation of refugee concerns since the establishment of the PNA are problematic. The PLO has been weakened and its role in the camps is challenged by other groups. At the same time, radical Islamic groups and the various Right of Return committees that have been set up in the host countries and the OPTs are not consulted by the leadership. In the Palestinian context, UNRWA could be considered as a possible vehicle for such representation, fulfilling the functions normally taken up by UNHCR in this regard, but such a politicization of the organization would not be welcomed by Israel or the host Arab states.

The third prerequisite concerns local and regional development. It is now increasingly accepted that, during their period of exile, refugees become economic actors in a broader regional market. There is much evidence that this has also occurred in the Palestinian case, with its transnational networks and their integration in the local economies of the OPTs, Jordan and Syria.[33] In addition, all the major Palestinian refugee camps are very close to either Israel or the OPTs. Thus, family, business and political networks that have developed in exile will encourage Palestinian returned refugees to remain part of the regional labour market. A Palestinian repatriation programme needs to be flexible enough to build on these networks as a developmental and integrative asset. This suggests that any agreement should include a transitional period in which a refugee is permitted to retain residency status in the host country for a number of years, in order that employment, accommodation and other services prospects in the place of destination can be explored. The Arab League may need to be asked to consider temporarily suspending clauses in its charter prohibiting its member states from offering dual citizenship.

There has been much discussion on the 'absorptive capacity' of a new Palestinian state to receive large numbers of refugees. Subsequent discussions focused more on the question of costs and tailoring a programme to both the regional and local economies, institutional capacities and financial regulatory mechanisms. Rex Brynen, as both academic and consultant to the World Bank, has offered a detailed critique of some of the more ambitious plans and has argued that there is no such thing as the 'absorptive capacity' of the OPTs to act as a brake on repatriation. Instead returnees should be assisted in voluntarily relocating in a way that minimizes bureaucratic intervention. He highlights housing finance initiatives as a critical element of any refugee absorption strategy, and that in this connection the Palestinian state ought not to construct housing for returnees or relocate refugees to evacuated Israeli settlements, or attempt to remove the existing refugee camps.[34]

A fourth prerequisite concerns measures that promote justice and reconciliation. Clearly, the measures proposed will be contingent on the details of a peace agreement. If the solution being suggested is what is known as a 'one-state solution' – that is, a single, bi-national state involving the return of some refugees to their homes in what is now Israel – then there are a series of essential reforms to be carried out and institutions to be established. One can envisage that, as part of a package of restorative justice measures, there would have to be a degree of land and property restitution, compensation for destroyed or appropriated land and property, and the repeal of laws which discriminate against non-Jews on the Israeli side and non-Palestinians on the Palestinian side. Other important measures would include a formal apology from both sides to the citizens of the other side for any harm and suffering experienced during the course of the conflict, the establishment of a truth commission, and commemorative activities such as national days of mourning, museums, public monuments, etc. to mark the refugee experience, dispossession, achievements in exile and the role of peacemakers and other leaders.

If the solution is based upon what is known as a 'two-state solution', in which a new Palestinian state would be established in the OPTs and it would be the main location for resettlement of refugees, then the reconciliation process which takes place is between two states rather than between two peoples within a single state. The process would be much more a top-down one, with the focus on more official institutions and formal programmes at the elite level. An essential first step would be an apology for the harm done and the suffering of the other side to be reciprocated by an amnesty for the criminal actions of political and military leaders. We can see that the first tentative steps for this were taken at the Taba talks, where the Israeli side suggested an expression of 'its sorrow for the tragedy of the Palestinian refugees, their suffering and losses', without accepting full responsibility. An important second step would be a limited restitution and compensation package which recognized both material losses and the trauma of dispossession and exile. It would also be important to include a programme of official visits for leaderships on both sides to express their remorse and commitment to coexistence at a suitable national site or monument. This would be supported by a range of measures to encourage harmonious interactions in trade, culture and education. There would need to be a review of legislation, media policy and curricula which promote contentious and negative images of the other side. Connected with this is the establishment of a truth commission to determine the events leading up to 1948 and to clarify the subsequent actions of the Israeli government and the PLO. As we have already mentioned, there may be considerable resistance to this idea, particularly from the Israeli establishment, but this should not deter Israeli and Palestinian civil society from taking the initiative with international support.

As indicated in the introduction to this chapter, the protracted nature of the Palestinian refugee situation can be partly attributed to the lack of agreement over the causes of the conflict. To the Palestinians, the conflict began once the Zionist settlers started colonizing Palestine in large numbers in the 1920s and 1930s, culminating in the *Nakba* of 1948. For the Israelis, the land acquired in 1948 is not an issue. The land acquired in 1967, the recognition of Israel as a Jewish state and the nature of the Palestinian state *vis-à-vis* Israeli security are the issues to be negotiated. We can see this dissonance very starkly in the proposals for dealing with the causes of Palestinian refugeedom. While clearly this began with the creation of the Israeli state in 1948, the Oslo Accords in 1993 refer only to 'persons displaced from the West Bank and Gaza Strip in 1967'.[35] It was only after the collapse of the Camp David talks in 2000 that it was clear to the Israelis that no deal was possible without a consideration of the rights of the refugees from 1948. This conflict, over what the conflict is about in the first place, is what Christine Bell has termed the 'meta-conflict'.[36] In Bell's view, unless this meta-conflict is resolved, peace

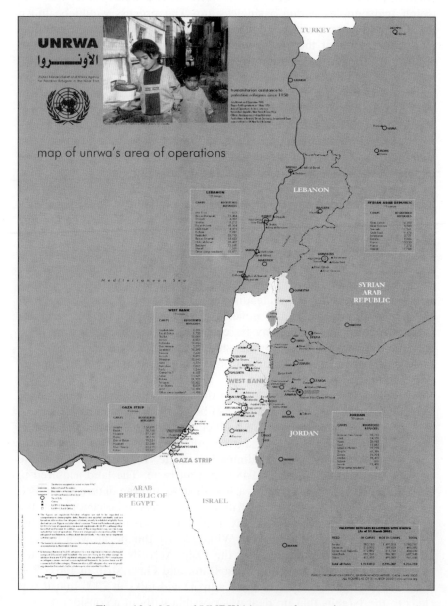

Figure 10.1 Map of UNRWA's area of operations

agreements and the post-conflict institutions they establish will not be addressing the heart of the conflict and will lead to a renewed cycle of disagreements and conflict.

Thus, under the one-state solution scenario mentioned above, there is one package of reforms to be carried out and institutions to be established. Under a two-state solution scenario, then another series of measures and institutions would need to be adopted.

Most analysts agree that an Israeli state which defines itself as a Jewish state would not contemplate an agreement that involved the large-scale repatriation of Palestinian refugees or the significant restitution of property. It would prefer to continue the present political impasse, with all its military and security burdens and costs placed on society, than to surrender or dilute the *raison d'être* of its existence – the creation of a Jewish state. In the light of this, there is no immediate solution to the Palestinian refugee case while the present regional and international balance of power provides Israel with the means to maintain its dominant position. In this context, any fundamental improvement in the refugee situation will occur only if there is a shift in the balance of power against Israel, or if the Palestinians signal that they would be willing to accept a two-state solution, with the OPTs being the location for the refugees. In some ways, this was beginning to take place during the 1990s but the combination of Israeli caution, Palestinian frustration, post-9/11 US interventions and the reconfiguration of regional alliances has reversed this trend. Nevertheless, if negotiations resume it is likely that they will resume more or less based upon the 'Clinton parameters', with great efforts exerted by the Palestinian leadership to inject a measure of property restitution into the proposals.

Notes

1. BADIL Resource Center for Palestinian Residency and Refugee Rights, *Survey of Palestinian Refugees and Internally Displaced Persons*, Bethlehem: BADIL Resource Center, 2003, p. 34.
2. See M. Dumper, ed., *Palestinian Refugee Repatriation: Global Perspectives*, London: Routledge, 2006, p. 5 ff.
3. The only equivalent cases can be drawn from other post-war partitions, such as the partition of Germany and India, where a political settlement has since been reached although individual refugees still nurse a sense of loss and grievance. In the case of Germany and German refugees from Eastern Europe, legal attempts at restitution are taking place.
4. See Table 1, UNHCR, *The State of the World's Refugees: A Humanitarian Agenda*, Oxford: Oxford University Press, 1997, p. 287.
5. BADIL Resource Center, *Survey of Palestinian Refugees and Internally Displaced Persons*, p. 34.

6. See my discussion in the Introduction of M. Dumper, *The Future of Palestinian Refugees: Toward Equity and Justice*, Boulder, CO: Lynne Rienner Publications, 2007, pp. 5–6.
7. M. Dumper, *Palestinian Refugee Repatriation*, p. 13.
8. Anglo-American Committee of Inquiry, *A Survey of Palestine*, Vol. 1, 1946, cited in Badil and Centre on Housing Rights and Evictions (COHRE), *Ruling Palestine: A History of the Legally Sanctioned Jewish-Israeli Seizure of Land and Housing in Palestine*, Geneva: Badil/COHRE, 2005, p. 20.
9. UN General Assembly Resolution 181.
10. UN General Assembly Resolution 194 III, 1948.
11. UNRWA Mandate 302 IV 'Assistance to Palestine Refugees 8 December 1949' is archived on the website of the UN Information System on the Question of Palestine (UNISPAL), available online at: http://domino.un.org/UNISPAL.NSF.
12. W. Lehn and U. Davis, *The Jewish National Fund*, London: Kegan Paul International, 1988, pp. 113–14, 109–11, 131, 133–4; see also Don Peretz, *Israel and the Palestinian Arabs*, Washington, DC: Middle East Institute, 1958, p. 158.
13. M. Fischbach, *Records of Dispossession: Palestinian Refugee Property and the Arab-Israeli conflict*, New York: Columbia University Press, 2003, p. 70.
14. Ibid., p. 246.
15. D. McDowall, *The Palestinians: The Road to Nationhood*, London: Minority Rights Publications, 1994, p. 38.
16. UNRWA statistics, available online at: http://www.un.org/unrwa/publications/index.html.
17. Palestine Liberation Organization, *The Palestinian Refugees Factfile*, Ramallah and Jerusalem: PLO (Department of Refugee Affairs), 2000, p. 8.
18. Convention relating to the Status of Refugees, 1951: Article 1.
19. D. Peretz, and G. Doron, *The Government and Politics of Israel*, Third Edition, Boulder, CO: Westview Press, 1997, p. 47.
20. D. McDowall, *The Palestinians*, p. 88.
21. L. B. Jacobsen, *Finding Means – UNRWA's Financial Situation and the Living Conditions of Palestinian Refugees*, Summary Report 415, FAFO, 2003, p. 13.
22. Ibid., p. 75.
23. Ibid., p. 10.
24. A. Plascov, *The Palestinian Refugees in Jordan 1948–1957*, London: Frank Cass & Co., 1981, p. 16.
25. Palestinian Central Bureau of Statistics, 2002: Summary, available online at: http://www.pcbs.gov.ps/DesktopDefault.aspx?tabID=1&lang=en.
26. N. Boqae'e, *Palestinian Internally Displaced Inside Israel: Challenging the Solid Structures*, Bethlehem: BADIL, 2003.
27. A. Jamal, 'The Palestinian IDPs in Israel and the Predicament of Return: Between Imagining the Impossible and Enabling the Imaginative', in A. M. Lesch and I. S. Lustick, eds., *Exile and Return: Predicaments of Palestinians and Jews*, Philadelphia: University of Pennsylvania Press, 2005.
28. J. Peters, *Pathways to Peace: The Multilateral Arab-Israeli Talks*, London: Pinter, 1996.
29. *The Geneva Accord: A Model Israeli-Palestinian Peace Agreement*, 2003; text available online at: http://www.geneva-accord.org/HomePage.aspx?FolderID=11&lang=en.
30. See also the work of the Civitas Project in articulating these views in K. Nabulsi, *Palestinian Register: Laying Foundations and Setting Directions*, Oxford: Nuffield College, University of Oxford, 2006.
31. See my discussion in M. Dumper, *Palestinian Refugee Repatriation*, chapter 14.

32. For a full discussion on the relationship between international law and the proposals put forward in the Middle East peace process, see M. Dumper, *The Future of Palestinian Refugees*, chapter 3.
33. S. Hanafi, 'The Sociology of Return: Palestinian Social Capital, Transnational Kinships and the Refugee Repatriation Process', in Eyal Benvenisti, Chaim Gans and Sari Hanafi, eds., *Israel and the Palestinian Refugees*, Berlin: Springer, 2007.
34. R. Brynen, 'Perspectives on Palestinian Repatriation' in M. Dumper, ed., *Palestinian Refugee Repatriation*, pp. 63–86; and R. Brynen, 'Refugees, Repatriation, and Development: Some Lessons from Recent Work', in R. Brynen and Roula el-Rifai, eds., *Palestinian Refugees: Challenges of Repatriation and Development*, London: I. B. Tauris, and Ottawa: International Development and Research Center, 2007, pp. 102–120.
35. Declaration of Principles on Interim Self-Government Arrangements, 1993, Article XII.
36. Christine Bell, *Peace Agreements and Human Rights*, Oxford: Oxford University Press, 2000, p. 15 and p. 316.

11
Somali refugees: Protracted exile and shifting security frontiers

Peter Kagwanja and Monica Juma

In the late 1980s, opposition to the despotic regime of Mohamed Siyaad Barre in Somalia gave rise to one of Africa's most profound and protracted refugee crises. In 1988, fighting between the regime and its challengers in the north-west led to the flight of some 400,000 registered refugees to Ethiopia and Djibouti. The fall of the regime in 1991 and the ensuing civil war in 1991–1992 displaced more than half of the Somali population. Even as more than 1 million refugees returned home voluntarily in 1992–2004 – 485,000 of these with UNHCR assistance – as of 2005, some 400,000 Somali refugees were still stuck in exile in Djibouti, Kenya, Ethiopia, Uganda, South Africa, the Middle East, North America, Europe and Australia. Ongoing insecurity resulting from the collapse of the state, instability and conflict, famine and diseases – especially in central and southern Somalia – has prevented the return of over 240,000 Somali refugees in the Horn. To compound their difficulties, they have themselves become identified with insecurity, making Somali refugees targets of new security regimes and restrictions aimed at containing them.

The security imperatives of the American-led global 'war on terrorism' in the countries of the Horn of Africa after 11 September 2001 have exacerbated this. Counter-terrorism strategies have spawned new dynamics of insecurity within Somalia, hampering voluntary repatriation as one of the durable solutions. The rise of Islamism as a political force in Somalia has been viewed through the lens of anti-terrorism. This has intensified xenophobia and reinforced long-standing negative perceptions and stereotypes associated with the Somali populations in the Horn of Africa.

Protracted refugee situations: Political, human rights and security implications,
Loescher, Milner, Newman and Troeller (eds),
United Nations University Press, 2008, ISBN 978-92-808-1158-2

Anti-Somali sentiments have weakened the protection of Somali asylum seekers and refugees, reducing the chances of local integration and resettlement in the West. This has taken a heavy toll on a range of asylum rights, protection and assistance.

Defining a 'Somali refugee' poses unique challenges, largely because the people generally identified as Somalis are spread across Djibouti, Ethiopia and Kenya. Ethnic Somalis from Ethiopia and Kenya have from time to time fled war, political repression and persecution. For the purpose of this study, however, the term 'Somali refugee' is generally applied to Somalis who fled insecurity and instability in their Somalia homeland after 1991. Only a fraction of 'Somali refugees' have found their way into UNHCR statistics, with many of them living as self-settled refugees in rural and urban areas across the Horn.[1] The identity of a 'Somali refugee' also poses its own conceptual challenges. Theorists of the refugee phenomenon have contested the overriding idea of a 'refugee' as a simple identity construct that emerges from the experiences of violence, war, persecution and the subsequent flight and displacement from the country of citizenship. 'Refugeeness', as Liisa Malkki points out, may be seen as 'the process of becoming ... a gradual transformation, not an automatic result of the crossing of national boundaries'.[2] Somali refugees went through the obvious path of identity formations in response to the concrete realities of their exile lives. However, in the Horn, the 'Somali refugee' quickly took on the negative stereotypical perceptions and marginality of the local ethnic Somalis, thus fashioning the dynamics of reception, protection and integration and hindering other forms of assistance. Specifically, the Somali refugees share the collective stigma attached to local ethnic Somalis in Kenya since the *shifta* (bandit) war in the 1960s.[3] The criminalization of the Somalis as a state security burden has presented unique and intractable difficulties to the protection of Somali asylum seekers and refugees.

This chapter focuses on the fundamental connection between shifting security dynamics in the Horn of Africa, especially after 9/11, and the prolonged Somali refugee crisis. The chapter analyses the causes of this crisis, delving into the responses of the Kenyan state and how this has impacted on the protection, assistance and experiences of Somali refugees. Finally, the chapter examines the measures taken at the national, regional and international level to strengthen the protection of Somali refugees and to expand space for durable solutions: repatriation, local integration and resettlement. The chapter argues for the stabilization of Somalia as the best option for ending the country's protracted refugee crisis, and it advocates a careful mix of 'hard' power options (military/peacekeeping) with the means of 'soft' power, such as diplomacy and dialogue, to stabilize Somalia, urging the toning down of the prevailing military-heavy style, driven by the dictates of the 'war on terrorism'.

Causes of Somalia's protracted refugee situation

Somalia's protracted refugee crisis has deeper historical roots than the events of the late 1980s that triggered it. At independence in 1960, the Somali nationalist elite put their country on the path to modernization but tragically failed to lay a firm foundation for a democratic dispensation. Foremost was the failure to strike a balance between the centralizing tendencies of the post-colonial state and the centrifugal trends within Somali society, traditionally based on diverse, fluid and competing clan networks and loyalties.

The dictatorship of Siyaad Barre (1969–1991) stretched this tension to breaking point by cynically utilizing clan identity and sensibilities as part of its divide-and-rule tactic. Chronic underdevelopment, economic decay and the regime's over-reliance on military aid and external debts alienated the state from the citizenry. They also reinforced popular public faith in clan and other primordial forms of identity. Despite the wind of change after 1989, a weak and divided opposition and an obdurate regime derailed Somalia's transition from despotism to democracy. The stage was set for the dramatic implosion of the 1990s.

State collapse and civil war

In 1988–89, the Barre regime's clamp-down on the rebel Somali National Movement (SNM) and opposition politics in the north-west created the 'Hargeysa Exodus', in which some 400,000 refugees fled to Ethiopia and Djibouti. The popular uprising in late 1990 and early 1991 forced Barre and his regime's stalwarts to cross the border to Kenya. The fall of the government in Mogadishu led to chaos and anarchy that pushed over half a million people into Ethiopia and some 200,000 into Kenya.[4] At the height of the civil war in 1991–1992, an estimated 2 million people were internally displaced, and another 800,000 forced to flee to neighbouring countries.[5] The bloody battle for the soul of Mogadishu[6] between Ali Mahdi – elected as President by a cluster of Hawiye politicians on 28 January 1991 – and General Mahammed Faarah Aydiid after 1992 turned Somalia into one of the leading refugee-producing countries in the world.[7] The impact of this violence was compounded by institutional decay, droughts, diseases, floods and other natural disasters. A crippling drought in 2005–2006 and crop failure in mid-2006, against the backdrop of escalating war, almost wiped out people's livelihoods in southern Somalia and drove many people to refugee camps in north-eastern Kenya.

Evasive peace

More than a dozen abortive peace deals were signed in the period 1992–2004. These failures fostered political uncertainty, impeded repatriation as a durable solution and spawned new refugee influxes. In October 2000, the Transitional National Government (TNG) was created by the Arta peace process brokered by the President of Djibouti, Ismail Omar Guelleh. However, the TNG managed neither to stamp its authority over the whole country, nor to win the backing of regional and international actors suspicious that it was under the strong sway of Islamic militants.[8] The promise of a more enduring peace came in October 2004, when the Somali Transitional Federal Government (TFG) was launched, with Abdullahi Yusuf Ahmed as President. Owing to insecurity, the TFG delayed relocation to Somalia, eventually setting its base in Baidoa in March 2005. This created a power vacuum in Mogadishu, which militants in the Islamic Courts Union exploited to seize power in mid-2006. Abortive peace pacts thwarted efforts to create a peaceful environment, shattering hopes for early repatriation of exiles.[9]

International intervention in Somalia in the 1990s went disastrously wrong, accentuating Somalia's prolonged refugee crisis. In 1992–1995, the UN established two operations in Somalia – UNOSOM I in 1992–1993 and UNOSOM II in 1993–1995 – to restore law and order. However, the intervention ended in a fiasco in May 1995, when the US pulled out its troops from Somalia following the killing of 18 of its soldiers in Mogadishu during the 'Black Hawk Down' incident. The failure of the so-called 'Operation Restore Hope' dashed hopes for significant repatriation through the Cross-Border Operations that UNHCR launched.[10] The collapse of this intervention was followed by a massive cut-back in Western support. Ironically, the entry of 'Western' forces in Somalia fanned the embers of Islamism, bolstering it as a real source of insecurity. As Roland Marchal rightly noted, 'Somali Islamist groups developed more during the international intervention than beforehand'.[11]

The Islamic Courts

The rise of the Islamic Courts was as dramatic as their fall. The two events, however, prolonged the predicament of Somali refugees. Islamist groups like al-Ittihad al-Islami, al-Tabliq and al-Islaah vied for power after the ousting of Siyaad Barre in 1991, leading to violent confrontations with secular warlords. In the period 1991–1992, fighting between al-Itihad militants and General Aydiid's forces over the control of the port towns of Kismayo, Baraawe and Merka forced some 25,000 refugees to flee to Kenya.[12] Similar clashes between al-Itihad fighters and the Somali National Front (SNF), in 1993–1996 in the Gedo region, devastated and displaced farming communities like the Gobweyn, who had already faced years of war and starvation.

In 1999, the Islamists formed a nationwide Islamic Courts Union (ICU) with armed militias and police to mobilize resources and power to promote their dual goals of pan-Somali nationalism, regionally, and Islamic jihadism, internationally. In June 2006, the ICU rode to power in Mogadishu after its dramatic victory over the US-backed warlords' Alliance for the Restoration of Peace and Counter-Terrorism (ARPCT). The ICU–ARPCT clash killed 400 civilians and pushed the total of internally displaced persons (IDPs) in Mogadishu to 250,000. To their credit, the Courts restored some order in the war-torn city and much of the south.[13] Anticipation of the imposition of *sharia* law on areas under its control forced thousands of refugees to flee into Kenya.[14]

An imminent armed showdown between the ICU militants and the TFG in Baidoa fostered a climate of uncertainty. As a result, 14,000 Somalis crossed the border into Kenya in September alone, bringing the total of new arrivals in 2006 to 34,000 refugees and raising the caseload of Somali refugees in the country to 190,000.[15] In December 2006, the ICU was defeated and removed from power by the TFG forces backed by Ethiopian troops. Despite the ICU's dramatic fall, the battles of Baidoa, Bay region, Bandiradley, Mudug, Beledwweyne and Hiran region displaced no fewer than 25,000 people in central and southern Somalia.[16] Thousands of refugees flocked to Kenya's border to escape the fighting as the Islamists abandoned Mogadishu and their last strongholds of Jilib and Kismayo in the Juba River valley. Wary of the security repercussions of Islamic militants and their supporters flocking into its territory, Kenya barred over 7,000 Somali refugees from crossing its border with Somalia, precipitating a profound humanitarian crisis which sparked an international outcry.[17] The rise of Islamic extremism also resurrected the problem of Somali irredentism, rekindling a perennial source of conflict in the region.

The 'war on terrorism'

The ascendancy of the Islamists as a political force threw Somalia into the spotlight of Washington's anti-terrorist strategy in the Horn of Africa. The strategies used to combat terrorism spawned security dynamics that, eventually, exacerbated the Somali refugee crisis. The US backed

Ethiopia to oust the Islamists in Mogadishu in late 2006. As the TFG and Ethiopian troops advanced on Mogadishu, it was the threat of terrorism, and not humanitarian considerations, which fashioned responses to Somali refugees. Kenya, another long-time US ally, deployed significant forces to seal its borders with Somalia. It also set up screening points to ensure that escaping Somali Islamic fighters and international terrorists did not use refugees as cover to enter its territory, leading to thousands of refugees being denied entry. On its part, the US launched military strikes against fleeing fighters of Shabaab and suspected al-Qaeda allies.

What has launched Somalia back onto the radar-screen of Washington's security priorities are the security concerns of the campaign against terrorism. During the strikes, the security priorities of crushing Islamic terrorism eclipsed the plight of Somali refugees fleeing the war. Indeed, the lens of terrorism has added a new security angle to the problem of the protection of nearly 240,000 Somali refugees in camps in the Horn of Africa. As the case of Kenya shows, Somali refugee protection has always been undermined by an overbearing emphasis on security and the view of Somalis as a security burden. These causal factors led to the continuation of the Somali refugee situation in the Horn of Africa. By February 2007, the caseload of Somali refugees in the region had increased to 240,550, with nearly 400,000 internally displaced; 188,868 Somalis were in hosted in Kenya.

Somalis in Kenya: from 'abdication' to 'intervention'

Between 1989, when the first group of Somali refugees arrived in Kenya, and 2007, the country's asylum policy went through two discernible phases: the era of 'abdication' and containment (1989–2002) and the age of 'interventionism' after 2002. Both phases reflected the shifting dynamics of democratization in Kenya, which saw the exit of Daniel Moi's authoritarian regime in 2002, ushering in the National Rainbow Coalition of Kenya (NARC) government of President Mwai Kibaki. Despite their different approaches, the two administrations viewed Somali refugees through a distinct security prism, with the dictates of state security eclipsing their humanitarian obligations.

Moi's 'abdicationist' state

As early as September 1989, when the Somali state started to unravel, 3,000 refugees fled to Kenya's north-eastern province through the border town of Liboi, settling near the current site of the Dadaab camps. Owing to relatively warm relations between Siyaad Barre and Moi, these refugees, mostly from clans opposed to the Barre regime, were received with hostility. In late 1989, after senior officials of the two governments met and agreed to cooperate, President Moi ordered Somali refugees to return to Somalia, in utter disregard of Kenya's humanitarian obligations to receive and protect these refugees.

Until the fall of the Barre government in 1991, the Kenyan military and police routinely sealed the border and the Kenyan coast to prevent Somali refugees from entering the country, beating and forcibly returning those who sneaked through the lines. Kenya's state security machinery also barred UNHCR, the Red Cross and other humanitarian agencies from assessing and providing assistance to these refugees. Even though the regime received and allowed government and military officials of the defeated Barre regime, who entered the country by air and booked themselves into Nairobi hotels, boats carrying thousands of refugees were pushed back and prevented from landing before the 'dam burst' in January 1991. An estimated 300,000 fled to Kenya in the aftermath of the fall of Siyaad Barre in 1991 and the subsequent power struggle in central and southern parts of Somalia. As a result, Kenya's refugee population shot meteorically from a low of 16,000 in March 1991 to 427,278 at the end of 1992. This overwhelmed and outstripped the government's capacity to manage and assist the avalanche of refugees.[18] The numbers stabilized between 225,000–245,000 in the next 10 years, largely because of shrinking resettlement opportunities and dwindling chances for repatriation as a result of instability at home.

The government established the National Refugee Secretariat within the Ministry of Home Affairs and National Heritage and the Office of the President in September 1992. The system, however, was too feeble to deal effectively with refugees. It thus responded to the influx by adopting the policy of confining refugees to isolated camps. In a drastic change of policy, the Moi government appealed for international aid and invited UNHCR to manage the refugee camps. In the intervening years, the government trod a decidedly 'abdicationist' line, arguing that 'refugees are the UNHCR's responsibility, not ours', and deliberately ceding refugee affairs to UNHCR.

The government also allowed the agency to take on a wide spectrum of responsibilities which are normally the preserve of the host state, giving it untrammelled powers in refugee administration.[19] By marginalizing itself from the refugee arena, the state contributed to the loss of the experience and capacities accumulated over the years, and stifled the emergence of new capacity for humanitarian intervention. Prior to the influx of Somali refugees, the Moi state drafted a refugee bill which could have augmented its capacity for refugee administration, but the draft was shelved in the 1990s. On their part, the agencies seized the chance to step into the breach created by the withdrawal of the state from refugee governance. Paradoxically, this turned the refugee arena into a sphere where UNHCR and its partners assumed untrammelled control over refugee affairs, without the corresponding power to protect refugees. This unintended blurring of the lines of authority over refugee matters became the bane of asylum in Kenya and the pitfall of refugee protection.

The policy of encampment

In 1992, the government opened seven camps which were later merged into two main camps. The first was the Dadaab cluster of camps – Ifo, Dagahaley and Hagadera – located in north-eastern Kenya, 80 kilometres from the border with Somalia and hosting 166,208 Somali refugees. The second was Kakuma camp in the north, near the border with Sudan, which hosted some 23,000 Somali refugees. Designed primarily to reduce the real or imagined threat that refugees posed to national security, the policy of encampment rested on two pillars: abdication of responsibility to humanitarian agencies, particularly UNHCR, and pushing refugees to the margins of society, away from the main economic activities in farmlands and urban areas.

The architects of the encampment policy may have imagined that remoteness, poverty, lack of services and insecurity in the bandit-prone, semi-arid sites of the camps would bar further refugee flows and force those in the country to voluntarily return. Refugees were forced to reside in the camps in order to qualify for assistance. Those found outside the camps were considered illegal aliens, with the possibility of deportation. Over and above the security concerns of the state and the burdens associated with the hosting of refugees, the scale and ethnic character (Somali) of the refugees provided a constant excuse for this draconian policy.

Camps consigned refugees to a status of limbo, with no legal status, no opportunity to integrate locally and the chances of repatriation appearing bleak due to insecurity in Somalia and narrowing resettlement openings. The protracted nature of the refugee situation saw international assistance – and therefore nutrition – decline. Shelter became dilapidated and hopelessness settled in. But the real threat to the rights of refugees has been acute physical insecurity. Refugee camps, especially the Dadaab camps which hosted 135,000 Somali refugees, became a haven for all sorts of violence, abuse and criminality.[20]

Widespread insecurity was attributed to the deteriorating political economy of the host state, the character of refugees, the poverty of the host communities and the failure of intervention measures by governments and agencies.[21] The mingling of fleeing refugees with armed extremists and combatants using camps as humanitarian shields exacerbated insecurity, posing a new challenge of unmasking and holding violators accountable and protecting refugees.[22] That said, the security situation has improved since 2001. Cases of violent crime dropped from 300 in 1998 to 36 in 2003 and reports of rape, murder and armed robbery have also been declining significantly.[23] However, bandits, armed gangs and criminals still stalk the camps and their environs.

Following the 1998 US Embassy bombings, analyses of insecurity in the refugee camps came to be framed within the discourse of international terrorism. The Dadaab camps came to be viewed as recruiting and training grounds for terrorist organizations. Terrorist imprints in the refugee camps were traced to the presence of Somali radical Islamist organizations, such as al-Ittihad al-Islami, with linkages to al-Qaeda and Islamists in Taliban's Afghanistan.

Festering frustration among the Somali refugees, arising from the protracted humanitarian crisis, created fertile ground for al-Ittihad and other militants to recruit refugees and local Somali youths.[24] Al-Ittihad exploited gaps and lapses in Kenya's banking system, utilizing the trust-based Hawilaad or Hudi banking system, which leaves no paper trail, to finance its activities among refugees in the camps. Somali warlords also used the Hawilaad system to buy weapons to sponsor war. Although UNHCR, the Kenyan government and NGOs disputed these charges, counter-terrorists' intelligence focused attention on terrorism in the country.[25]

On 7 November 2001, the US forcibly shut down al-Barakaat offices worldwide, confiscated their assets and cut telecommunication lines as part of the 'global war on terror', alleging that tens of millions of dollars a year were moved from al-Barakaat to al-Qaeda. Hawilaad was the pipeline for the transfer of funds from Somalis in rich Western diaspora to poor refugees, in order to supplement the evidently meagre assistance from UNHCR and make a living in northern Kenya's extremely hostile environment. As Cindy Horst and Nick van Hear have shown, the shutting down of Hawilaad offices in Dadaab dealt a devastating blow to the Somali refugee economy and livelihoods.[26] As expected, this action stirred up intense anti-Americanism, which has sustained Somali's al-Qaeda cell.

Also accused of feeding the cannons of terrorists are Islamic charities operating in refugee camps. Efforts to curb terrorism in Dadaab focused on the al-Haramain Islamic Foundation, the sole Muslim charity in Dadaab. Besides running religious schools and social programmes, the charity even began distributing rice, sugar and other commodities that the World Food Programme did not provide. The charity won the hearts and minds of many Somali Muslim refugees by offering up camel and goat meat during the holy month of Ramadan.[27] Internationally, US counter-terrorist operatives had already placed the charity on their list of organizations propping up terrorism. Our own research from 1997 to 2006 in Dadaab found that al-Haramain was working closely with al-Ittihad agents in Dadaab to provide training and political education to Somali refugees, along the lines of Pakistan-style madrassa classes, to prepare them to 'defend Islam and the Somali nation'.[28]

However, al-Haramain denied these accusations, arguing that it is not a radical group, and posted a message on the charity's website asserting that 'we separate ourselves from [extremists] because our deen [Islamic law] is not one of extremes'. Despite this, the Kenyan government banned al-Haramain in the immediate aftermath of the 1998 bombings, accusing it of 'working against the interests of Kenyans in terms of security'.[29] Although the government deported the charity's Sudanese director, Sheikh Muawiya Hussein, in January 2004, the charity's work continued among Kenyan Muslims.

Local Somali hosts

Somali refugees were received by their kith and kin in the northern region, enabling a high degree of compassion and sacrifice among Kenya's ethnic Somali population to accommodate and assist them. However, undercurrents of historical animosity and prejudices against Somalis among Kenya's non-Somali population provoked resentment and xenophobia. It did not help that the vast majority of poor Somali refugees fled to the country's north-eastern province inhabited by Kenya's ethnic Somalis, who existed on the margins of Kenyan society as subjects of resentment and hostility.

The government embarked on long-running military campaigns against the *shifta* (bandit) insurgents of the Somali Youth League, who shared the irredentist goal of 'Greater Somalia' with other pan-Somali nationalists. The region became a frontier of human rights violations, and there were massacres of sections of Kenya's ethnic Somalis. The government of President Jomo Kenyatta imposed a state of emergency on the entire north-eastern province, which was only lifted in 1991. The public, however, remained hostile to refugees and Kenyan Somalis, leading Sydney Waldron and Naima Hasci to the conclusion that 'Somali refugees have received little "African hospitality"'.[30]

The construction of a monolithic Somali identity that blurred the citizenship divide between Kenyan Somalis and Somali refugees undermined refugee protection. Despite this, the location of the Dadaab camps in semi-arid and economically marginalized northern Kenya exposed refugees to conflict with the locals over resources, which characterized relations among Kenya's Ogadeni clans long before the arrival of refugees. In the study cited earlier, Crisp identified rivalry between and within clans and sub-clans over scarce resources as a significant source of insecurity within and around the camps.[31]

Although Crisp stresses that 'in Dadaab the distinction between "refugees" and "local population" is in many ways a fuzzy one', the protracted nature of the refugee situation began widening the identity gap between refugees and Kenyan Somalis. Interviews with refugees and local Somalis revealed strained relations, leading Kenya's Somali community to amplify its Kenyan identity to give it an edge in the fierce intra-Somali competition for control over trade and scarce resources. 'They are Somali. We are Kenyan. We are different', is a popular line one hears from community leaders and politicians from Kenya's ethnic Somalis.[32] Aggravating the local/refugee divide was the perception of refugees by the former as a privileged group and a threat to the fragile economy – a card that local Somali politicians played with glee. This forced agencies to design programmes in ways that would reduce hostilities between refugees and the local population.

Somali urban refugees

Of the 300,000 Somali refugees who fled to Kenya after 1991, some moved into Kenyan cities and towns, swelling the caseload of Kenya's self-settled urban refugees from 14,400 in 1990 to an estimated 55,000–150,000 by 1998.[33] Thousands of Somali refugees fled to the Kenyan coast, settling in Hatima, Jomvu, Marafa, Swaleh-Nguru and Utange on the outskirts of Kenya's port city of Mombasa. While some of these refugees were resettled in the West, thousands of them were relocated to the Kakuma camp in northern Kenya in line with the government's encampment policy. Others, however, migrated to Nairobi, joining the larger population of Somali refugees living in the suburb of Eastleigh, traditionally inhabited by Kenya's ethnic Somalis from colonial days.

Here, Somali refugees faced xenophobia and police harassment. With serious glitches in the processing of asylum claims by host states or UNHCR, refugees and asylum seekers found themselves without proper documentation, and thus exposed to the vagaries of extortion in the form of bribes and harassment by corrupt security officers.[34] As noted earlier, Kenya's security operatives exploited the new discourse of counter-terrorism to step up repression against the overwhelmingly Muslim Somali refugees in Nairobi and Mombasa.[35] Somali refugees continued to be singled out for police harassment, even after Kenya's return to democracy following the opposition's victory over the Moi regime in the December 2002 elections. For example, they were targets of swoops against the refugees in Nairobi in June 2003.

Somali refugees were also victims of repressive legislation, introduced in the aftermath of 9/11 as curbs against terrorist incursions. These anti-terrorist laws failed to strike a balance between the security imperatives of combating terrorism and the need to protect the values of democracy and human rights. This made refugees easy targets for unlawful harassment, arrest and prolonged detention on suspect grounds, in violation of the 1951 UN Refugee Convention. Refugees also became targets of the special counter-terrorist unit that Kenya formed in February 2003, consisting of officers picked from the police force with insufficient knowledge of refugee entitlements or identity documents. Although lobby groups for refugee rights, such as the Refugee Consortium of Kenya (RCK), launched courses to train the police on refugee rights, arbitrary arrests of refugees in the anti-terrorist campaigns increased.[36] The discourse of counter-terrorism gave new impetus to the state-centric view of security prevalent in the Cold War era, with the role of the National Intelligence Services in refugee matters increasing significantly.

Kibaki's interventionist state, 2003–2007

Upon coming to power in January 2003, the new government of Mwai Kibaki adopted an 'interventionist' approach to refugee matters, abandoning the 'abdicationist' policy of the Moi era. The new interventionist policy was driven more by the new administration's security concerns, spawned by the assumed link between refugees and the terrorist menace, than by a desire to genuinely take up its responsibilities under the 1951 UN Convention. The Kibaki government reintroduced the policy of giving identity cards to refugees, which was used in response to Ugandan refugees in the 1970s and 1980s, but terminated in the 1990s.

After more than 13 years of delay, on 3 August 2006, the Kenyan government tabled a revised Refugee Bill in parliament. The Bill passed into law towards the end of 2006, giving Kenya for the first time in history a legal instrument to deal with refugees and providing for the recognition, protection and management of refugees. The new law provides for a Refugee Status Determination Committee, an inter-ministerial committee serving as an adjudication channel through which refugees can appeal to the Refugee Appeal Board and then to the High Court of Kenya if their application is rejected. It also creates a legal framework of refugee protection, in line with the 1951 UN Refugee Convention and the 1969 Organisation of African Unity Refugee Convention.

In February 2003, the government announced its plan to review the policy of keeping the refugee population in camps in the insecure and arid north of the country.[37] However, Kenya's new refugee law entrenched the government's encampment policy by empowering the Immigration Minister to designate certain areas as refugee camps. This potentially hampered refugee integration into Kenyan society, restricting opportunities for refugees to become economically self-reliant. Curiously, the Refugee Bill was tabled in parliament at the same time as the revised version of the controversial Suppression of Terrorism Bill, earlier rejected by parliament in 2004. This fuelled concerns among refugee lobbies that the new refugee law was no more than a curtain raiser for the more draconian Terrorism Bill, which targets refugees, among other aliens.[38]

In addition to the new legal edifice, the establishment of a National Refugee Secretariat in 2006, under a brand new Ministry of Immigration and Registration of Persons, which replaced the refugee section at the Ministry of Home Affairs, looked set to boost the system of refugee governance. In March 2006, the Refugee Secretariat launched an alien registration exercise, focusing almost exclusively on the Somali suburb of Eastleigh, revealing the security concerns driving Kenya's new refugee architecture. The officially stated objective of the exercise was to conduct a baseline survey to determine the total number of refugees living in urban areas and to identify refugees capable of working or doing business so as to give them legal status and Class 'H' permits. The head of the Refugee Secretariat, however, let the cat out of the bag by disclosing the primary purpose of the registration as being to 'weed out illegal aliens' and to give the government an opportunity to determine the number of people who reside in the country, legally and illegally.[39]

This move created a stampede of refugees to UNHCR offices seeking to register claims for asylum. Emphasis on security in the registration exercise may have had an adverse effect on the rights of Somali refugees. Kenya has grudgingly accepted to serve as a holding ground for refugees.[40] The government and local actors, however, expressed the fear that the West might be pushing this policy to offload the problem of refugees onto the state and walk away from their international obligations under the burden-sharing responsibilities.[41]

Enchained 'good Samaritan'

With its ability to take charge of refugees outstripped, the government surrendered its role to UNHCR and NGOs. Although this gave the agency unprecedented sway in the refugee arena, it was ill equipped to respond to the refugee influx effectively. Its small capacity, comprising 40 staff by July 1991, was only sufficient to address the needs of a relatively small urban caseload that never exceeded 10,000 refugees. This staff, however, had risen to 162 by July 1992, while the creation of a fully fledged emergency response capacity at UNHCR Headquarters in Geneva quickened the process of delivery of human resources and materials.[42]

With absolutely no experience in working in the hostile environment of northern Kenya, UNHCR was unable to deliver assistance on time or to stem the security threat to refugees and aid workers posed by the incursions of Somali militias and bandits. This contributed to 'one of the worst results in years anywhere in Africa', with malnutrition reaching a peak of 54% among refugee children, death rates at 100 a day per 100,000 refugees, and epidemics like malaria, pneumonia and tuberculosis, as well as water shortages and gun fights, taking a heavy toll on refugees' lives.[43]

In May 1992, UNHCR introduced a range of programmes to curb malnutrition and to promote a healthy environment at an estimated cost of US$37.5 million. International attention given to the Somali crisis in the early 1990s favoured UNHCR's efforts to mobilize resources to provide assistance to refugees. In April 1992, the UN Security Council passed Resolution 794, declaring the 'magnitude of human suffering in Somalia' as a threat to international peace and security. The Security Council, working within chapter VII of the UN Charter, also sanctioned the use of 'all necessary means to establish as soon as possible a secure environment for humanitarian relief operations in Somalia'.[44] This enabled UNHCR to initiate its Cross-Border Operations and almost 300 Quick Impact Projects (QIPs) with a budget of US$35 million in southern Somalia. As argued earlier, the success of these initiatives was limited by continued insecurity in Somalia itself. Realizing that its management of refugees was a long haul, UNHCR shifted its focus from stabilizing southern Somalia to stabilizing the situation in the Kenyan camps.

The UN refugee agency introduced a number of initiatives aimed at reducing the high level of physical insecurity for refugees. One of these was the 'security package', including equipment, facilities and incentives to beef up the capacity of the Kenyan police to ensure law and order in and around refugee camps. The agency also introduced mobile courts that enabled the sitting of the district court in Dadaab to enhance refugee access to the judicial process for victims of rape and other crimes.

A firewood project was launched in 1998 with a special contribution of US$1.5 million from the United States to UNHCR to curb sexual violence against refugee girls and women while foraging for firewood. The project provided 30% of refugee firewood needs, making a significant boost to refugee security. An environmental angle was added to the firewood project to, first, rehabilitate the environment and, second, reduce the resource-based conflict pitting refugees against the local population.[45] Cumulatively, these programmes brought down the number of incidences of crime and violence from 300 in 1998 to 36 in 2003, with Dadaab touted as 'the safest region in the entire north-eastern province of Kenya'.[46]

Despite this initial success, the capacity of UNHCR to provide assistance and protect refugees was undermined by the plummeting levels of assistance to refugee camps across Africa. The protracted stay of refugees in host countries appeared to trigger donor fatigue. The budget for food in Dadaab shrank repeatedly by 20% in 1999, 2000 and 2001, creating critical interruptions in the food supply line and a dramatic rise in debilitating malnutrition.[47] With dipping levels of assistance, opportunities for resettlement also diminished.

Resettlement through security lenses

The resettlement programme for Kenya has been described as 'the largest in Africa'.[48] Between 1991 and 1998, almost 35,000 refugees were resettled to third countries from Kenya, with more refugees resettled from Kenya in 2000 than from any other African country. However, the slow pace of resettlement to third countries has undermined the principle of burden-sharing, forcing the Kenyan government to express concern over the burdens involved in the protracted stay of refugees.

In a strange twist, corruption scandals that pervaded Kenya's social fabric also dramatically affected the scale of resettlement. The UN Office of Internal Oversight Services disclosed in January 2002 that 'a criminal enterprise' had infiltrated the refugee status determination and resettlement process in Nairobi in the late 1990s, extorting bribes from refugees seeking resettlement in third countries. As a result, three UNHCR staff and two members of an affiliated NGO were arrested with four others. Sadly, the investigations leading to these arrests resulted in the temporary suspension of the resettlement process, momentarily closing one major door for a durable solution.[49] Coming on top of the suspension of the US Resettlement Program in the aftermath of 11 September 2001, the freeze in UNHCR-referred cases seriously undermined the burden-sharing process.

Even though a number of Somali refugees were resettled, the perceptions of security spawned by 11 September 2001 reduced the chances for resettlement of Somali refugees. The US drew a sharp divide between 'war-like' or 'aggressive' nomadic Somalis and the peace-loving, non-violent, farming or trading ethnic Somalis, using this ideological classification to allocate resettlement opportunities. Based on this criterion, from the spring of 1996, about 3,000 Benadir, a Somali ethnic group from the Benadir region of Somalia in the southern coastal region, including

Mogadishu, with a long history as urbanized merchants and artisans, were resettled in about 20 sites throughout the US.

The Benadir were resettled because they are 'devout Sunni Muslims who are well known for their peace-loving, non-violent ways, targets of jealousy and animosity from nomadic Somali clans whose homes and businesses were destroyed, women raped in front of male relatives, and countless slaughtered'.[50] Similarly, over 12,000 ethnic Somali Bantu refugees were settled in the United States between 2002 and 2004. US Christian organizations such as the Church World Service stressed their 'distinctness, physical differences from Somalis, disadvantaged status as second-class citizens, which cut them away from access to education, land ownership and political representation'.[51]

Out of a total of 6,000 refugees resettled from Kenya in 2004, only 800 were Somali refugees from Dadaab, signifying the bleak future of their resettlement. When interviewed in Kenya, even civil society organizations involved in resettling refugees conceded that Somali refugees have fallen off their resettlement register. Resettlement refracted through the prism of security sidelined 'pastoral' Somali refugees trapped in a protracted stay in Kenya. With reduced chances for repatriation, settlement and local integration, attention shifted to the role of refugees and how to deepen it to strengthen protection within the host countries.

Back to the boon/burden balance-sheet

Despite the bleak picture of protracted refugee situations as a financial and security burden to the host country, prolonged stays of refugees have made a remarkable difference in the social and economic lives of host populations, as well as from the perspective of security. UNHCR and NGOs have established infrastructure and social amenities that have boosted economic development in and around the camps. Admittedly, the presence of refugees has heightened conflict over scarce resources, but it has also brought some benefits to refugees as well as local populations. The UNHCR and partner agencies in the area have created employment opportunities, provided healthcare and education, boosted security, and dug and maintained boreholes. These developments have outweighed the burdens associated with refugees.

The presence of refugees in Kenya ensured the flow of billions of US dollars as foreign currency to fund humanitarian activities at a time when the country was reeling under economic sanctions from its external donors in the 1990s. Refugees boosted the populations of towns like Dadaab, accelerating the pace of urbanization of marginal areas. Dadaab's population grew from 3,000 in 1993 to over 15,000 in 2004, opening employment opportunities and stimulating the dramatic expansion of infrastructure, such as roads and telecommunications, and vital services, such as access to newspapers and radio stations. Since 1995, UNHCR ploughed in US$446,000 to improve roads and airstrips in Dadaab, constructed 30 water boreholes for the local population, and installed water and electricity in the local high school. Donor countries like the US and Japan also invested in improving local schools. In 2002, Japan built a library and laboratory in the Dadaab Secondary School as part of the Local Assistance Project.

In February 2004, Kenya's leading mobile telephone company, Safaricom, extended its services to a previously remote frontier. Availability of water and resources has promoted environmental regeneration in Dadaab, with nearly 1.5 million seedlings planted between June 1994 and December 2003, with a survival rate of 60%. Activities of humanitarian agencies have boosted the economy, providing a tax base for local authorities and the central government. For instance, UNHCR's firewood project enabled the Kenyan government to levy tax revenue of 2.5–3 million Kenyan shillings per year while creating employment for some of the local youths who would have otherwise been forced into banditry to make a living.

However, as it has expanded on the margins of Kenyan society, Dadaab has attracted illegal trade and trafficking in small arms, contributing to insecurity in north-eastern Kenya. Today, for an AK-47 rifle that would have cost 60 heads of cattle in the 1960s, one can pay as little as a chicken, making gun-running a real threat in the region.

Regional responses

Although the TFG entered the Somali capital of Mogadishu in December 2006, for the first time since its formation in October 2004, it was not clear at the end of 2006 that the problem of refugees overstaying in Kenya was ending soon. In fact, the number of refugees increased and the crisis deepened, pointing to the need for a comprehensive approach to end insecurity in Somalia and improve human rights conditions in Kenya's refugee camps. Resolving the problem of protracted refugee situations in the Horn, especially in regard to Somali refugees, demands that responses endeavour to strike a balance between security and human rights needs at the local, national, regional and international level.

EAC and IGAD

The East African Community (EAC), comprising Kenya, Uganda and Tanzania, may become increasingly involved in Somalia's complex crisis, with Uganda having promised 1,000 peacekeeping troops and Tanzania joining the International Contact Group (ICG) on Somalia. The EAC's treaty, which highlights peace and security and good neighbourliness as touchstones of the Community, might provide the guiding norms in the regional efforts to resolve the Somali security and refugee emergency.[52] The EAC has been largely powerless to protect refugees owing to scarcity of resources and operational capacity and its reluctance to boldly confront refugee situations in the region. In 2003, the EAC's Sectoral Committee on Cooperation in Defence and Inter-State Security proposed a common refugee registration mechanism. If fully put into action, such a mechanism has the potential for improving the protection of Somali refugees in Kenya, opening opportunities for regional burden-sharing within the EAC's five member countries.

Another regional organization that has played a pivotal role in resolving the Somali dispute is the Intergovernmental Authority on Development (IGAD), consisting of Djibouti, Eritrea, Ethiopia, Kenya, Somalia, Sudan and Uganda. IGAD has pursued dialogue as the best way to stabilize Somalia and, in October 2004, it brokered the peace agreement that led to the formation of the Somali Transitional Federal Government. Dialogue is IGAD's best option as a tool for securing a durable solution to the dilemma of the protracted Somali refugee crisis. Long-running tension between three of IGAD's members – Kenya, Ethiopia and Djibouti – and pan-Somali nationalists undermines the prospect of a concerted regional military approach to the Somali crisis. Furthermore, IGAD's members are also divided on this front. On the one hand there is Ethiopia, which has been backing a military approach to the extremists and throwing its support behind the TFG and President Abdullahi. At the other extreme is Eritrea, which has backed Islamic extremists. While both the Moi and Kibaki administrations in Kenya have been inclined to support the Ethiopian camp, they have pursued a policy of cautious optimism in Somalia, defined in Nairobi's official circles as 'good neighbourliness'.

IGAD has moved to bring the TFG and Somalia's Islamic Courts to a dialogue to avoid an Iraq-like scenario evolving in Somalia, with dire consequences for refugees. On 2 December 2006, the IGAD Secretariat signed a communiqué with the Islamic Courts addressing many of the security concerns in Somalia. The Courts pledged to respect the territorial integrity of neighbours and refrain from interference in their internal affairs; asserted they would deny sanctuary to 'any forces which are intent on undermining the security of IGAD member states'; and condemned all acts of terrorism.[53] On its part, the IGAD Secretariat expressed appreciation for the efforts the Courts were making to restore peace and stability in Somalia, and called for the withdrawal from Somalia of all foreign [Ethiopian] troops. The communiqué largely reflected the efforts of IGAD's current chair – Kenya – to tread a neutral path on the Somali conflict, while expressing the basic unease of Ethiopia and the

West with the Courts. However, IGAD lacks a strong regional policy and a legal and institutional framework to harmonize strategies and approaches to refugee protection. Resource constraints have also hindered its capacity to sufficiently address insecurity relating to refugees.[54]

The AU stabilization force

The African Union (AU) has viable instruments to address the Somali refugee crisis. One of these is the Convention Guiding Specific Aspects of Refugee Problems in Africa, adopted in 1969 by its predecessor, the Organisation of African Unity (OAU). The Convention, however, must be revised to take into account the security challenges posed by the post-9/11 refugee protection environment.

The AU also has the Convention on the Combating and Prevention of Terrorism (adopted in 1999). The exigencies of the 'war on terrorism' are increasingly blurring the line between refugees and 'terrorists'. This has enabled states to encroach upon the rights of refugees and has eroded Africa's reputation as one of the most hospitable regions towards refugees in conformity with African traditions.[55] The AU must insist that refugees are not terrorists, clarifying the definition of terrorism to ensure that the agenda to protect refugees is not eclipsed by the discourse on counter-terrorism.[56]

On 6 December 2006, the UN unanimously adopted Resolution 1725 (2006), approving an AU protection and training mission in Somalia acting under chapter VII of the Charter.[57] This stabilization force has a strong mandate allowing it to monitor progress by the Transitional Federal Institutions and the Union of Islamic Courts in implementing agreements reached in their dialogue; ensure the free movement and safe passage of all involved with the dialogue process; and maintain and monitor security in Somalia. The AU force has improved prospects for creating a congenial environment to allow refugees to voluntarily repatriate.

Putting soldiers on the ground is, however, proving a tall order. Of the expected 8,000 troops, by February 2007 only about 4,000 forces had been pledged by Uganda, Burundi, Malawi and Nigeria. Although the AU Peace and Security Commissioner, Said Djinnit, is reportedly upbeat about the deployment of a stabilization force to replace the Ethiopian troops, countries such as Nigeria, South Africa, Mozambique, Angola, Tanzania, Rwanda and Tunisia, which have been approached to contribute troops, have been slow to respond. The International Contact Group meeting in Dar es Salaam on 9 February 2007 called on the world to provide backing and funds to the AU mission in Somalia.

The AU force is also faced with a difficult task. In February 2007, hundreds of Somali residents of Mogadishu took to the streets to protest against the plan, burning the flags of the United States, Ethiopia, Kenya, Uganda, Nigeria and Malawi.

The Security Council and the AU summit in January 2007 emphasized the need for continued credible dialogue between the Transitional Federal Institutions (TFI) and the Union of Islamic Courts. The AU should urge the TFI to engage in dialogue with broad sections of society to calm growing tensions and prevent new destabilization, especially by Islamists. Delayed dialogue is starting to undermine stability in Somalia, threatening the emergence of an Iraq-type situation in which local extremists allied to international terrorist elements terrorize the country and its neighbours.

The unresolved Somaliland question

The future of the self-declared Republic of Somaliland is another sticking point, with far-reaching implications for the Somali refugee crisis. Between 1983 and 1991, fighting between Somaliland's separatists and the government claimed some 50,000 lives and spawned nearly 1 million refugees and IDPs. Emboldened by the collapse of the Somali government in Mogadishu, Somaliland separatists declared 'independence' from Somalia on 18 May 1991. In December 2005, the President of Somaliland, Dahir Rayale Kahin, applied for Somaliland's membership to the African Union. The application has been opposed by pan-Somali nationalists, including those within Somaliland.

Somaliland's relative stability has evidently allowed a degree of repatriation of refugees. In June 2004, UNHCR and the government of Djibouti repatriated some 521 refugees to the territory. Another 1,454 refugees in Ethiopia's Hartisheik camp also returned to Somaliland.[58] In early 2005, UNHCR assisted some 240,000 refugees to spontaneously return to their homes in Somaliland. This brought the total number of returnees in Somaliland to some 700,000 by mid-2005.[59] But the unresolved issue of Somaliland's sovereignty has deadly potential for conflict and displacement. Despite calls in some international circles for the African Union to take a position that endorses Somaliland's separatism, this is a difficult issue that demands serious thought.[60]

International responses

Since the failure of the UN intervention in Somalia in 1992–1995, international engagement with the Somalia refugee crisis has been half-hearted, ad hoc and often uncoordinated. Despite this, Somalia's protracted problem of displacement has been central in framing international responses. Most of the international decisions, agendas and programmes adopted since 2000 to strengthen the protection of refugees and enhance durable solutions have also focused on Somali refugees. The best known of these approaches at the international level include: the Agenda for Protection (2001), the Convention Plus (2003), the Comprehensive Plan of Action for Somali Refugees (2004) and the UN Joint Needs Assessment and the Somali Reconstruction and Development Programme (2005). The resurgence of Islamism and the consequent escalation of armed conflict led to the formation of the Somali Contact Group in mid-2006, designed to serve as a diplomatic vehicle for promoting stability in Somalia. This section focuses on the implications of these interventions for Somalia's protracted refugee situation.

The Agenda for Protection (2001)

One of the frameworks guiding the UNHCR programmes aimed at protecting Somali refugees and ending the protracted crisis is the Agenda for Protection. In late 2000, UNHCR launched the Global Consultations on International Protection to engage states and other partners in a broad-ranging dialogue on refugee protection. The discussions culminated in the adoption of the Agenda for Protection by the 53rd Session of UNHCR's Executive Committee in December 2001 and it was welcomed by the United Nations General Assembly during 2002. The aim of the agenda is to explore how best to revitalize the existing international protection regime while ensuring its flexibility to address new problems.

The Agenda for Protection has six main goals:
- Strengthened implementation of the 1951 Convention and 1967 Protocol;
- Protecting refugees within broader migration movements;
- Sharing of burdens and responsibilities more equitably and building of capacities to receive and protect refugees;
- Addressing security-related concerns more effectively;
- Redoubling the search for durable solutions; and
- Meeting the protection needs of refugee women and children.

NGOs' critique of the Agenda for Protection centred on what they saw as its silence on certain vulnerable but sizeable groups such as urban refugees, and critical issues like the detention of asylum seekers.[61] Furthermore, it emphasized security as likely to impact negatively on Somalis who have long been viewed as a security risk by most host countries. The high point of the Agenda is the field-based mainstreaming of the Agenda for Protection by UNHCR. This has seen the setting up of an independent forum for protection and parallel protection fora in UNHCR field and branch offices in collaboration with other stakeholders, including NGOs and refugees. UNHCR, host governments and NGO partners should galvanize the fora to improve the protection of Somali refugees.

The Convention Plus (2003)

Another initiative with a profound impact on the Somali refugee situation is the Convention Plus. UNHCR launched the Convention Plus in 2003 as a strategy to improve refugee protection worldwide and to facilitate the resolution of refugee problems through multilateral special agreements.

The Convention Plus focuses on three core concerns:
- The strategic use of resettlement;
- Addressing irregular secondary movements of refugees and asylum seekers; and
- Targeting development assistance to achieve durable solutions.[62]

NGO partners have directed their criticism of the Convention Plus at the issue of development aid.[63] NGOs stressed four issues: harmonization of long-term poverty reduction strategies, humanitarian aid and the right to asylum with development assistance to refugees and host countries; centring of issues of justice and the rule of law in the search for sustainable peace and an environment of adequate protection for citizens, refugees and asylum seekers; adoption of regional perspectives in planning and implementing assistance programmes for refugees and host countries; and refugee participation in the development of plans and their implementation through their own organizations and groups. These issues have framed academic and policy debates on the protracted Somali refugee situation[64] and the Convention Plus has galvanized some response to Somali refugees. However, this response is increasingly refracted through the security lens in the Horn of Africa, thus rolling back the gains made in refugee protection.

The Somali CPA (2004)

A third international framework is the Comprehensive Plan of Action (CPA) for Somali Refugees. The CPA provides perhaps one of the most ambitious responses to the problem. The CPA was initiated in mid-2004 as a partnership between UNHCR, the European Commission, Denmark, the United Kingdom and the Netherlands in collaboration with the Somali authorities and regional host states. Its aim is to provide space for effective protection of Somali refugees and internally displaced persons through an integrated approach to the three durable solutions to refugee displacement: repatriation, local integration and resettlement.[65]

The CPA focuses on four asylum countries – Ethiopia, Djibouti, Kenya and Yemen – as well as Somalia itself. The relative stability in Somaliland

(1991) and Puntland (1998) and the launching of the Transitional Federal Government (October 2004) set the stage for the CPA. The CPA process started off with a high-profile stakeholder consultation in Kenya on 17–18 May 2005, and in other countries in the Horn whose outcomes fed into a larger conference.[66] The final outcome of the CPA will be a framework to comprehensively address the range of challenges of Somali refugees from protection to solutions, continued displacement both into Kenya and globally, and factors inhibiting return.

Continuing political instability in central and southern Somalia has impeded voluntary repatriation and implementation of durable solutions for the internally displaced. Furthermore, efforts to improve Somali refugees' access to local integration has had extremely limited success. Despite growing interest in resettlement schemes in several European Union states with significant Somali populations, the scope of resettlement has also been narrow. In the light of shrinking opportunities for durable solutions, the CPA seeks to improve the prospects for refugee self-reliance pending return. In this regard, it seeks to enhance the level of protection and assistance available to refugees in Djibouti, Ethiopia, Kenya and Yemen. The implementation of the CPA's specific programme areas and projects by UNHCR and its partners got under way in 2006, targeting issues with a high chance of being effectively implemented, irrespective of the direction of the war and ongoing peace process in Somalia. In the context of growing xenophobic perceptions of refugees of Somali origin regionally and globally, the CPA is increasingly being viewed as a tool to contain Somalis in the area of their origin – the Horn of Africa. Regional governments are raising the alarm that this policy will undermine burden-sharing, leaving responsibilities for refugees in the hands of the poorer states like Kenya.

UNHCR needs to focus on the threat to asylum by states, which are utilizing terrorism in an instrumental fashion to crack down on 'warehoused' refugees in camps and on urban refugees, often on the basis of religion. The best case is security measures taken to contain Somali refugees in camps and in urban enclaves in the Horn of Africa. One dilemma is how to separate extremist elements from bona fide refugees in camps. This will remain a highly dangerous task that host states and the international community must jointly undertake. The exclusion clauses in international law used against militias in the Great Lakes and West Africa may provide an instrument to deal with extremist elements in camps.

Somali JNA-RDP

Achieving sustained reconstruction and development and deepening the peace process in Somalia are recognized as vital to securing one of the durable solutions to the country's protracted refugee situation: repatriation. In 2005, the TFG and members of the international community asked the United Nations Development Group and the World Bank to co-lead a post-conflict needs assessment for Somalia. This led to the launching of the UN Joint Needs Assessment/Somali Reconstruction and Development Programme (JNA/RDP).

The main objective of the JNA process was to assess needs and develop a prioritized set of reconstruction and development initiatives to support Somali-led efforts to deepen peace and reduce poverty.[67] The JNA has been an inclusive and participatory process involving teams of Somali and international technical experts working together to assess needs and develop prioritized strategies. The implementation of proposals resulting from needs assessment is expected to lay solid foundations for the establishment of an effective, participatory and transparent system of governance within Somalia. It is also expected to achieve sustainable recovery, reconstruction and development to reverse regression from the Millennium Development Goals and to advance socioeconomic development for all Somalis.

The JNA process in central and southern Somalia focused on several clusters: governance, safety and the rule of law; macroeconomic policy framework and data development; infrastructure; social services and protection of vulnerable groups; productive sectors and the environment; and livelihoods and solutions for the displaced. Three cross-cutting issues were also examined: peacebuilding, reconciliation and conflict prevention; capacity building and institutional development; and gender equity and human rights. On 25 May 2006, the UN and the World Bank provided a brief to the Somali Transitional Federal Parliament in Baidoa on the Somali Joint Needs Assessment. The TFG undertook to endorse the final document of the Reconstruction and Development Programme, making the Somali Joint Needs Assessment a national document.

The RDP, itself founded on the outcomes of the JNA, is another framework for ensuring durable solutions to Somalia's protracted refugee crisis. It is designed as a pro-poor instrument for mobilizing, distributing and coordinating international recovery assistance. The RDP is organized around three pillars: deepening peace and strengthening institutions of governance; investing in people; and establishing an environment for rapid poverty-reducing development. Careful implementation of the three pillars is central to meeting the priority needs and core reconstruction and development objectives.

Part of the problem facing the JNA is how to bring home over 70,000 children conscripted into Somalia's fighting factions and separated from their families.[68] Also impeding quick take-off of the JNA is the absence of a decision on the status of Somaliland state, which has declared its independence from the larger Somalia. The international community has also been reluctant to provide funding and political support to lay the foundations of a comprehensive humanitarian and development programme in Somalia. The implementation of the JNA has been slowed down by the failure of the Transitional Federal Government to assert its authority over much of Somalia, and the continuing challenge posed by Islamists. Like the CPA, the JNA involves as its stakeholders Somali counterparts, the UN, NGOs, returnees and IDPs. Consequently, these two processes must be mutually reinforcing and closely coordinated. The successful implementation of the JNA/RDP process has the potential to resolve Somalia's protracted refugee crisis. Its success will, however, depend largely on diplomatic breakthroughs in finding a peaceful solution to the dispute between the TFG and the Islamists.

The International Contact Group

Another international instrument involved in stabilizing Somalia is the International Contact Group (ICG) on Somalia. The group held its inaugural meeting in New York on 15 June 2006. Those invited to the inaugural meeting were Norway, the US, the UK, Tanzania, Sweden, Italy, the European Union, and representatives from the African Union and the UN, including Jan Egland and Francois Fall, who is the Special Representative of the Secretary-General for Somalia. Norway was appointed the lead country to develop the group.

Its purpose is to coordinate a common policy on Somalia so that we can support the Transitional Federal Institutions. Its agenda reflects the larger context of the war on terrorism. Although it stresses social and economic development and humanitarian assistance to Somalia, the US Assistant Secretary of State, Jedayi Fraser, underlined its main objective as working with Somali parties 'to prevent the country from becoming ... a haven of terrorism and instability in the region and in Somalia itself'. The Dar es Salaam conference on 9 February 2007 stressed four objectives for the ICG: support for the TFG, support for aid to the people of Somalia, support for the efforts to prevent terrorism – terrorist acts using Somalia as a base and those carried out in Somalia – and support for the stabilization of the entire region.

The ICG has backed the Transitional Federal Government and its institutions but has warmed to the idea of a dialogue with the Islamic Courts Union to promote lasting peace, stability and development in Somalia. The meeting in Dar es Salaam urged the world to back and fund an African mission to stabilize Somalia after the fall of the Islamists from power. It also called for the urgent dispatch of an African peacekeeping force to stabilize the country after a December offensive by Ethiopian and government troops drove rival Islamists from territory they had captured since June. The group can now serve as a force of unification, emphasizing and supporting inclusive dialogue with the various segments of the Somali community.

Facing the future

The future of the Somali refugees will continue to look bleak unless drastic measures are taken at the regional and international level to ensure stability in Somalia to enable the return of refugees, and also to enhance the protection of Somali refugees across the region. Reform of the refugee protection system in Kenya is central to ensuring protection and the delivery of assistance.

Strengthening refugee protection capacity

Although Kenya adopted a long-delayed Refugee Bill in 2007, a national refugee policy is also needed to complement the legislation. The policy should clarify the government's strategy and commitments with regard to refugee protection. Members of the international community, UNHCR and local experts should step in to assist the government in building the capacity of the National Refugee Secretariat at all levels of administrative capacity.

The new Ministry of Immigration and Registration of Persons should establish a system of gathering and managing data relating to refugees in both the camps and urban areas, including deploying staff for registration in Kakuma and Dadaab camps. The government should ensure that all refugees and asylum seekers are issued with identification documents (cards) to reduce the risk of harassment of refugees and the potential for *refoulement*, and also to monitor criminal elements using refugee cover. A clear budget should be worked out to support the implementation of the new changes in refugee management systems and governance. The government should also expand space for partnerships between its officials, NGOs and refugee communities to strengthen protection machinery and facilitate training for refugees.

Specifically, building the capacity for the National Refugee Secretariat is critical to the protection of refugees and maintaining oversight over the delivery of assistance. The government should increase the personnel of the National Refugee Secretariat, and mobilize resources to ensure its efficient functioning. Because UNHCR and NGOs are currently involved in protection delivery, the government should also introduce measures to strengthen their capacity. One way of doing this is to look into their human resources constraints with a view to building a sustainable local humanitarian capacity linked to UNHCR.

The Kenya stakeholders' consultation in May 2005 urged the government to recruit camp liaison officers and support staff in the main refugee camps: Dadaab and Kakuma. One of the direct positive spin-offs of building such capacity is that successful asylum seekers would be issued with the relevant documentation granting them convention refugee status at the camp level without having to travel to Nairobi.

Central to strengthening refugee protection is the provision of training in refugee law, protection, and refugee governance and management to the Refugee Secretariat staff and staff of the relevant ministries and departments and NGOs. The Ministry of Immigration and Registration of Persons needs to consult with UNHCR, local human rights and refugee NGOs and refugee centres to put in place such a training programme. The government should work with local and international experts at the Centre for Refugee Studies, Moi University, and the Institute of Diplomatic Studies, Nairobi University, to develop the contents of the training, including in-depth induction on key protection issues, status determination, registration processes and data management. UNHCR also requires adequate staffing resources in order to properly manage the influx of asylum seekers and to reduce refugee status determination waiting times to a reasonable level.

Security and judicial capacity in camps

The original rationale for the encampment policy was to facilitate refugee access to protection and assistance. In practice, the policy has inhibited refugees' potential to contribute to local development, severely restricted their freedom of movement, and limited their access to markets, employment and opportunities for self-reliance. In a word, warehousing of refugees behind barbed wire has taken away the right to asylum. An urgent review of the policy is, therefore, required to strengthen the assistance and protection goals of asylum. This necessitates the issuing of work permits and licences to refugees to enable them to access employment and promote their self-reliance. Similarly, the host communities should have access to camp-based services.

Ensuring refugees economic access and rights is also an integral part of the revitalization of the protection system within the camps and urban areas. As part of empowering refugees in public decision-making and participation, it is imperative that the government and Kenyan civil society factor in refugees in national poverty reduction and development strategies and bring refugee-hosting areas to the centre of National Development Plans. In an effort to provide more coherent development,

District Development Plans should promote sustainable development for refugees and shift from the prevailing preoccupation with short-term relief. Refugee participation in economic exchanges with local communities requires greater freedom of movement than is currently allowed, opening access to business and trading licences.

Although robust policing and community involvement in security has led to significant drops in the incidence of serious crime in the Dadaab and Kakuma camps, security problems continue, including banditry, rape and murder. UNHCR and donors should work in partnership with the government to increase the number of trained security personnel assigned to the camps. The security personnel also need to be provided with adequate equipment, including telecommunications, transportation and other logistical means to effectively undertake their roles. Many of the security incidents in the camps are intra-communal, which demands conflict prevention strategies rather than criminal procedures. The government, agencies and NGOs should consider appropriate training on conflict management for refugees and the host communities to ensure peace and enhance refugees' personal safety and security.

Mobile courts have improved access to justice in the camps but cases have been slow, and witness protection not guaranteed. Camps have also been hit by a shortage of police and magistrates, leading to administrative delays and prolonged pre-trial detention. This requires a substantial increase in the number of mobile courts and police deployed in the camps, and measures to provide them with additional training in investigative techniques and documentation of crime scenes. The government, in partnership with UNHCR and donors, should provide financial and technical assistance to strengthen judicial capacity to effectively prosecute perpetrators of crime in the refugee-hosting areas. There is a need to cultivate closer coordination between law enforcement personnel and refugee communities to develop common strategies to address crime.

Protecting urban refugees

The government, UNHCR and NGO partners must prioritize a baseline survey of urban refugees in all cities and towns to obtain accurate information about the urban caseload in Kenya. It is also necessary to take a tally of agencies providing assistance and protection to urban refugees with a view to sharing resources, burdens and responsibilities. The idea of identifying police personnel to serve as refugee focal points at police stations has been mooted.[69] Such a policy, it is argued, will enable the relevant stakeholders to easily contact specific police officers to facilitate protection. The focal point will also know who to contact when faced with refugee issues.

Dealing with xenophobia and violence against refugees, especially women and children, holds the key to effective protection of urban refugees as well as those in camps. There is a need for extensive use of existing communication channels present in government, UNHCR, NGOs, civil society, advocacy groups and the media to counter xenophobic and negative perceptions and attitudes towards Somali refugees. Refugee women and children constitute a category that remains vulnerable to violence, abuse and exploitation. The ineffectiveness of existing responses to violence and abuses in camps requires vigorous implementation of strategies that have proven to work, including awareness creation targeting traditional and religious authorities; provision of firewood and alternative forms of energy to keep women and girls away from bandits and criminals; investment in law enforcement capacity such as deploying more female police; monitoring and reviewing adherence to the Code of Conduct by law enforcement officials; and investigating and prosecuting those believed to have abused women or children. In addition to strengthening judicial capacity, it is necessary to review the traditional forms of justice practised in the camps with a view to improving them.

Conclusion

The collapse of the Somali state entered its sixteenth year in 2007, amid spiralling violence and intensification of the country's protracted refugee situation. The causes of Somalia's refugee crisis are deep, complex and diverse: a legacy of authoritarian manipulation of clannism, state collapse, failed peace processes, a history of irredentism, the rise of political Islam and Islamic jihadism, and the security consequences of the US 'war on terrorism'. Cumulatively, these factors forced no fewer than half a million registered refugees to flee their homes for safety in the neighbouring countries in the Horn of Africa – particularly Kenya, Ethiopia and Djibouti. The legacy of conflict and irredentism produced stereotypes and xenophobic perceptions of Somali refugees, creating a distinctly negative security lens through which they have been viewed. These securitized negative perceptions of Somali refugees were also reinforced by the presence of a marginalized, local ethnic Somali category, especially in Kenya. The refugee protection regime that emerged from the confluence of these factors impeded the protection of the rights of Somali refugees as well as the system of delivery of assistance. Similarly, the security prism also produced a segregated system of resettlement that criminalized and discriminated against 'pastoral' Somalis. Collectively, political instability within Somalia, a security-driven and repressive refugee protection regime and discriminative resettlement patterns hampered the search for corresponding durable solutions: repatriation, local integration, and resettlement. This has prolonged and deepened the Somali refugee emergency. Comprehensively addressing these causal factors is critical to effectively tackling the Somali refugee problem. In the last seven years, positive steps have been taken at the regional and international level to provide frameworks for addressing this problem comprehensively.

Repatriation: Somalia has bounced back onto the radar-screen of the regional and international community, raising prospects for the country's return to peace and the voluntary return of exiles from Kenya, Ethiopia and Djibouti. In December 2006, the UN Security Council adopted Resolution 1725 (2006), authorizing a strong IGAD/AU protection and training mission in Somalia. The force is designed to monitor progress by the Transitional Federal Institutions and the Union of Islamic Courts in implementing agreements reached in their dialogue; ensure the free movement and safe passage of all involved with the dialogue process; and maintain and monitor security within Somalia. A more optimistic view is that the effective implementation of the mission might begin creating opportunities for a durable solution to Somalia's protected refugee crisis. Sustainable peace enabling the repatriation of Somali refugees must also be based on comprehensive dialogue. In this regard, the UN also stressed the centrality of continued credible dialogue between the main protagonists in Somalia – the Transitional Federal Government and the deposed Union of Islamic Courts.

Strengthening protection: Local integration of Somali refugees has proven difficult in the Horn of Africa. The alternative is to bolster the local protection regime to ensure refugees access to their basic rights, including asylum rights, and economic and employment opportunities as well as protection from abuse, violence and discrimination. The passing of the long-delayed Refugee Bill in Kenya and the creation of a new ministry dealing with refuges have expanded the space for refugee protection. This space is, however, restricted by a number of factors, such as weak security and judicial capacity in camps coupled with a frail refugee governance and management capacity at all levels of government, including the Refugee Secretariat, the relevant ministries and departments and NGOs. This calls for comprehensive capacity building with the government, NGOs and international agencies and at the camps to strengthen the protection of refugees. Constructive partnerships are required to secure the relevant capacity, training and resources to achieve this objective. However, partnership models which are likely to undermine the principle of burden-sharing between the wealthier and poorer countries must be discarded.

Resettlement: The security environment after September 11 coupled with corruption scandals dramatically reduced the scale and opportunities for resettlement, especially in the West. A number of Somali refugees were resettled in the West, but selection was based on skewed and ideologically biased criteria that denied opportunity to the vast bulk of Somali refugees. Some 12,000 ethnic Somali Bantus were settled in the United States in 2002–2004 as a vulnerable group because of their 'physical differences from Somalis, disadvantaged status as second-class citizens.'[70] Similarly, 3,000 Benadir resettled to the US under similar circumstances. Despite constituting the largest caseload of refugees in Kenya, of the 6,000 refugees resettled in Kenya in 2004 only 800 were Somalis. While resettlement opportunities have declined globally, available opportunities should be given to all without discrimination based on ethnic, racial, religious or occupational criteria. 'Vulnerability' must continue to serve as a criterion for resettlement. That said, discriminative patterns must be avoided to level the playing field for all needy refugees, including Somalis.

Notes

1. Somali refugees range from those recognized by UNHCR, asylum seekers, rejected applicants, those whose files have been closed, and even some who, for a variety of reasons, have not gone through the process of applying to UNHCR for asylum.
2. Liisa Malkki, *Purity and Exile: Violence, Memory, and National Cosmology among Hutu Refugees in Tanzania*, Chicago: University of Chicago Press, 1995.
3. Peter Kagwanja, 'Unwanted in the "White Highlands": The Politics of Civil Society and the Making of a Refugee in Kenya 1902–2002', PhD Dissertation, University of Illinois, Urbana-Champaign, 22 December, 2003.
4. United Nations High Commissioner for Refugees (UNHCR), *Somalia: A Situation Analysis and Trend Assessment*, Nairobi: UNHCR, August 2003, p. 3; Joakim Gundel, 'The Migration-Development Nexus: Somalia Case Study', *International Migrations* 40, no. 5, 2002, p. 264.
5. Marc-Antoine de Motclos and Peter Kagwanja, 'Refugee Camps or Towns? The Socio-Economic Dynamics of the Dadaab and Kakuma Camps in Northern Kenya', *Journal of Refugee Studies* 13, no. 2, pp. 205–222.
6. Mogadishu was split into two parts – the North (Karaan) controlled by Ali Mahdi and the South (excluding Medina) controlled by General Aydiid, fostering an atmosphere of insecurity.
7. Somalia was amongst the top 10 countries of origin for asylum applications to the European Union during the 1992–2001 decade, and amongst the top 20 countries of origin for asylum applications in 28 mostly industrialized countries from January to June 2002. See Forced Migration Organization, *Country Report: Somalia*, 2003, p. 3.
8. Roland Marchal, 'Islamic Political Dynamics in the Somali Civil War: Before and After September 11', in Alex de Waal, ed., *Islamism and its Enemies in the Horn of Africa*, London: Hurst and Company, 2004, p. 137.
9. UN Security Council, 12 February 2004.

10. By the end of 1993, 320 Quick Impact Projects were initiated as part of the multi-million dollar Cross-Border Operations (CBOs) set up by UNHCR Geneva funded from the Special Emergency Fund for the Horn of Africa, including US$13 million UNHCR mobilized for CBOs. By June 1993, only 30,000 of the refugees had returned home from Kenya. US Committee for Refugees, *World Refugee Survey*, Washington, DC: Immigration and Refugee Service of America, 1998, p. 73.
11. Roland Marchal, 'Islamic Political Dynamics in the Somali Civil War', p. 130.
12. Ibid., p. 123.
13. The ICU organized a clean-up campaign for the streets of Mogadishu in July 2006, the first time rubbish had been collected since the collapse of the government in 1991.
14. Birgit Michaelis, 'How Much More Suffering for Somali People?', *The Reporter*, 21 October 2006.
15. Authors' interview with the Refugee Consortium of Kenya (RCK), September 2006; speech by UNHCR spokesperson Ron Redmond at the press briefing, on 15 September 2006, at the Palais des Nations in Geneva.
16. 'UNHCR Sends Supplies, Emergency Teams to Somalia and Ethiopia', Reuters, 12 January 2007.
17. Boniface Mwangi and Victor Obure, 'Kenya: Misery of Refugees Stranded at Liboi', *The East African Standard* (Nairobi), 7 January 2007.
18. UNHCR, Branch Office for Kenya, *Information Bulletin*, Nairobi, January 1993.
19. These were the authors' earlier findings. See Peter Kagwanja, 'Strengthening Local Relief Capacity in Kenya: Challenges and Prospects', in Monica Juma and Astri Suhrke, eds., *Eroding Local Capacity: International Humanitarian Action in Africa*, Nordic African Institute, 2002, p. 105.
20. Jeff Crisp, 'Forms and Sources of Violence in Kenya's Refugee Camps', *Refugee Survey Quarterly*, vol. 19, 2000, pp. 54–70; also 'A State of Insecurity: The Political Economy of Violence in Kenya's Refugee Camps', *African Affairs* 99, no. 397, 2000, pp. 601–632; Peter Kagwanja, 'Ethnicity, Gender and Violence in Kenya', *Forced Migration Review*, no. 9, December 2000.
21. Jeff Crisp, 'A State of Insecurity: The Political Economy of Violence in Kenya's Refugee Camps'.
22. LCHR, *Refugees, Rebels and the Quest for Justice*, New York: The Lawyers Committee for Human Rights, 2002.
23. UNHCR, Sub-Office Dadaab, 'Assessment of Security Situation in Dadaab from 1998 to 2002', Dadaab: UNHCR, December 2002. In authors' files.
24. ICG, 'Somalia: Countering Terrorism in a Failed State', *Africa Report* no. 45, 23 May 2002, p. 11.
25. Author's interview with a UNHCR official, Nairobi, September 2006.
26. Cindy Horst and Nick van Hear, 'Counting the Cost: Refugees, Remittances and the "War Against Terrorism"', *Forced Migration Review* 14, July 2002.
27. Author's interview with an NGO official, Dadaab Camp, northern Kenya, August 2002.
28. Authors' interviews with Dadaab local authority officials, April 2002; and with RCK officials in Nairobi, August 2006.
29. Cited in Mohammed Salih, *Islamic NGOs in Africa*, Nairobi: UNHCR, p. 175.
30. Sydney Waldron and Naima Hasci, *Somali Refugees in the Horn of Africa: State of the Art Literature Review*, Studies on Emergencies and Disaster Relief, Refugees Studies Centre, University of Oxford, Report no. 3, 1995, p. 13.
31. Jeff Crisp, 'Forms and Sources of Violence in Kenya's Refugee Camps'.
32. This became clear from the authors' interviews, confirmed by Cindy Horst's own research in Dadaab.
33. Peter Kagwanja, 'Strengthening Local Relief Capacity in Kenya'. In authors' files.

34. Interview with Rose Kimotho, RCK Advocacy Officer, 1 September 2006.
35. Dave Rowan, 'Kenya: Crackdown on Refugees Following Hotel Bombing,' *World Socialist Website*, 14 December 2002.
36. Interview with Judy Wakahiu, Executive Director of the Refugee Consortium of Kenya, Nairobi, 1 September 2006.
37. Interview with Prisca Kamungi, a researcher on refugees, Nairobi, 3 September 2006.
38. Interview with Rose Kimotho, RCK Advocacy Officer, Nairobi, 31 August 2006.
39. See the transcript of the RCK interview with Mr. Peter Kusimba, Head of the National Refugee Secretariat, *Refugee Insights*, 10, 2006, pp. 12–14.
40. Ibid., p. 13.
41. Author's interview with RCK official, Nairobi, 29 August 2006.
42. UNHCR, Inspection and Evaluation Service, *A Review of UNHCR's Women Victims of Violence Project in Kenya*, Geneva: UNHCR, March 1996, p. 6.
43. US Committee for Refugees, *World Refugees Survey*, Washington, DC: USCR, 1993, p. 63.
44. United Nations Security Council Resolution 794 Granting the Secretary-General Discretion in the Further Employment of Personnel of the United Nations Operations in Somali, 1992, S.C. res 794, 47 UNSCOR at 63, UN Doc. S/RES/794.
45. The environmental project was spearheaded by the German agency, the Gesellschaft für Technische Zusammenarbeit (GTZ). By 2004, the project had provided 8,000–10,000 metric tons of firewood a year to the refugees, bringing US$600,000 as earnings to the local population. See UNHCR, *Evaluation and Policy Analysis Unit*, June 2001, p. 1.
46. Author's interview with an NGO aid worker, Nairobi, 4 August 2006.
47. UNHCR, 'Kenya', in *Mid-Year Progress Report*, Geneva: UNHCR, 2001, p. 72.
48. UNHCR, Branch Office for Kenya, *Annual Protection Report 2002*, Nairobi, 2003, p. 11.
49. UNOIOS, *Report of the Office of Internal Oversight Services on the Investigation into Allegations of Refugee Smuggling at the Nairobi Branch Office of the United Nations High Commission for Refugees*, 21 December 2001, UN Doc. A/56/733.
50. E. Mypist, 'Notes on Benadir Refugees', Mombasa: United Nations High Commissioner for Refugees, 1995; United States Catholic Conference, 'Benadir Refugees from Somalia', Washington, DC: Refugee Information Series, Migration and Refugee Services, 1996.
51. Interview with NGOs, Nairobi, 2 September 2006.
52. 'EAC to Hold Joint Counter/Anti-Terrorism Exercise', press release, the East African Community Secretariat, Arusha, 22 August 2005.
53. Communiqué issued at the end of consultations between the delegation of the Somali Council of Islamic Courts and the IGAD Secretariat, Djibouti, 1–2 December 2006.
54. See Peter Kagwanja, 'Counter-Terrorism in the Horn of Africa: New Security Frontiers, Old Strategies', *Africa Security Review* 15, no. 3, 2006, pp. 72–86.
55. A. Katz, 'Terrorism and Its Effects on Refugee and Extradition Law', in Jackie Cilliers and Kathryn Sturman, eds., *Africa and Terrorism: Joining the Global Campaign*, Institute of Security Studies, Monograph Series no. 74, Pretoria, 2002. See http://www.iss.co.za/PUBS/MONOGRAPHS/No74/chap4.html.
56. Refugee International, *No Power to Protect: The African Union Mission in Sudan*, 9 November 2005. See http://refugeesinternational.org.
57. United Nations, *Security Council Approves African Protection, Training Mission in Somalia, Unanimously Adopting Resolution 1725 (2006)*, Security Council, 5579th Meeting, 6 December 2006, SC/8887. See http://www.un.org/News/Press/docs/2006/sc8887.doc.htm.

58. OCHA, *Affected Population in the Horn of Africa*, United Nations Office for the Coordination of Humanitarian Affairs, Central and East Africa, May 2004.
59. UNHCR press briefing by UNHCR spokesperson Ron Redmond, Palais des Nations, Geneva, 27 May 2005. See http://www.unhcr.org/news/NEWS/429703574.html.
60. International Crisis Group, 'Somaliland: Time for African Union Leadership', *Africa Report* no. 110, 23 May 2006.
61. 'NGO Statement on the Agenda for Protection', Executive Committee of the High Commissioner's Programme Standing Committee 24th Meeting, 25 June 2002. See http://www.icva.ch/doc00000657.html.
62. UNHCR, High Commissioner's Forum, Progress Report: Convention Plus, 20 February 2004. See http://www.unhcr.org/protect/PROTECTION/403cd06a4.pdf.
63. 'Convention Plus Work on Targeting Development Assistance to Achieve Durable Solutions: Statement on Behalf of NGOs', United Nations High Commissioner for Refugees' Forum, 12 March 2004. See statement at http://www.icva.ch/doc00001127.html.
64. Swiss Forum for Migration and Population Studies, 'The Path of Somali Refugees in Exile: A Comparative Analysis of Secondary Movements and Policies', *SFM Studies* 46, 2006.
65. See UNHCR, *The State of the World's Refugees 2006*, chapter 2, Safeguarding Asylum: Box 2.5 – The Comprehensive Plan of Action for Somali Refugees, Geneva: UNHCR, 2006.
66. UNHCR, *Strengthening Refugee Protection, Assistance and Support to Host Communities in Kenya and Comprehensive Plan of Action for Somali Refugees*, Report on Stakeholder Consultation, Nairobi, Kenya, 17–18 May 2005.
67. JNA/RDA, *Somali Reconstruction and Development Framework: Deepening Peace and Reducing Poverty, Volume 2: South-Central Somalia*, October 30, 2006. See http://www.somalia-un.org/jna.html.
68. 'Somalia: Government Calls for Assistance to Rehabilitate Child Soldiers', IRIN, Nairobi, 2 February 2007.
69. Authors' interview with RCK officials, Nairobi, September 2006.
70. Authors' interview with a human rights lawyer, Nairobi, 2 September 2006.

12

Sudanese refugees in Uganda and Kenya

Tania Kaiser

Sudan's conflict is infamous on the continent for its longevity and brutality, and for the numbers of people it has displaced. The conflict in the South alone has created over 4 million internally displaced people (IDPs) and 100,000s more are refugees in neighbouring countries.[1] Forced migration has been explicitly part of the conflict dynamics, with military actors on both sides forcing or exploiting the displacement of civilians to further war ends. Successive, repeated and protracted displacement has been a common feature of the conflict in Southern Sudan since the late 1980s, as it has so obviously also been in the more recent and ongoing Darfur conflict.

The respective responses of host states Uganda and Kenya to the protracted exile of the Sudanese are instructive; there are similarities and differences in the two governments' approaches as well as in their motivating concerns and objectives. Both host states have favoured some form of encampment for long-term refugees, but have adopted contrasting styles of management. Different issues have arisen in each country, demonstrating the importance of considering environmental and socio-political conditions in any single host country. In Uganda in particular, ongoing internal conflict across the refugee-hosting area has presented serious security challenges to refugees and those who attempt to assist them.

Refugees in Kenya have been more directly affected than those in Uganda by a post-9/11 security clampdown which has clearly affected state attitudes to refugees and asylum seekers. This is probably not

Protracted refugee situations: Political, human rights and security implications,
Loescher, Milner, Newman and Troeller (eds),
United Nations University Press, 2008, ISBN 978-92-808-1158-2

surprising given the experience of the Nairobi US Embassy bombing and especially reflects concerns about possible links between Somali nationals in Kenya and the al-Qaeda network.

In Uganda, 'terrorism' has been a more provincial phenomenon but responses to the Lord's Resistance Army (LRA) have also been made by the government using the language of the 'war on terror'. Huge numbers of IDPs have been forcibly encamped, reportedly 'for their own protection', by the government in the context of the long-lasting internal conflict.[2] The existence of this population has undoubtedly lowered the profile of the Sudanese refugee population in the country.

Uganda, which neighbours Sudan and the Democratic Republic of Congo (DRC) on its north and north-western border, has a long history of cross-border forced migration in both directions. Uganda hosted large numbers of Sudanese and then Zairian refugees in the 1960s, while many Ugandans fled the West Nile to those countries at the end of the 1970s. Uganda has again hosted Sudanese since the late 1980s, and increased numbers of Congolese refugees since the 1990s. Almost all Ugandans living in the northern border region either have been refugees themselves or have hosted refugees, at some point in their lives.[3]

The complexity of the forced migration picture in Sudan itself, and in Uganda and Kenya as major refugee-hosting neighbours, warrants further consideration of the impacts of protracted exile.[4]

Background – Sudanese refugees in Uganda and Kenya

The immediate cause of the presence of Sudanese refugees in neighbouring countries including Kenya and Uganda is the long and bloody conflict between the government of Sudan and various Southern Sudanese rebels groups, first and foremost amongst which has been the Sudan People's Liberation Movement/Army (SPLM/A, and frequently 'SPLA' when the army specifically is being referred to). The conflict – the second in Sudan since it won independence from the British in 1956 – (re)commenced in 1983 and has often been described, over-simplistically, as being characterized by a series of oppositions: Northerners against Southerners, Arabs against Africans, Muslims against Christians and 'animists'. It is certainly more useful – not least given the recent conflict history of Sudan, in which violence has broken out in both the west of the country (in Darfur) and the east – to see it as one of a series of conflicts between the Islamist, Khartoum-based government of Sudan and various marginalized groups who feel themselves to have been exploited and oppressed by a political system from which they have been largely excluded. Douglas Johnson points to the legacy of colonial and pre-colonial political structures and relations, the role of oil and other natural resources, and the involvement of Cold War actors in perpetuating the conflict.[5]

At the time of writing, almost two years after the signing of the Comprehensive Peace Agreement (CPA) between the government of Sudan and Southern rebel groups led by the SPLM/A in January 2005, tripartite agreements on repatriation have been signed between the governments of Sudan and Uganda/Kenya and UNHCR. UNHCR has been registering refugees for voluntary repatriation since shortly after the signing of the CPA, although any immediate moves to begin the process were halted temporarily after the sudden death of John Garang, leader of the SPLM/A, and following a deterioration of security in the South of the country. The organized repatriation began in December 2005 and UNHCR had assisted with the return of close to 20,000 refugees by December 2006. Of these, 1,500 came from Kenya and 4,500 from Uganda. More than 70,000 others have returned from neighbouring countries without UNHCR assistance.[6] Many more still await the return of conditions in their areas of origin that would make it possible for them to go home.

A massive post-conflict package of aid has been pledged to Sudan for the reconstruction and rehabilitation of the South and for its integration into the unitary state that Khartoum hopes will be the end result of a referendum provided for in the peace settlement. The implementation of the peace agreement has not been without problems, and some commentators note that Sudan's wealth and recent development continue to be concentrated in the North, while the South remains relatively neglected.[7]

The conflict and forced migration

This conflict has been notorious for its impact on civilian populations who have suffered death, disease, displacement and impoverishment as a result. At least 1.5 million people have been killed and millions more displaced from their homes in the short and long term. For most of the nearly 200,000 Sudanese refugees now living in settlements in Uganda's Arua, Moyo, Koboko, Yumbe, Adjumani, Masindi and Hoima Districts and Kenya's Kakuma camp, exile has been not only repeated but also protracted. Some Sudanese have lived in a succession of transit camps, settlements and non-settlement locations, depending on changes in security conditions, government policy, personal circumstances and the availability of assistance. In addition, a large number of refugees live in urban areas of both countries, in great uncertainty due to their precarious legal status, especially in Kenya.

Douglas Johnson has argued strongly that, far from being treated as neutral victims by conflict actors, forced migrants have been positioned politically and manipulated for conflict objectives by both the government of Sudan and the SPLA. For the government of Sudan, the forcible resettlement of civilian populations has served to ensure an easy supply of labour where this has been required by the government, while its capacity to move them has demonstrated its control of populations in the conflict context. For the SPLA, control of populations has also been of importance – this was perhaps most clearly evidenced in relation to its manipulation of Southern Sudanese refugee populations in Ethiopia before 1991, but is also visible in its later attempts to persuade refugee populations to return to the 'liberated areas' under its control.[8] In brief, the displaced have become part of the conflict and, as such, solutions to their problems which do not involve a decisive end to the conflict itself have seemed very distant throughout the war.

As this chapter will argue, the governments of both Uganda and Kenya have always made it clear that their preferred 'durable solution' for the Sudanese caseloads in their countries is voluntary repatriation. The intractability of the conflict in Sudan, the infrastructural devastation wrought by the fighting, the landmines despoiling large and often unmarked areas of territory, and continuing insecurity in Southern Sudan have made any early repatriation an impossibility for the vast majority. For years, many young Sudanese males feared conscription by the SPLA, while at other times it has been fear of LRA activity in Sudan which has made return a distant hope. Meanwhile, local integration as a durable solution has been almost entirely ruled out by the governments of Uganda and Kenya, while resettlement places – although recently increased in the Kenyan context in particular – are still so few that they do not represent a significant possibility for the majority.

Causes of flight

Sudan first experienced internal conflict even before the country's independence from the British in 1956. The first Sudanese civil war only came to an end with the signing of the Addis Ababa Agreement in 1972 which made provision for greater regional autonomy for the South. During the time of the first Sudanese civil war, 1955–1972, UNHCR estimates that 86,000 Sudanese sought refuge in Uganda,[9] at which time many of them were able to exploit local connections and self-settle in the Ugandan border area. Many returned to Sudan in the 1970s, only to return again to Uganda in the late 1980s after the resumption of conflict between the newly formed SPLA and the government of Sudan following the introduction of *sharia* law and what was perceived by Southerners as a betrayal over the autonomy of the South by the Khartoum government.[10]

In Kenya, only very small numbers of Sudanese arrived before the 1990s, partly because many from south-eastern Sudan fled instead to Ethiopia where SPLA forces were in good standing with the government. After the fall of President Mengistu in 1991, however, the SPLA and large numbers of refugees were ejected back to Sudan and after this point larger numbers of Sudanese began arriving in Kenya, where they were accommodated in the newly established Kakuma camp.

All the Sudanese in Uganda and Kenya were forced to leave their homes directly or indirectly as a result of fighting between the government of Sudan and various Southern factions including the SPLM/A. While some fled from the government of Sudan forces or local defence militias, others fled from the SPLA itself.[11] As well as fleeing violence caused by fighting between the Northern and Southern actors, refugees were also forced to leave due to inter-factional fighting within some parts of the South, especially after the SPLM/A split in 1991. Finally, refugees were continuing to arrive in Uganda up to 2006 to avoid attack by the LRA, which had been pushed by the Ugandan military over the border into Sudan.

While many people escaped from outright fighting or its consequences, over time others were affected by combinations of fighting, loss of local infrastructure and services, including health and education facilities, poor environmental conditions and conflict-induced challenges to livelihoods. When the term was fashionable, Sudan was regarded as a classic 'complex political emergency' and the refugee populations in Uganda and Kenya broadly reflected this in that many fled for a complicated mixture of reasons.

The conflict context and the failure to find peace

Since the failure of the 1972 Addis Ababa Agreement and the return to war in Sudan in the early 1980s, a large number of internal and external peace initiatives have been tried and failed in Sudan, leading up to the success of the Intergovernmental Authority on Development (IGAD)-sponsored talks culminating in the Comprehensive Peace Agreement (CPA) in 2005.[12] While a number of regional actors have been involved at various moments, peacebuilding efforts for Sudan have been notable for the absence of any heavyweight international involvement until relatively recently. The recent interest and involvement of the United States has been explained by some as a response to its perception of an Islamicist threat emerging from Sudan, particularly after 11 September 2001.

The regional dimension has evidently been important to the pursuance of the war and hence the perpetuation of the refugee situation. In the late 1980s the SPLA held secure bases in Ethiopia and received support from its government, which increased their capacity in military terms. The Ugandan government – with which Khartoum suspended diplomatic relations for many years – has also been embroiled in the Sudanese conflict via its direct and indirect support for the SPLA. The government of Sudan, in return, for many years sponsored Uganda's own insurgent group, the LRA, which has in turn been positioned as an enemy of the SPLA. The government of Sudan has regarded refugees as partisan – as rebels rather than refugees, assuming that flight means guilt as far as association with the SPLA is concerned. Ironically, some refugees have also suffered the disapprobation of the SPLA, which considered them unsupportive of the cause in the early stages of the conflict.

Uganda's 'war in the North' has been inextricably interconnected with the Sudanese conflict through the course of both. Catastrophic insecurity within the Ugandan refugee-hosting area during much of the 1990s represented one of the worst threats to the security of nationals and refugees alike.[13] Arguably this interconnectedness of conflicts suggests that any 'durable solutions' for the Sudanese in Uganda must involve durable peace in both countries. The fact that the current ongoing peace negotiations between the government of Uganda and the LRA are sponsored by the new government of Southern Sudan is highly significant.[14]

The government of Kenya has not been directly involved in the Sudanese conflict in the same way as has the Ugandan government, but it has had an interest and has won applause for its steering of the IGAD process, which eventually led to the CPA. Otherwise, the government of Kenya has regarded the Sudanese conflict primarily as a security issue – especially since the late 1990s. While its concerns spring fairly clearly from risks associated with the Somali presence and concerns about their assumed links with al-Qaeda or other terrorist groups, there is little indication that a distinction has been made between the Somali and Sudanese populations in terms of refugee policy.[15]

Asylum in Uganda and Kenya

Who are the refugees?

The Sudanese conflict has generated 693,267 refugees,[16] many of whom have experienced protracted exile. They are to be found in Uganda, Kenya, the DRC, Ethiopia, Central African Republic and Chad, as well as resettlement countries including the US, Australia and the UK. Uganda currently hosts 269,800 refugees, of whom 228,700 are Sudanese who arrived in large numbers in the early 1990s.[17] Refugees in Uganda come from various ethnic groups, from all over Southern Sudan, especially from areas contiguous with the Ugandan border. In some cases they have co-ethnic Ugandan hosts such as Kakwa, Acholi, Madi and other groups. Kenya currently hosts 243,320 refugees, of whom 69,000 are Sudanese.[18] Most of them arrived after 1991 when Sudanese refugees in Ethiopia were largely forced to flee back to Sudan where no security was then to be found. Multi-ethnic in character, these groups nevertheless probably include a higher concentration of majority Nuer and Dinka refugees than does the refugee population in Uganda.

The legal basis for asylum in Uganda and Kenya and refugee conditions

In both cases, refugees arrived *en masse* at the border and were accepted by the Ugandan and Kenyan states on a *prima facie* basis, making it unnecessary for them to make individual applications for asylum. Both Uganda and Kenya have had draft refugee legislation in progress for many years. In Uganda, the Refugees Act finally became law in 2006, thus replacing the much more draconian Control of Alien Refugees Act, which dates back to 1960. In Kenya, the new Refugee Bill passed its third reading in parliament in November 2006 and awaits being signed into law by President Kibaki, at which point it will replace the Aliens Act. Both governments are signatories to the 1951 UN Refugee Convention and its 1967 Protocols. Under recent restrictive, domestic legal frameworks, both states have had considerable freedom to manage their respective refugee situations as suits their wider political agendas.

Through the period of hosting Sudanese refugees, both Kenya and Uganda have made clear that their eventual preference and priority is the repatriation of refugees to their country of origin. This is clearly reflected in their hosting arrangements, which emphasize the containment and control of refugees in camps and settlements and decline to support the temporary or permanent integration of refugees in any meaningful sense.

The fact that refugees in neither Kenya nor Uganda are free to choose their place of residence according to employment, family or other factors denies their freedom of movement under the 1951 UN Convention and other human rights instruments. This is only one of many ways in which the warehousing of refugees in camps and settlements has been shown, in these countries and elsewhere, to undermine the rights of refugees.

Linked to the restrictions on movement and extremely limited opportunities for economic activity in the Kenyan case in particular, refugees in both countries find themselves living in conditions of impoverishment without any remedial action available to them. In Kakuma, for example, they have no option but to rely on inadequate and sometimes unreliable food rations provided by the World Food Programme (WFP), any substantial agricultural activity being ruled out by an absence of available land and water resources, as well as by the government's own prohibition on such activity. In Uganda, agricultural activity is encouraged, but faces a number of serious constraints, not least of which is remoteness from markets, inadequacy of plots provided as families increase, and exhaustion of the soil. In both contexts, security is far from assured, with challenges emanating variously from within the refugee communities themselves, from alienated hosts and from military actors from within and without the host countries.

In both countries but to a greater extent in Kenya than in Uganda, refugees risk harassment and discrimination at the hands of the security forces if they are caught without documentation outside of camp areas. As the Refugee Consortium of Kenya put it, describing abuses against refugees in the course of round-ups and other security operations, 'harassment, extortion and intimidation of refugees by police officers continue unabated'.[19]

The impact of protracted exile

Impact of the protracted refugee situation on refugees and states

Sudanese refugees in Uganda have mainly had little reason to fear systematic or deliberate persecution from the government of Uganda since they began to arrive in large numbers in the late 1980s and early 1990s. As assumed clients of the SPLM/A, a close ally of Uganda's governing National Resistance Movement (NRM), no obstacles were put in the way of refugees seeking entry to Uganda and they were granted refugee status on a *prima facie* basis, obviating the need for individual status determination procedures. Although it was the case that some refugees fled explicitly from the SPLA itself, and that the SPLA viewed certain of the refugee camps and settlements in Uganda in the early to mid-1990s as 'friendly' or 'hostile', engaging with them accordingly, there is little evidence to suggest that such subtleties affected the government of Uganda's attitude to or interaction with refugees in specific locations in northern Uganda.

In practice, although it has of course been of significant benefit to the refugees that over the duration of their protracted exile in Uganda the head of state has been someone who is politically well disposed to them as constituents of the SPLM/A, it is not correspondingly the case – as is demonstrated throughout this chapter – that the state has ensured the rights of Sudanese refugees in Uganda. On the contrary, the state itself has been a notable abuser of refugee rights in various ways, as will be explained in more detail below. But it ought to also be recognized that, as time has passed since the arrival of the Sudanese in Uganda, they have been very far from being the biggest population or perhaps the most difficult population of forced migrants with whom the Ugandan government has had to deal. In the same period, a vast number of Ugandan citizens have been forcibly displaced and confined by the government to IDP camps 'for their own protection' and this group – for political and conflict as well as for humanitarian reasons – has increasingly occupied the attention of the government and humanitarian actors. This chapter does not argue that this has been a mistaken emphasis – the suffering of the Ugandan IDPs has been extreme and appalling. It remains a pity, however, that the Sudanese and responses to their plight have to some extent been regarded as less pressing and serious towards the end of the 1990s and early 2000s, in the face of such a comparison.

Uganda has been relatively indulged in its prosecution of the northern war by an international community that wants it to 'remain' a secure haven in a regional context of precarious states and insecurity. When aid conditionalities have been applied to the government of Uganda, for example by the Department for International Development in 2005, this has been in relation to its slow progress on political liberalization rather than in response to its treatment of either refugees or IDPs in northern Uganda.

While settlements in Uganda were still being at least partly serviced by the implementing partners of UNHCR, refugees represented neither threat nor competition to local populations. As has been noticeably the case in relation to the Kiryandongo settlement in Masindi, as time has passed and services have been withdrawn from the refugee settlement, local hosts have become increasingly aware of the refugees' need to leave the settlements to compete for resources in the local environment. In line with the analysis of Crisp and Slaughter (this volume) regarding the extent to which refugee responses are conditioned by the respective attitudes of UNHCR and host states to look to the former rather than the latter to meet their needs, it is notable in this case that, when refugee settlements are fully serviced by the international community, refugees are also less likely to be perceived as a burden by local hosts. When such support is withdrawn in protracted exile, refugees remain restricted in settlements but are forced to leave them to compete for subsistence opportunities and natural resources outside them and hosts are more likely to feel resentful or burdened by them.

Sudanese refugee communities in Uganda have had a positive impact on the local economies in areas where they have been permitted to make a contribution via agricultural activity or business. Merkx notes highly increased economic activity around remote refugee settlements in the border area,[20] and this is also immediately obvious in Masindi, where, for example, the Bweyale trading centre near the Kiryandongo settlement has more than doubled in size and importance since the arrival of both the Sudanese refugees and also significant numbers of Ugandan IDPs fleeing fighting in Gulu, Kitgum and Pader districts.

In Kenya one could argue that the economic impact of the protracted refugee presence has been less positive – a clear result of government refugee policy. With the exception of 'refugee assets' in terms of aid flows which have made a significant contribution to the economy of hosting areas which are already impoverished, refugees are not able or entitled to assert themselves and thus make an economic contribution.

In Kenya, the predicament of the Sudanese can generally also be argued not to have been the main driver of the two states' response to refugees. All refugees are restricted to camps, where they live in very difficult conditions and where hostile relations with the local host population are virtually guaranteed by the structural relationship they find themselves in. Regular round-ups take place of any refugees, only a small minority of whom are Sudanese, who have managed to slip through the net and remain in Nairobi.

The location of refugee camps and settlements in remote, politically marginal border areas in both refugee-hosting states reflects governments' desire to maintain the separation of refugee populations, positively preventing integration with local populations. In Kenya, this process has been managed in such a way that local hosts have struggled to assert themselves in any substantial way, with local political representatives largely unable to convert local pressure for recognition into political or economic capital with respect to the refugee presence. In Uganda, the situation has varied from one refugee-hosting district to another, with the emergence of entirely new administrative and political districts tightly linked to successes in the making of political claims by hosting populations and their more successful political intermediaries. The creation of the new districts of Adjumani, Yumbe and Koboko, in the heart of the refugee-hosting area, is indicative of these kinds of dynamics. Similarly, districts such as Arua and Yumbe have proved quite effective at targeting 'refugee assets' or the aid associated with a refugee presence in ways that imply long-term benefits for the local population. Jacobsen has noted that refugee assistance programmes may be used as a way for states to 'broadcast' power to remote border areas and, while the NRM government in Uganda may be atypical in that its system of participatory local government has increased its political reach to the very boundaries of the country, it remains the case that the presence and management of refugees at the margins has offered a further domain of activity and power for them.[21]

Refugee security

The most important questions relating to refugees and security in Uganda refer to refugees' own experience of insecurity caused by LRA (and other rebel groups') attacks and activities, rather than any insecurity caused by the presence of the refugees. For example, refugees in transit camps in Adjumani district were attacked by the LRA in 1991, and were moved to a new settlement at Kiryandongo by the Ugandan government and UNHCR as a result. LRA attacks have continued with varying frequency in this and other locations throughout northern Uganda until the present. Similarly, refugees accommodated in West Nile districts (namely Arua) experienced serious assaults by the West Nile Bank Front and related groups during the middle of the 1990s. While areas have passed from being more to less secure over time, it is probably fair to say that at no time has the entire refugee-hosting area been secure.[22] This has been as true for nationals as for refugees, who on numerous occasions have been forced to flee together. At the Achol-pii settlement in Pader district, refugees and nationals joined together to form security patrols to attempt to ensure the security of all. There is also evidence that refugees have been targeted directly because of their status as Sudanese and reputation as clients of the SPLA and the NRM. A series of attacks on Achol-pii, culminating in massacres at the settlement in 1996, were attributed by refugees and others partly to the fact that Sudanese refugees from that settlement had been apprehended by the LRA in Sudan, in possession of refugee identity cards which showed their origin in Achol-pii. Wider security impacts of this situation are discussed separately below.

Responses to protracted exile; the camp/settlement system in Kenya and Uganda

Although strategies and practices have varied widely, both Kenya and Uganda have from an early stage resorted to the 'warehousing' of refugees in camps and settlements.

In the early 1990s, both countries saw a significant upsurge in refugee arrivals, with Kenya especially receiving as many as 420,000 Somali and Sudanese at around the same time. While the vast majority were Somali, refugees of both nationalities were initially mainly encamped in the coastal region before their removal to the UNHCR-run camps at Dadaab and Kakuma. The expansion of UNHCR's role in refugee management and assistance in Kenya at that time led to an increased reliance on the use of camps.[23]

In Uganda, Sudanese refugees were initially accommodated in transit camps in close proximity to the border until their relocation to agricultural settlements on land provided by the government. As in Kenya, land offered was not always of good quality, or close to markets and services.

Arguments about the disadvantages for refugees of encampment, as well as the extent to which and ways in which, as a system, it undermines refugee rights to freedom of movement and related rights, are compelling and these represent the point of departure for a consideration of situations of protracted exile. Even relatively sophisticated livelihood models have not paid much attention to the effects of the passage of time on refugee economies.

In Kenya it may be the case that the assistance model employed means that there has been less variation over time. Since no livelihood opportunities at all are allowed to refugees in Kakuma – no significant farming is permitted and business opportunities are highly restricted by lack of access to any serious markets and the hostility of the local population to the gathering of natural resources by refugees – 'full' (although frequently inadequate) food rations continue to be delivered to Kakuma residents by the WFP. Opportunities for any other income-generating activity – restricted as they may be – have ebbed and flowed over the years but the structural limitations have remained broadly the same. Even the providers of assistance to refugees in Kakuma express concerns about the extreme impoverishment of refugees living in the camp and it seems clear that over an extended period even basic needs have not been met there.[24]

In Uganda there has been a greater shift over time. In the Ugandan settlement system, refugees are also required to remain resident in settlements allocated by the government. Here, though, they are able to practise subsistence agriculture and the objective is that they should produce a substantial enough surplus of cash crops that they can sell them and cater for their own non-food needs. There are large discrepancies between settlements regarding the extent to which it has been possible for refugees to achieve this kind of 'self-sufficiency' over the years. Critical variables have included the local security conditions (which have varied widely), access to markets, soil and climactic conditions, and relations with local populations. These have been important both as an indicator of the likelihood that refugees have been able to acquire more land, and in relation to employment opportunities for refugees (formal or informal) outside settlements.[25]

However, there are evidently limits to the period of time that one can expect a farmer to re-use the same piece of land without having the opportunity to leave it fallow, before yields start to fall dramatically. This, together with the fact that no account has subsequently been taken by settlement managers of the fact that, while plots were allocated on the basis of family size, many if not all refugee families have grown in size since their arrival in Uganda over a decade ago, has led to a situation more difficult in the later stages of protracted exile than in the early days. It should also be noted that external inputs to most settlements have also declined over time as refugee communities have been evaluated or assumed to be able to provide themselves with a range of services, etc. Thus, in the Kiryandongo settlement in Masindi district, for example, support for community associations (Widows and Orphans Association, Disabled Association) and occupational associations (bee keepers, poultry keepers, Cooperative Society), which was offered in the early 1990s, had run out by the late 1990s, whereupon such institutions largely collapsed. Later still, further inputs were removed or abolished due to repeated and catastrophic cuts in UNHCR's budget for Uganda's settlements. In early 2006, for example, Kiryandongo had already lost both its UNHCR-funded agricultural and environmental extension programmes as budget cuts made them unfeasible from UNHCR's point of view. These cuts, in combination with the generic factors mentioned above and unfavourable climactic conditions in late 2005/early 2006, led to a disastrous harvest and severe hunger in the settlement later in the year. It is important to note that it has been in the context of UNHCR's inability to continue supporting such programmes in settlements that the government of Uganda has considered scrapping the settlement system in favour of encampment.[26]

As appears to be frequently the case in situations of protracted exile, the predicament of the Sudanese refugees in both Uganda and Kenya can be argued to have become more rather than less difficult as their stay in the host country has become protracted. The international humanitarian system is organized in such a way that funding and attention for specific refugee or other crises may vary significantly depending on their perceived geo-political significance. For very many years, Sudan's conflict was not high on the agenda of international political actors and responses to humanitarian needs in and outside the country were consistently inadequate. In relation to the refugee populations in particular, as time passed, more pressing needs elsewhere increasingly took precedence. One clear indicator for this is that UNHCR has routinely and consistently had trouble in adequately resourcing its refugee assistance programmes in Uganda and Kenya. The problem of attracting funding to protracted refugee situations has been thoroughly dealt with in internal and external accounts and analyses.[27] Here it is relevant to note by way of example the fact that UNHCR's Uganda programme in 2005 had a total budget of zero to support crop production, fisheries, livestock and forestry.[28] Without substantial evidence to demonstrate that district-level inputs were meeting refugee livelihood needs to an adequate level, this fact is disappointing, to say the least.

DAR/SRS in Uganda: Failures of development in the absence of integration

In Uganda, unlike in Kenya, the government's willingness to provide land for agriculture to refugees and its collaboration with UNHCR on a so-called 'Self-Reliance Strategy' (SRS) for refugees have led it to participate in one of UNHCR's new flagship initiatives under the Convention Plus initiative. Development Assistance for Refugees (DAR) is one of UNHCR's responses to the challenges of protracted refugee situations and, as its name suggests, involves adopting a significantly more developmental approach to refugee assistance and management than has often previously been the case.[29] In Uganda, this strategy has evolved reasonably naturally out of the SRS (which has its roots in discussions between enlightened UNHCR staff and the government since 1996 and which was formally launched in 1999, 10 years after some of the first Sudanese refugees arrived in Uganda).

While SRS programming relied heavily on refugees and their labour and exhorted them to achieve self-sufficiency within the constraints of the settlement system, under DAR, UNHCR's remaining refugee safety net is all but removed. Under the SRS/DAR strategy, responsibility for the delivery of services and support to refugees shifts mainly to district authorities, who are expected to have budgeted for refugees in their district development plans. The laudable intention is to avoid the creation and maintenance of parallel services for refugees and rather to integrate services for refugees into district delivery. UNHCR and its implementing partners take a minimal or non-existent role under the scheme, which seeks to support refugee-hosting areas including refugees rather than target refugees directly and explicitly. The consequences of this 'beneficiary blindness' in terms of refugee protection have apparently been relatively little considered. The practical consequence of the implementation of the DAR programme in Uganda so far appears to be a substantial reduction in services provided to refugees, with a corresponding increasing requirement that they share costs for those that are available. In the absence of any meaningful income-generating opportunities and given the economic constraints of life in a refugee settlement, these requirements are contradictory.

Role of UNHCR/NGOs/civil society organizations

UNHCR's laudable attempts to generate a new developmental agenda in refugee assistance are hobbled by the fact that an adequate budget to support such programming is not available to it in the Ugandan case. It has neglected to address with the government the fact that any meaningful developmental activity may require a more substantial commitment to the integration of refugees than the government appears willing to contemplate. If refugees were at liberty to settle freely and negotiate access to land in places where conditions for agricultural activity and trade were favourable, their potential contribution to the economy of Uganda could well increase markedly. Under current conditions and for the reasons outlined above, the longer they are restricted to under-serviced settlements without being enable to achieve economic autonomy, the worse off they are likely to become. An additional concern is that the closeness of the working relationship between UNHCR and government officials at the field level leaves some room for doubt as to UNHCR's capacity to deliver protection in the current circumstances. In Rhino camp settlement, for example, refugees testify to the fact that they are required to acquire a 'chit' from a government official before they are entitled to approach the UNHCR field officer with any protection problem.[30] Clearly this process is not conducive to a safe environment within which abuses can be raised and discussed safely.

This links to Crisp and Slaughter's wider point (made in their chapter in this volume), namely that UNHCR and the government of Uganda in this case have carved out an implicit working arrangement whereby UNHCR, rather than the host state, is ultimately seen as responsible for refugee well-being. This can be seen most explicitly in the government's effective refusal in most cases to take on financial responsibility for refugees as was envisaged under the SRS, and is also reflected in the government's reliance on UNHCR funding for it to carry out basic refugee-related functions.

A fundamental concern in the case of both Uganda and Kenya is that the predicament of refugees is constructed as an apolitical issue which is best addressed with humanitarian – or perhaps developmental – tools. The political roots of the problems which refugees in both countries face, and the undoubtedly political nature of any solution to them, are rarely raised or discussed.

Having positioned itself as a humanitarian/developmental actor, and due to its mandate which requires its actions to be non-political, UNHCR is limited in the extent to which it can address such questions. Nevertheless, it seems clear that claiming refugees as its 'territory' and then failing to address crucial political questions is not a sufficient answer to the challenges it faces. UNHCR's implementing partners – often international NGOs with their own funding sources outside the UN system – have also shown remarkably little inclination to address these questions in their own right. Where are the voices of protest about the conditions in Kakuma in which refugees have been forced to live for so many years? Why are NGOs working in Uganda not protesting more loudly at the conditions of insecurity in which refugees have been obliged to live throughout the 1990s? While in both countries some international and national refugee advocates have been heard, they have not been powerful enough to sway governments and change policy. UNHCR, arguably, has been unable or unwilling to act as a strong refugee advocate for fear of compromising its own operational position. Its need to maintain good working relations with governments has sometimes led to a situation where it has been unable to be forceful in relation to protection issues. In some cases, when it has attempted to be clear in its support for refugee rights, the government has reacted strongly. In 2004, UNHCR's country representative in Uganda was made *persona non grata* by a government that did not appreciate his loud opposition to the forcible relocation of refugees from Masindi to Arua district – a location considered unsafe by the refugees in question.

The involvement of development actors including the United Nations Development Programme (UNDP) and national and international NGOs in refugee-hosting areas has been minimal. In most cases, refugee areas are excluded from development interventions taking place in the wider refugee-hosting area on the grounds that refugees do not yet fall within a developmental remit. This remains true even in areas where significant numbers of self-settled refugees are living illegally and without external support in border areas and elsewhere. Notably, the Ugandan communities supporting these refugees are also not targeted for help by either development or humanitarian actors. It is also the case that refugees in Uganda tend to be settled in the heart of the conflict-affected area, which has itself suffered from years of developmental neglect regardless of the presence or absence of refugees. In Kenya, the situation has been somewhat different in that pressing security problems resulting from tensions between refugees and hosts have been at least partly addressed via supportive interventions for host populations.

Local, national and regional security challenges; direct and indirect threats

In addition to the socio-political and economic dimensions discussed above, the protracted refugee situations in Uganda and Kenya inevitably also raise questions of security at the local, national and regional levels. As will have become clear from the above, most of the direct security threats relate to the protracted refugee situations in combination with other regional or international risks. In Uganda, the most important dynamic refers to the interaction of the Southern Sudanese and northern Uganda conflicts and resulting relations between respective governments and rebel movements, as well as to the international relations between the governments of Sudan and Uganda themselves. In Kenya, refugee security is viewed through the prism of the government's fears about extremist political Islam, and such fears are focused most clearly on its Somali refugee caseload. Nevertheless, all refugees in the country are dealt with similarly, regardless of how close to this threat they reach.

Direct threats, therefore, included serious tensions during the 1990s between the governments of Sudan and Uganda, each of which supported insurgent movements in the other's country. The commitment of the NRM to the SPLM/A was clearly based on a strong historical and ideological link, while the collaboration of the fundamentalist Sudanese Islamic government with an ostensibly Christian rebel group was less obvious, except as a retaliatory measure. Diplomatic relations between the two states were reduced to a minimum during the period of hostility and military activity did occasionally transpire. Sudanese military Antonov aircraft bombed Ugandan border towns including Adjumani in 1998 while the Ugandan military was observed supporting SPLA operations in Southern Sudan, including the SPLA offensive in early 1997.

The capacity of the LRA, supported by the government of Sudan, to inflict damages on northern Uganda, including targeting refugees, has already been described. Until the recent ceasefire between the government of Uganda and the LRA, their continued activity in Southern Sudan put pressure on Sudan's CPA[31] and was one of the principal obstacles to repatriation for those refugees originating in the parts of Eastern and Western Equatoria where they had established bases.

The LRA and the government of Sudan were not the only threats to security to refugees in settlements in northern Uganda. Refugees in Ugandan border camps and settlements and in Kakuma in Kenya were at risk from the 1990s from conscription by the SPLA, which was well known to use some settlements as 'R & R' bases when it was not on active duty.[32] Many SPLA soldiers kept their families in settlements in Uganda and Kenya, visiting them when they were able to. It is hard to insist on the civilian nature of settlements when this is taking place.

Further security threats to refugees in Uganda in particular emanated from the host government itself. In September 2003, thousands of Sudanese refugees were forcibly relocated by the Ugandan military and other forces to settlements north of the Nile, where they insisted they would be at risk from both Ugandan rebel groups and the SPLA. Despite representations made by UNHCR and other actors, the government showed little restraint and refugees were forced at gun point and with the aid of tear gas onto trucks for their removal. Refugees' accusations of deaths and injury during the process have been little considered.[33]

Indirect security threats resulting from the presence of a protracted refugee presence in Uganda and Kenya refer to tensions and conflict between refugees and host populations, as well as to violence within the refugee communities themselves. Some analysts have argued that refugee camps and their frequent proximity to borders themselves invite or exacerbate insecurity. In the case of Kakuma, located in a very poor and semi-arid part of Kenya, refugees outnumber the local Turkana, who feel in competition with them for crucial natural resources. There exists no cultural or ethnic affinity between the Turkana and the refugees they host and, although UNHCR does provide some infrastructural support for nationals, they do not receive food rations as the refugees do. The area as a whole is insecure partly because of its endemic poverty, and partly because of its proximity to the borders of Sudan and Uganda, which have both been countries at war. Competition between the refugees and hosts over employment and resources causes political instability.[34] In northern Uganda too, tensions have been expressed between refugees and their hosts – often, although not exclusively, over the land which they have been allocated or otherwise accessed for agricultural purposes.[35]

Conflict is not infrequently also to be found within refugee populations or communities. In Kakuma, violence has broken out both between Sudanese and members of other national groups (e.g. Somalis or Ethiopians) and between sub-groups of single national groups. Crisp notes that in 1997 several incidents took place involving intra- and inter-group fighting – in this case he specifies violence between Nuer and Dinka refugees – which led to more than 200 causalities.[36] Such conflicts are often linked to the ongoing conflict in the country of origin, in this case Sudan.[37] Kiryandongo refugee settlement in Uganda has also seen inter-ethnic violence between refugees in which deaths occurred and huts and crops were razed.[38] Here too, explanations as to the cause refer to arguments within the rebel movement in Southern Sudan.

Finally, indirect security threats must also include the risk of power abuses by refugee leaders and other powerful figures within refugee communities. Harrell-Bond and Verdirame have pointed out that, in both Ugandan and Kenyan camps and settlements, evidence has been found of incidences of customary justice and abuses in contravention of national legislation.[39]

What future for the Sudanese?

Immediate prospects

The most obvious and decisive conditions for the resolution of protracted refugee situations are typically seen to emerge when the root causes of the refugee-generating situation are resolved and the reasons why people became refugees in the first place are removed. This most obviously points to a repatriation model as far as solutions are concerned. Even if it is the case that the majority of people – when such conditions are achieved – will be able and willing to go 'home', it is also understood that there will always be sub-groups within the refugee population for whom this is not a desirable outcome. This is why it is so important that a mixture of durable solutions is offered, even when the root causes of refugee-generating conflict have been addressed. It is critical that we remember that people's needs and aspirations may have changed during their period of protracted exile, transforming previous expectations about durable solutions for them.

In the case of Sudan, 2005/6 saw a dramatic transformation in the prospects for an end to the protracted refugee situation of its nationals in Uganda, Kenya and other host countries in the region and beyond. The signing of the 2005 Comprehensive Peace Agreement in Sudan, in combination with significant progress towards a peace agreement in Uganda, constitutes the best hope in nearly 20 years of a large-scale repatriation to Sudan.

For any large-scale repatriation from Uganda and Kenya to be feasible, several assumptions need to hold and several conditions met. Firstly, and most substantially, the peace in Sudan needs to hold under the government of national unity despite destabilization risks in the short and medium term from the ongoing conflicts in the east and west of the country and from potential renewed conflict over oil resources, and from the future risk of Southern factional fighting in the future. Significant recent risks in this respect have included Southern factional fighting, especially at Malakal, and the increasingly appalling situation in Darfur, where hopes that the active involvement of the Southern leadership in the national government might have a positive effect on negotiations for international intervention appear to have been unfounded. Secondly, the Ugandan peace talks ongoing in Juba (as of December 2006) need to reach a successful conclusion so that the ceasefire agreement is maintained. Under these circumstances, refugees in Uganda and Kenya need not fear that they are to repatriate only to continue suffering from insecurity at the hands of the re-mobilized LRA. It should be noted that the government of Southern Sudan's sovereignty, and therefore its capacity to act with strength in the government of Sudan, has been somewhat compromised by the presence on its territory of the LRA and the Uganda People's Defence Force, in pursuit of the rebels under an agreement with the government of Sudan. The embedding of peace in Sudan is, therefore, to some extent dependent on a resolution to the Ugandan conflict as well.[40]

Prospects for the completion of the large-scale repatriation to Sudan, which began in December 2005 and had by December 2006 assisted close to 20,000 refugees to return to Sudan, depend to a large extent on confidence-building measures within the refugee population. The sudden cooling of enthusiasm for return after the sudden death of John Garang in 2005 demonstrates how a willingness to gamble on and invest in what must be an uncertain future is influenced by confidence-shattering events such as this. In addition, a comprehensive return programme depends on the presence of an adequate level of security and the availability of funds from the international donor community. Neither of these was assured during 2006 and progress towards a substantial return has been relatively slow as a result.[41]

The resolution of a protracted refugee situation has to incorporate activity in a range of locations and by a varied cast of actors. If repatriation is to be the solution of choice at this time for Sudan, massive efforts will be needed to transform the conflict-ravaged country to which refugees are to return. In interviews carried out in settlements in Arua in 2004 and in Masindi in 2006, refugees argued that before they were ready to move back to Sudan they wanted to see substantial developmental progress made in their home areas. Minimally, mention was made of a need for the clearance of landmines, (re)construction of roads and infrastructure, the (re)establishment of social services including schools and health facilities, and the maintenance of a level of security to enable ordinary living to take place there. It is clear that the rehabilitation and development challenges in Sudan are immense, but they must be understood to be a critical component of any durable solution for returnees from other parts of Sudan or from neighbouring countries. What this implies, at the very least, is that it will not be sufficient to expect UNHCR alone to deal with returnees in isolation from wider development efforts. Plans are already being put in place for an integrated reconstruction and development effort which includes UNHCR and its 4Rs programme. The fact that many of the funding pledges made to the government of Sudan have yet to be honoured and that civilians in the South complain that little development activity is visible away from Juba and other urban centres, however, is not a promising sign at this critical moment.

The scale of reconstruction needs in Southern Sudan represents a massive challenge to the post-conflict transformation which needs to take place in the country if peace and the reintegration of the displaced population are to be achieved. Furthermore, it is politically crucial that progress is made quickly, both so that exiles can be persuaded to return in time to be registered for the population census which will form the basis for the electoral register to be used in the eventual referendum on secession in 2011, and so that meaningful consolidatory state-building processes can be established in the meantime. Not only does the 2007 UN Work Plan for Sudan anticipate a large increase in spending on recovery and development, 'for the first time in Southern Sudan recovery and development programming (US$350 million) exceeds humanitarian activities (US$280 million)'.[42] As far as the resolution of the protracted refugee situation is concerned, it is clear that the implementation of a strong developmental intervention in response to refugees in their countries of exile could have reaped benefits in terms of their ability to contribute substantially to post-conflict development processes in Sudan. As things stand, however, the limits which have been set on refugee development by the policies and assistance practices of the host states and UNHCR are also likely to have negative consequences on return to Sudan in this respect.

Other dimensions of a solution

In the event that a mass return to Sudan becomes unfeasible in the short to medium term, what prospects exist for a durable solution to the protracted refugee situations in Uganda and Kenya? What are the alternative options for refugees who are unwilling or unable to repatriate even if the majority of their peers do? The answers to these questions must lie in the two remaining durable solutions – resettlement and local integration – neither of which can be excluded as a viable part of a solution to the problem of Sudanese refugees.

In 2005, after the signing of the CPA, UNHCR assisted in the resettlement of 4,600 Somali and Sudanese refugees from Kenya and a further 600 from Uganda, demonstrating that even in the presence of a new peace agreement there were Sudanese refugees for whom a return to the country of origin was not considered a viable option. Some individuals have protection concerns about returning to Sudan which only a durable solution in another region can resolve. This being the case and given the evident development needs of Sudan even were repatriation to go ahead, this chapter argues that resettlement should continue to be considered a vital part of the search for durable solutions for the Sudanese. The developmental advantages for resettled individuals and their country of origin are likely to be great when their likely activity with respect to remittances, skills exchanges and the benefits of international business or trading links is considered.

Secondly and more controversially, it is clear that further consideration could be given to the solution of local integration. After the resolution of the first Sudanese conflict in 1972, significant numbers of former refugees remained in Uganda while their countrymen repatriated. These were mainly people who had integrated informally into the local populations and were able to negotiate or avoid the administrative expectation that they would return to Sudan. Although substantial numbers of Sudanese have also 'self-settled' in Uganda on this occasion, the changed political landscape at the local level (i.e. the existence of the local council system[43] which monitors residents in any given village or urban quarter) has made this kind of spontaneous settlement more difficult to negotiate. This being the case, it may be that, in the absence of government authorization for the local integration of any 'residual' refugee caseload, remaining in Uganda may be less straightforward in 2006 than it was in 1972.

Various issues arise. Firstly it should be noted that there was some indication from government of Uganda officials in refugee-hosting areas of northern Uganda from around 2004 onwards that the long-term local integration of a minority of Sudanese would not be viewed with alarm in the event that the majority returned home. This view was not shared by officials of central government who insisted, as they had always done, on a policy of repatriation. Perhaps more importantly, both Uganda and Kenya have implemented refugee policies through the period of protracted exile which have positively prohibited the integration of refugees in the national society. While some limited integration of refugee services has been the objective of the Ugandan SRS/DAR strategy, the socio-economic integration of refugees themselves has been actively resisted by the government.[44] Kenya's position has been even clearer and any attempt to educate the governments of these two host states on the benefits of local integration as a durable solution would clearly have to overcome the inbuilt difficulties associated with their previous rejection of it, even as an interim strategy.

Almost entirely absent from the general discourse on refugee protection and assistance in Uganda and Kenya is any sense that policy thinking has incorporated the need to consider the eventual requirements of any durable solution in the short to medium term. The logic of refugee assistance programming in both countries has been predominantly concerned with the containment of refugees and with states' fears that they will not be adequately recompensed by the international community for the 'burden' which refugees are assumed to represent. In fact, and as we have seen, the international donor community has indeed been averse to generously funding supportive interventions for refugees experiencing protracted exile. The result has been that in both Uganda and Kenya the limited discussions regarding durable solutions for refugees have tended to assume repatriation as early as possible, while refugee assistance programming in the meantime has not engaged directly with the implications of this fact. Thus, it has been a constant struggle for any developmentally oriented organization seeking to support long-term investment in refugee communities to secure funding or even agreement that this is an appropriate response to refugees in exile. As such, interventions including capacity building, peace and civic education programmes, training and higher educational supports have been few and far between, much to the frustration of refugees stranded in camps and settlements. As has been noted above, the consequences of this short-sightedness will be felt most acutely when refugees are able to return home and precisely those skills with which they have not been equipped in exile are most urgently needed.

Addressing the human rights and security challenges of the Sudanese refugee situations in Uganda and Kenya

As this chapter has argued, Sudanese refugees do not currently enjoy their rights in either Uganda or Kenya and this is the first failing of the current asylum and refugee regime in these host states. Urgent action is required in both Uganda and Kenya to ensure that the more generous provisions of new refugee legislation are implemented as quickly as possible. In addition, action is required from diverse actors in the short, medium and long term to support a comprehensive solution to the Sudanese protracted refugee situation. Such action should include, but is not limited to, the following:

- Ongoing support from the international community for the implementation of the Sudanese CPA, including full support for the funding and implementation of the rehabilitation and reconstruction effort
    - requires the collaboration of peacekeeping, political, diplomatic, development and humanitarian actors including the United Nations Mission In Sudan (UNMIS), UNDP, World Bank, UNHCR and national and international NGOs
    - requires continuing confidence-building measures among refugee, returnee and 'stayee' populations as well as their involvement and participation in decision-making and implementation
- Support for the current peace negotiations between the government of Uganda and the LRA
    - requires continued involvement of the government of Southern Sudan
    - requires continued political pressure by international political actors on both warring parties with respect to the rights of conflict-affected civilians including refugees
- In the lead up to any large-scale repatriation, humanitarian aid for refugees in Uganda and Kenya where this is required, as well as more extensive preparatory developmental inputs from specialist development actors (national, international, governmental and international organizations)
- Active consideration of alternative durable solutions for refugees for whom repatriation to Sudan is unsuitable at this time. Work towards implementation of local integration as an interim solution in the event that repatriation programming takes time

- requires involvement of resettlement states, UNHCR/International Organization for Migration.
- local integration work will require the involvement of district authorities in both host states, UNHCR, UNDP and (international) NGOs after authorization of host states has been achieved.

Policy lessons for the region and beyond – preliminary conclusions from the Sudanese case study

A major step in the direction of the enjoyment of refugee rights would be for states to remove the requirement that refugees live only in designated camps and settlements, rather allowing them to integrate freely with the local populations in each respective country. It is clear that this happy situation is very far from being a reality in either of the case study states, for a number of more or less compelling reasons. In the first instance, it has not been clear that the international donor community is serious about committing itself to a genuine and extensive process of burden-sharing with respect to refugee populations in Uganda and Kenya. This is a function of the messages European and other industrialized states send out via their own restrictive policies and practices of asylum, and also results from their unwillingness to provide the funds for even basic refugee programming in East Africa. As such, the political atmosphere has not been generated within which East African states are ready to work with international actors to overcome their concerns about the burden that an integrated refugee population might represent on their already overstretched resources.

Even in the absence of a durable solution to the Sudanese refugee problem, a shift in policy on the part of the Ugandan and Kenyan governments could result in a transformation of the experience of exile for refugees. States need to take a more enlightened approach to the distinction between forced and other migrants, allowing a situation where professional refugees are enabled to continue their careers in exile much more easily, where adequate support is given to those without special skills or resources and where, in general terms, the differentiated needs of a disaggregated population are responded to with more subtlety with benefits for all.

This approach implies a much more extensive application of the developmental approach with which UNHCR and the government of Uganda have had so little luck in that country. A developmental approach predicated on the removal of the constraints of settlement living and the meaningful integration of refugees, in conjunction with a realistic budget in the short to medium term, could lead to very different outcomes than we have seen thus far in Uganda. Such an approach would recognize the 'normality' of cross-border movements in this region, regardless of the conflict context, and would build on these opportunities.

The existing division of labour in relation to refugee management and assistance has been found lacking in the Ugandan and, arguably, in the Kenyan cases in that defining certain kinds of aid as 'refugee aid' and failing to link or relate it to other supportive interventions in a refugee-hosting region has shown a limited impact. The conflict between the refugees and hosts in Kakuma is one indicator of this failure. The fact that in Adjumani district in Uganda a small number of refugee-oriented organizations provide support in settlements without reference to local development planning or activity and that local planners are still reluctant to fully involve refugee populations in their calculations is a further case in point.

What would be required for these more enlightened arrangements to be put in place? In the first instance, governments must be persuaded that they run no greater risk than was previously the case, either in terms of security or in relation to the economic burden they are to carry. With confidence in these facts, states are more likely to be willing to support the integration of refugees into local populations. Secondly, assistance organizations, both local and international, need to find ways to work creatively with both refugee and non-refugee groups as a matter of course. This will require a working style which is more collaborative than competitive. Finally, it requires funding arrangements which are more conducive to mixed-mandate operations at the international and national levels.

If developmental approaches are to be attempted, the Ugandan case provides a warning as to what will result if efforts are not wholehearted and adequately funded. UNHCR's experience in both Uganda and Kenya in recent years shows clearly that attempting to run any kind of programme in the context of continual financial crisis and endless mid-year budget cutting is likely to result in poor-quality programming of whatever kind. UNHCR staff should not be left in a situation as in Kenya where they are obliged at year-end to report that 'due to budgetary constraints, minimum standards of protection and assistance in Kenya were not reached in 2005'. Similarly, attempts to support the much-advertised SRS in Uganda in the same year were devastated by the suspension of 'crop production, environmental work, animal husbandry and income generation' due to budget cuts.[45] Hardly surprising that self-reliance was not achieved by refugees in the second decade of their exile.

Emphasis is required on the critical linkages between the experience of protracted exile for refugees, the approaches adopted by host states and assistance providers and any eventual set of durable solutions to their problems. This chapter has sought to show that de-linking assistance modes and durable solutions is counter-intuitive and counter-productive in the context of such prolonged exile. Only by re-linking these 'phases' of the exilic experience will each of the stakeholders be able to exploit the continuities which exist between them. As such, this chapter has argued that only developmental approaches to asylum and refugee management in Uganda and Kenya provide an optimal preparation to each and all of the three durable solutions of repatriation, local integration and resettlement in the context of protracted exile. The major obstacles to such an approach being adopted remain the inability or unwillingness of international donors to strongly support and contribute to it (perhaps because it so clearly contrasts with their own actions in industrialized countries) and the extreme nervousness of host states about the economic, security and social consequences of so doing. In response to the latter it is worth emphasizing the extent to which the Ugandan and Kenyan case studies point to the inherent insecurity of refugee camps and settlements, which are frequently themselves accused of bringing insecurity to refugee-hosting regions.

Notes

1. Of the 5 million IDPs in Sudan as of August 2006, 1.8 million come from the western Darfur region while the majority of the remainder come from the South. It is estimated that 1–1.2 million Southerners had returned home since the signing of the Comprehensive Peace Agreement (CPA) in January 2005. See Internal Displacement Monitoring Centre (IDMC) web page: http://www.internal-displacement.org/8025708F004CE90B/ (httpCountries)/F3D3CAA7CBEBE276802570A7004B87E4?OpenDocument&expand =3.2&link=45.3.2&count=10000#45.3.2.
2. The IDMC estimates that, as of June 2006, between 1.7 and 2 million people remained internally displaced due to conflict in northern Uganda. See http://www.internal-displacement. org/8025708F004CE90B/(httpCountries)/04678346A648C087802570A7004B9719 ?OpenDocument&expand=2&link=49.2&count=10000#49.2.
3. T. Kaiser, L. Hovil and Z. Lomo, '"We Are All Stranded Here Together": The Local Settlement System, Freedom of Movement, and Livelihood Opportunities in Arua and Moyo Districts', Refugee Law Project, Working Paper no. 14, Kampala, 2005.
4. Any analysis of the dimensions of forced migration in Uganda and Kenya needs also to take into account Allen's warning; 'Research findings made it clear that in north-east Africa, formal distinctions between refugees, internally displaced, returnees and stayees violate the realities of circumstances on the ground'. Tim Allen, ed., *In Search of Cool Ground: War, Flight and Homecoming in Northeast Africa*, London: James Currey, 1996, p. xii.
5. Douglas Johnson, *The Root Causes of Sudan's Civil Wars*, Oxford: International African Institute in association with James Currey, 2003.

6. UNHCR, 'UNHCR Resumes Repatriation of Sudanese from Ethiopia's Bonga Camp', 14 December 2006, see http://www.unhcr.org/news/NEWS/4581752611.html; UNHCR, 'South Sudan: Return Convoys from Uganda Suspended Following Attacks', 20 October 2006, see http://www.unhcr.org/news/NEWS/45389c577.html; UNHCR, 'UNHCR Resumes Repatriation of Sudanese from Kenya's Kakuma Camp', 22 November 2006, see http://www.unhcr.org/news/NEWS/4564600e4.html.
7. *The Economist*, 'An Island Unto Itself', 5–11 August 2006, p. 47.
8. Douglas Johnson, *The Root Causes of Sudan's Civil Wars*, pp. 145–6.
9. J. O. Akol, 'A Crisis of Expectations: Returning to Southern Sudan in the 1970s', in Tim Allen and Hubert Morsink, eds., *When Refugees Go Home*, Geneva: UNRISD, 1994, cited by J. Merkx, 'Refugee Identities and Relief in an African Borderland: A Study of Northern Uganda and Southern Sudan', *New Issues in Refugee Research*, Working Paper no. 19, Geneva: UNHCR, 2000, p. 13.
10. Early arrivals in Uganda were accommodated mainly in Adjumani district, while subsequent arrivals in the early 1990s were divided between there and Arua, being placed in transit camps until they could be relocated to agricultural settlements in line with government policy at that time.
11. J. Merkx, 'Refugee Identities and Relief in an African Borderland', p. 11.
12. International Crisis Group, 'Sudan's Comprehensive Peace Agreement: The Long Road Ahead', Africa Report No. 106, 31 March 2006.
13. West Nile was plagued by attacks by the West Nile Bank Front from 1996 onwards, which led to further displacement of refugees and others. On the East of the Nile, attacks by the LRA that disrupted refugee life included numerous attacks on refugee settlements in Moyo and Adjumani districts in the late 1990s.
14. A ceasefire agreement between the government of Uganda and the LRA came into effect on Tuesday 29 August 2006, with its most recent extension reaching until the end of February 2007.
15. 'As Peter Kagwanja has explained, Kenya's long-standing apprehension with regard to large refugee influxes is the result of several factors: a chronic shortage of arable land, which comprises only three per cent of the country's territory; … and a more general concern that the arrival of refugees will lead to the spread of firearms, increased levels of crime and social unrest. As a result of these concerns, Kagwanja argues, the colonial and post-colonial Kenyan states have sought to limit the number of refugees on the country's territory and have consistently rejected any suggestion that exiled populations be given land and allowed to settle in the country on a long-term basis'. Cited in J. Crisp, 'A State of Insecurity: The Political Economy of Violence in Refugee-populated Areas of Kenya', *New Issues in Refugee Research*, Working Paper no. 16, 1999, p. 17.
16. UNHCR, *2005 Global Refugee Trends: Statistical Overview of Populations of Refugees, Asylum-seekers, Internally Displaced Persons, Stateless Persons, and Other Persons of Concern to UNHCR*, Annex 2. Figures as of June 2006.
17. These planning figures date from January 2006 and are likely to be an overestimate for officially registered refugees. *UNHCR Global Appeal*, Geneva: UNHCR, 2006.
18. Planning figures as of January 2006. *UNHCR Global Appeal*, Geneva: UNHCR, 2006.
19. Refugee Consortium of Kenya, 'Refugee Management in Kenya', *Forced Migration Review* 16, 2003, pp. 17–19. For further details on abuses of refugee rights in predominantly urban settings in Kenya and Uganda see Human Rights Watch, *Hidden in Plain View: Refugees Living Without Protection in Nairobi and Kampala*, 2002, http://www.hrw.org/reports/2002/kenyugan/.
20. J. Merkx, 'Refugee Identities and Relief in an African Borderland', p. 6.
21. K. Jacobsen, 'Can Refugees Benefit the State? Refugee Resources and African State-building', *Journal of Modern African Studies* 40, no. 4, 2002, pp. 577–596.

22. T. Kaiser, 'Experience & Consequences of Insecurity in a Refugee Populated Area in Northern Uganda', *Refugee Survey Quarterly* 19, no. 1, 2000.
23. G. Verdirame, 'Human Rights and Refugees: The Case of Kenya', *Journal of Refugee Studies* 12, no. 1, 1999, p. 57.
24. A. Jamal, 'Minimum Standards and Essential Needs in a Protracted Refugee Situation: A Review of the UNHCR Programme in Kakuma, Kenya', UNHCR, Evaluation and Policy Analysis Unit, EPAU/2000/05, 2000.
25. T. Kaiser, L. Hovil and Z. Lomo, 'We Are All Stranded Here Together', p. 18.
26. UNHCR, *UNHCR Global Report 2005*, Geneva: UNHCR, 2006.
27. J. Crisp, 'No Solutions in Sight: The Problem of Protracted Refugee Situations in Africa', *New Issues in Refugee Research*, Working Paper no. 75, Geneva: UNHCR, 2003.
28. UNHCR, *UNHCR Global Report 2005*, Geneva: UNHCR, 2006.
29. UNHCR, *Handbook for Planning and Implementing Development Assistance for Refugees (DAR) Programmes*, Geneva: UNHCR, 2005.
30. Research interviews by the author with refugees in Rhino camp settlement, 2004.
31. Internal Displacement Monitoring Centre, *Only Peace Can Restore the Confidence of the Displaced. Update on the Implementation of the Recommendations made by the UN Secretary-General's Representative on Internally Displaced Persons Following His Visit to Uganda*, March 2006.
32. L. Hovil, 'Refugees and the Security Situation in Adjumani District', Refugee Law Project, Working Paper no. 2, Kampala, 2001.
33. See T. Kaiser, 'Participating in Development? Refugee Protection, Politics and Developmental Approaches to Refugee Management in Uganda', in *Third World Quarterly* 26, no. 2, 2005, pp. 351–367.
34. J. Crisp, 'A State of Insecurity', p. 20; E. Aukot, '"It Is Better to Be a Refugee Than a Turkana in Kakuma": Revisiting the Relationship between Hosts and Refugees in Kenya', *Refuge* 21, no. 3, 2001, pp. 73–83.
35. See T. Kaiser, L. Hovil and Z. Lomo, 'We Are All Stranded Here Together'; and L. Hovil and E. Werker, 'Refugees in Arua District: A Human Security Analysis', Refugee Law Project, Working Paper no. 3, Kampala, 2001.
36. J. Crisp, 'A State of Insecurity', p. 9.
37. Ibid., p. 23.
38. A. Tumusiime, 'Riot Police Rushed to Quell Refugee Camp Riots', *The New Vision*, Wednesday 29 May 2002.
39. G. Verdirame and B. H. Harrell-Bond, *Rights in Exile: Janus-faced Humanitarianism*, New York: Berghahn Books, 2005, pp. 121–23.
40. One possible spoiler in relation to the latter is that no resolution has yet been found to the fact that, while the government of Uganda has offered an amnesty to the LRA and its leadership, the International Criminal Court has issued indictments for the arrest of the leadership which it still insists must be implemented.
41. UNHCR, 'South Sudan Funding: UNHCR Thanks US and Canadian governments', 22 September 2006, see http://www.unhcr.org/news/NEWS/4513b48f16.html; UNHCR, 'UNHCR Resumes Repatriation of Sudanese from Kenya's Kakuma Camp', 22 November 2006, see http://www.unhcr.org/news/NEWS/4564600e4.html.
42. UNMIS News Bulletin, 19 December 2006, see http://www.unmis.org/english/2006Docs/PIO-UNMISbulletin-dec19.pdf.
43. The local council system is closely associated with Uganda's national resistance movement, having grown out of 'resistance councils' which were established as the National Resistance Army took territory by force in Uganda in 1986. Both council systems represent a form of local government and governance, operating at different levels from

village to district level. Officers are elected by their peers and wield considerable influence, even at the very local level. One of their responsibilities at village level is to record the arrival of any stranger to the village.

44. T. Kaiser, 'Between a Camp and a Hard Place: Rights, Livelihood and Experiences of the Local Settlement System for Long-term Refugees in Uganda', *Journal of Modern African Studies* 44, no. 4, 2006, pp. 597–621.
45. UNHCR, *UNHCR Global Report 2005*, Geneva: UNHCR, 2006.

13
Bhutanese refugees in Nepal

Mahendra P. Lama

Lhotsampas refugees from Bhutan, numbering over 105,600, have been in seven relief camps in eastern Nepal for the last 17 years.[1] Whilst claims and counter-claims by the Bhutanese and Nepalese governments exist over the nationality of these refugees, they represent a critical example of a protracted forced migration situation. This chapter deals with the case of Nepali-speaking Bhutanese – Lhotsampas refugees – and is divided into four sections. The first section deals with issues related to nationality and state behaviour in explaining the mass exodus of people from Bhutan. The second section examines the issues of refugee management in the camps in Nepal. The third section looks into recent developments including the effectiveness of the ongoing repatriation initiatives and emerging security dynamics. The final section presents a sketch of the road ahead.

Explaining the mass exodus from Bhutan

At the very outset four major questions need to be raised, because these issues have continued to both confound and constrain negotiations for the repatriation of the Lhotsampas to Bhutan. Firstly, are these Lhotsampas both legally and emotionally Bhutanese nationals? Secondly, aside from the technical question of the nationality of these people, what lies behind their expulsion from Bhutan? Thirdly, does the behaviour of the Bhutanese state and its neighbours have implications for international norms relating to forced migration and the management of refugees?

Protracted refugee situations: Political, human rights and security implications,
Loescher, Milner, Newman and Troeller (eds),
United Nations University Press, 2008, ISBN 978-92-808-1158-2

And, finally, what are the likely ramifications for regional security and for the welfare of the refugees if the situation remains protracted?

Questions of nationality

The question of the nationality of the Lhotsampa refugees has been extremely sensitive. Bhutan maintains that they are illegal immigrants and its government has marginalized these people over many decades through restrictive nationality laws, even though official documentation of citizenship has never been routine in the country. In fact, the prime reason behind both their forced expulsion and voluntary movement was the 1985 Citizenship Act and the census exercise of 1988. However, there has been a history of restrictive and discriminatory practices. For example, the government imposed strict adherence to nationality clauses for representation in the *Tshogdu* (National Assembly) set up in 1953 and the right of the Nepali-speaking Bhutanese in southern Bhutan to own property in the 1950s. This was heightened by political initiatives taken by the Bhutan State Congress (BSC) set up in Assam in India to bring a democratic system to Bhutan in the wake of India's independence in 1947. The visit of Indian Prime Minister Pandit Nehru to Bhutan in 1958 and pressure from the BSC largely prompted the Bhutanese authorities to enact the 1958 Citizenship Act that decided to grant citizenship to all ethnic Nepalese domiciled in the country before 31 December 1958. This law declared that any person can become a Bhutanese national if the father is a Bhutanese national residing in Bhutan. In addition to this, if any foreigner who has reached the age of majority and is otherwise eligible presents a petition to an official appointed by His Majesty the Druk Gyalpo and takes an oath of loyalty according to the rules laid down by the official, he may be enrolled as a Bhutanese national provided that the person has been a resident of the Kingdom of Bhutan for more than 10 years and owns agricultural land within the Kingdom. Furthermore, if a foreigner has reached the age of majority and is otherwise eligible, and has served satisfactorily in government service for at least five years and has been residing in the Kingdom of Bhutan for at least 10 years, he may receive a Bhutanese nationality certificate through the same process. Once the certificate is received, such a person has to take the oath of loyalty according to rules laid down by the government.[2]

The oath of loyalty in the Bhutanese politico-cultural context meant something much larger than a simple commitment to the constitution. It meant an emotional integration with the mainstream and ruling Drukpa community, both in terms of accepting their politico-economic superiority and in strictly abiding by their sociocultural norms. This also indicated an acceptance of the consolidation process of the only surviving traditional Buddhist monarchy. These perceptions, as we shall later realize, have in fact been the guiding principles of Bhutanese nationality. In other words, whoever is found to violate these norms in the view of the ruling Drukpas would be liable to penalties, including forfeiting his/her Bhutanese nationality. Among many such grounds of this forfeiture, the three most striking are: a person is registered as a Bhutanese national but has left his agricultural land or has stopped residing in the Kingdom; if a nationality certificate is obtained on presentation of false information, or the omission of information; and if any citizen or national is engaged in activities against His Majesty the Druk Gyalpo or speaks against His Majesty, or the people of Bhutan.

The majority of those who were given citizenship in 1958 were notified by royal proclamation, which was not accompanied by any special certification process, and there is little evidence that the enactment of the 1958 law made any real practical difference to the population.[3] This absence of any paper documentation in the citizenship declaration to each individual or family has resulted in considerable controversy.

The Bhutan Citizenship Act, 1977

The Bhutan Citizenship Act of 1977 made the eligibility criteria for citizenship more stringent. The integration of Sikkim as the 22nd state of India in 1975 and the widely perceived role of the ethnic Sikkimese Nepalese (who constituted over two-thirds of Sikkim's population) in aiding the process of joining the Indian democratic system, thereby abruptly ending the Chogyal's rule,[4] shocked the ruling elites in Bhutan. This was because Bhutan also had more or less the same type of socioeconomic structure as Sikkim, with the ethnic Nepali-speaking Bhutanese playing a significant role. Here also, India had been supporting Bhutan's development and strategic interests significantly and, like the Treaty with Sikkim of 1950, it had a 1949 Treaty with Bhutan which regulated the latter's foreign relations. Against this backdrop it was quite natural for the Bhutanese absolute monarchy to panic. In fact, this was manifested in many ways, including Bhutan's new directions in its relations with countries other than its closest neighbour, India.[5]

Tough conditions were imposed for citizenship, including knowledge of the Bhutanese language and history. But even those who fulfilled these conditions were not assured of citizenship, as the new Act also mentioned that the power to grant or reject an application for citizenship rests solely with the Royal Government of Bhutan (RGB). Hence, not all applicants who fulfilled the formal conditions were necessarily eligible for citizenship.[6]

The oath of loyalty clauses were made more transparent and equivocal this time with the inclusion of clauses such as: 'I owe allegiance only to His Majesty the Druk Gyalpo of Bhutan, I shall abide by and observe the Laws and Rules and regulations of the Royal Government with unswerving reverence, I shall observe all the traditions and customs of the people of Bhutan, I shall not commit any act against the *tsa-wa-sum* of Bhutan (the country, the people and the King) and as a citizen of Bhutan, I hereby take this oath in the name of Yeshey Goempo and undertake to serve the country to the best of my ability'.

Even the citizenship forfeiture clause was made more unambiguous. It stated that 'anyone having acquired Bhutanese citizenship if involved in acts against the King or speaks against the Royal Government or associates with people involved in activities against the Royal Government shall be deprived of his/her Bhutanese citizenship'.[7] On account of geographical and cultural factors, these new criteria requiring knowledge of the Bhutanese language and history in particular, combined with unflinching loyalty to the King, and a literal ban on any democratic rights and liberty were very difficult to follow for the southern Bhutanese.

Bhutan Marriage Act, 1980

With the promulgation of the Bhutan Marriage Act of 1980 (regarding marriage with a non-Bhutanese) the loyalty issue in the concept of nationality deepened further. Bhutanese who were married to non-Bhutanese were now subject to stringent laws relating to their personal lives. These laws prohibited them from employment in the National Defence Department or the Ministry of Foreign Affairs and limited their access to various public services. Again, these were primarily targeted at Indians and some recently migrated Nepalese who otherwise could have gained citizenship by virtue of their Bhutanese spouses. This was a major disincentive for the Bhutanese to marry people from outside their culture.[8] This was particularly a warning to the Lhotsampas in southern Bhutan, for whom seeking a spouse from across the border in India and further beyond in Nepal had become an established practice.

The possibility of some people acquiring citizenship in a fraudulent manner was indicated in the 1958 Law and 1977 Act. The latter mentioned that all *kashos* (royal edicts) with the people which were not granted by His Majesty the King will be investigated by the Home Minister and reported to the RGB. This was even mentioned to the author by the Bhutanese monarch himself during an audience with him in July 1993.[9]

Bhutan Citizenship Act, 1985

The most significant act came in 1985 in the form of the Bhutan Citizenship Act. This was the direct sequel of a three-year census carried out by district officials and village headmen following which 'formal' citizenship identity cards began to be distributed. This census was criticized for its unscientific approach and methods,[10] and was supposed to have given a rough idea to the Bhutanese authorities about the number of 'illegal immigrants', particularly in southern Bhutan.

This new act bestowed citizenship on three grounds: by birth, by registration and by naturalization. In order to qualify for the first case, both parents had to be citizens of Bhutan.[11] However, in the second category of citizenship by registration – to which most of the Indians and many of the Lhotsampas belonged – they had to be permanently domiciled in Bhutan on or before 31 December 1958 and their names had to be present in the census register maintained by the Ministry of Home Affairs. This ministry itself did not exist until 1968 and the records were generally held by the village headmen, although they were not comprehensive and accurate.[12] This implied a sudden and retroactive cut-off year of 1958, a full 27 years back, when the certification procedures were highly unscientific and informal. This inevitably meant mandatory showing of 30-year-old land tax receipts. Interestingly, the cash tax payment system was only introduced in 1964.[13] This clause delegitimized the immigration into Bhutan that occurred between 1958 and 1985. What really astonished the now expelled population was the fact that despite the long-practised conscious policy of strictly regulating immigration through measures like project-specific labour imports through its Public Works Department, a strict monitoring and reporting system run by the National Assembly members, and the strict Acts of 1958 and 1977 related to citizenship, 'illegal immigrants' of this staggering magnitude were detected, identified and later expelled in the early 1990s.

In the third case of citizenship by naturalization, the conditions that at least one parent must be a citizen of Bhutan, the 15–20 years of recorded residence in Bhutan by the person seeking citizenship there, and the solemn oath of allegiance to the King and people of Bhutan are mandatory. What is more seriously discriminating is that the person has to demonstrate proficiency in Dzongkha and knowledge of the culture, customs, traditions and history of Bhutan. The non-Drukpas from both southern and eastern Bhutan regarded these conditions as highly discriminatory. The forfeiture of citizenship laws was also made very stringent. Under them, a person who is deprived of Bhutanese citizenship must dispose of all immovable property in Bhutan within one year, failing which the immovable property shall be confiscated by the Ministry of Home Affairs on payment of fair and reasonable compensation.

Census of 1988

Another census was carried out in 1988 exclusively in the southern districts of Bhutan, which declared almost one-sixth of the population of Bhutan as 'illegal immigrants'. They were mostly Nepali-speaking people. In the eyes of the expelled populace, the now infamous 1988 census was only a ploy that was in effect forcibly used to impose the cut-off year of 1958 for legal immigration.[14] The very *modus operandi* of the census operation was highly questionable on the grounds of its intrinsic objectives, the absence of sound census criteria and the way enumeration was conducted. Before entering into the field of census operation, seven categories of people were identified who were supposed to have been living in southern Bhutan. These categories were defined as: genuine Bhutanese citizens, returned migrants, drop-out cases, non-official women married to Bhutanese men, non-official men married to Bhutanese women, adoption cases and non-nationals. Except the first, all the remaining categories had strong inbuilt provisions for disqualifying a Bhutanese resident from citizenship and declaring him or her as an 'illegal immigrant'. Since the census operation specified 1958 as the cut-off year it was natural for many genuine and long-settled Bhutanese nationals not to have the exact proof the census officials were purportedly looking for. The census officials came down heavily even on the genuine Bhutanese. Many Lhotsampas who were declared as citizens by district officials under the provisions of the 1958 Nationality Law found these declared null and void unless they could produce documents proving at least residence and often land ownership prior to 1958.[15] There are rampant allegations of aggressively partisan behaviour by the census officials, who were mostly from the north.[16]

Though it remains inexplicable by any demographic extrapolations, the results of the census exactly matched the expectation of the Bhutanese authorities that there had been a planned and systematic infiltration by the immigrant Nepalese.[17] Bhutan found that these 'illegal immigrants' used multiple methods of infiltration including matrimony; reverse adoption; acquisition of land and housing; working as orange/cardamom porters and farm hands; falsification of documents; displacement; enrolment in schools, and intimidation, bribery, force, etc.[18] This census left a large number of genuine Bhutanese citizens also in the category of 'illegal immigrants'.

A petition to the King by two eminent Lhotsampas (both members of the Royal Advisory Council) in April 1988 to review the entire operation under the 1985 Act and 1988 Census was declared seditious.[19] This led to widespread protests and demonstrations and many violent incidents in southern Bhutan. They were crushed with unprecedented brutality by the Bhutanese armed forces. All these dissident activities were declared 'anti-national' and the participants branded as 'terrorists'.

Beyond the issue of nationality

There must be something deeper and more malignant than the issue of nationality. Otherwise at least the genuine Bhutanese Lhotsampas would not have been expelled or left the country on their own. Besides the much-trumpeted nationality issue, which supposedly remained weak and ineffective even after so many amendments, the real issues behind these mass expulsions are located in the very structure and orientation of the present-day political economy of Bhutan.

The costs of an open border

Historically, Bhutanese Nepalese have been primarily a farming community with their demographic settlement concentrated in the southern districts. The fertile southern districts have been the backbone of Bhutanese economic progress. These districts have other distinct advantages in the form of better transport and communications, easy accessibility to the large markets in the border hinterland of 720 kilometres and the ready availability of a cheap labour force in the labour-scarce economy of Bhutan. These are the districts where most of the hydropower potential is concentrated.[20] All the major commercial centres and industries such as hydro-, mineral- and timber-based are located in these southern districts and have contributed hugely to national income.[21]

Three new trends emerged in the economic profile of southern Bhutan. Firstly, the traditional barter trade with bordering Indian townships largely acquired the form of a robust economic exchange with ever-expanding markets. Secondly, the concept of income and employment concentration crept in among the main farmers and the land-owning class. And thirdly, the demand for cheap labour increased. All the major actors in these newly emerging trends were either Lhotsampas, recently immigrated Nepalese or Indians from across the border. The other Bhutanese nationals, particularly from the north, had a very marginal role in the enactment and sharing of this economic take-off.

Here lay the economic genesis of the conflict. The intimately established economic exchange between the southern Bhutanese and the people on/across the border in West Bengal and Assam had given the farmers direct exposure to market forces. 'Owing to Bhutan's free access to the nearby large Indian market ... 94% of Bhutan's trade is with India but due to an open frontier and inadequate customs administration there is no accurate account for flow of goods and services between the two countries'.[22] In the process, they had rendered the entire state apparatus of procurement, supply and stocking redundant, as with the State Trading Corporation and the other export channels of Bhutan. This had serious long-term political implications and this process had to be reversed. The RGB thus gradually tightened its policy and compulsorily diverted economic exchanges through its State Trading and Food Corporation. The main farmers, along with their Indian intermediary partners, were hit hard and they thus started nursing a revengeful grudge against the system.[23] It is in this very context of fragmenting and de-concentrating the burgeoning power equations in the south that a proposal to resettle Lhotsampas in the northern *dzongkhags* (districts) was discussed as early as the 67th Session of the National Assembly in 1988.[24]

When feudal interests clash

Meanwhile, in a perpetual feudal mindset where clan ascription and the order of heirs predominate, the concentration of wealth outside the ruling Drukpa clans of the west became simply intolerable. Naturally, here the feudal remnants from the west were in direct collision with the new and upcoming symbols of wealth concentration in the south. This was manifestly clear in other sectors. Bhutan, where the private sector simply did not exist, had to accommodate increasing numbers of educated people from the south in the government sector only. As a result, the composition of elites in Thimphu started changing, which threatened the traditional elites, particularly the inward-looking Drukpas.

Some of the Drukpas whom the author interviewed were quite perturbed by the manipulative skills of upper-caste Lhotsampas (mainly Bahuns and Chettris) in crucial social and political arenas. At a particular point in time, the Bhutanese economy (a least developed country) witnessed a scenario of too many people chasing too few goods. This was where the issues of nationality and loyalty were mobilized in a more crystallized form so that the elites with most-favoured nationality status could be defined on a singular basis of the son-of-the-soil principle.

Illegal immigrants: a reality

For poverty-stricken and environmentally satiated rural Nepal (and also for states like Bihar, Orissa, West Bengal and North East India), outward migration has been a tremendous relief. This has been matched by a fertile and increasingly accessible southern Bhutan that needs a cheaper and more amenable workforce for the commercialization of agriculture. The Nepali immigrants, with similar socio-anthropological features, were further assisted by the provision of unrestricted cross-border movement in the India-Nepal Peace and Friendship Treaty of 1950. However, the Bhutanese Foreign Minister has remarked: 'But the majority of the illegal immigrants entered Bhutan after we started the process of socioeconomic development in 1961, when we launched our first five year plan ... We needed a lot of labour and Nepalese labour was easily available and very cheap. Unfortunately, many of them decided to settle down in Bhutan and that is the genesis of the present problem'.[25]

This brought about three sources of instability. Firstly, since the legal procedure to absorb immigrants as citizens was tedious and cumbersome, most of these illegal immigrants started ignoring it. Secondly, a new migrant settlement emerged which was socioculturally and politically alien to mainstream Bhutan. Thirdly, a clash of interests occurred both between the Bhutanese Nepalese and 'illegal immigrants', and also between patrons of illegal immigrants within Bhutanese Nepalese and the Drukpa elites in Thimphu.

Attraction of democracy

For an ever inward-looking nation like Bhutan with a historically institutionalized monarchy, the Indian form of democracy always looked unsuitable and threatening. However, for a socioeconomically dominant section of southern Bhutanese who invariably interacted with the people across the border, the vibrant-looking democracy has always been a lure. This took the form of community articulation and more conspicuous political assertion among the southern Bhutanese. The success of regionalized movements across the Indian border in Assam by the All Assam Students' Union and in Darjeeling by the Gorkha National Liberation Front and the smooth transition of Sikkim to the Indian brand of democracy further encouraged southern Bhutanese to assert the values of pluralism and democracy.[26] The proverbial last straw was provided by Nepal becoming a multi-party democracy in 1990 after three decades of Panchayati autocracy.

Sociocultural degeneration

Behind the façade of cultural conservation, a steady degeneration in both the natural and cultural heritage of Bhutan has also been recorded.[27] Modern communication techniques and the sudden exposure to alien culture made ordinary Bhutanese search for more liberal social and political norms, despite the rigidity and fastidiousness of the authorities. In order to restrict the further slide the state sponsored a cultural resurgence through the policy of *driglam namzha* (traditional etiquette) which mainly intends to inculcate 'respect for authority and a hierarchy that promotes the interest of the society and the nation'.[28]

For the southern Bhutanese it was suffocating and impossible to wear *kho* and *kira* in summer temperatures of 30–38 degrees Celsius.[29] The police clampdown on the offenders further embittered the good intentions behind *driglam namzha*. But this was just a practical/social resistance; the political resistance was more acute as it consciously promoted exclusionist nationalist ideology. Their apprehension was further confirmed and consolidated by many other state actions such as the withdrawal of the Nepali language from the school curricula in the south in 1989, and the mandatory requirement for Lhotshampa to hold a 'No Objection Certificate' (issued by the police to confirm that the bearer has no involvement in anti-national activity) for purposes such as school admission, employment, training, travel documents, trade and industry licences and for farmers to access their earnings from the cash crop.[30]

Issues of refugee management in the camps in Nepal

Categories of refugees

Though the RGB has its own version of the exodus, accounts given by refugees, interviews with refugees' leaders, reports by various human rights organizations and media reports lead us to conclude that the exodus of refugees from Bhutan has been broadly in three categories:

(i) The expulsions of illegal immigrants and even genuine Bhutanese Lhotsampas who 'did not qualify' under the 1988 Bhutan Citizenship Act. A large number of refugees were coerced into signing so-called 'voluntary migration forms', which stated that they were selling their land and leaving the country of their own free will. This was done with the threat of large fines or imprisonment if they failed to comply. This led to people fleeing their villages after a distressed sale of their properties.[31]

(ii) The expulsions of Lhotsampas who were supposed to have been involved in activities that were against the *tsa-wa-sum* of Bhutan. They are often referred to as *ngolops* (anti-nationals) in official statements and documents. In some instances, whole village blocks with a number of recognized Bhutanese families were forced out *en masse*, apparently in retaliation for a robbery or an attack on a local government official by 'anti-national elements'.[32]

(iii) Those who escaped ethnic violence under perceived fear of persecution by the state machinery. 'We were confronted with a stark choice. Either stay in Bhutan and die at the hands of the King's army or flee' said a refugee.[33] On the other hand, the RGB brazenly alleged that many innocent and illiterate farmers were persuaded or forced to leave because of the 'attractive international hospitality' in the camps and threats by terrorists.[34] In totality, it now appears that Bhutan accomplished the goal of 'ethnic cleansing'.

Phases of exodus

The first phase of 1990–1992 witnessed constant violence and counter-violence, including arbitrary arrests, torture, rape, killings, and detentions in horrendous conditions without trial. People were uprooted regardless of their status, mainly depending on predetermined geographical targets and localities. This phase produced the maximum number of refugees. This phase, however, came to an end when in January 1992 a royal decree was issued making it a criminal offence to force any genuine citizens to leave the country. In the second phase of 1993–1995, the exodus took a slightly subtler form, with the security forces and local authorities resorting to forcible eviction, destruction of houses, denial of public services and harassment and intimidation on a selective basis. A large number of senior Lhotsampa civil servants resigned. In the third phase of 1996–1997, the emphasis was on consolidating the gains of 1989–1995 by strengthening the methods of forcibly injecting fear and submissiveness. This phase also witnessed the intimidation of the remaining dissidents, by way of various administrative punitive actions and sociocultural restrictions, including continuing census operations in southern Bhutan with the same intentions as previously, the relocation of certain Lhotsampas and the settlement of northern Bhutanese in the south, restrictions on the physical movement of Lhotsampas and deploying senior Lhotsampa civil servants to humiliating posts and compulsorily retiring them.

Camp situations: major concerns

When the refugees first arrived in Nepal, they spontaneously located themselves in Maidhar, Timai and Sanischare. The Nepalese authorities housed them in harsh and overcrowded camps. Nepal had had no previous experience of refugee influxes except a few thousand Tibetan refugees who arrived in the 1960s and 1970s. The overall responsibility for camp administration has been with His Majesty's Government (HMG). The National Unit for the Coordination of Refugee Affairs (NUCRA) and the Refugee Coordination Unit (RCU) in Jhapa district was set up by the Ministry of Home Affairs function under HMG. UNHCR was invited to assist HMG in September 1991. The Lutheran World Service (LWS) was already there in the field, and the World Food Programme (WFP) commenced its operation for the distribution of food assistance in early 1992. By mid-1992 a large number of NGOs, both local and international, were aiding the refugees.

There are several issues that have acquired serious dimensions both within and outside the camps. Firstly, the warning from the UNHCR that 'there is no guarantee of continued assistance indefinitely since UNHCR is totally dependent on voluntary contributions' has steadily raised the level of gloom, despondency and frustration in the camps.[35] Secondly, the camps are cramped and crowded. A small hut accommodates at least eight people. One refugee commented: 'We have to survive on a small amount of food provided by various organizations ... due to widespread frustration and depression in the refugee camps, a number of young women have committed suicide while a number of young men have lost their mental balance ... we are prevented from working outside the camp. We are kicked out from the job as soon as the employer comes to know that we are Bhutanese refugees alleging that we are snatching away the job opportunity of the Nepalis'.[36] A sharp increase in domestic violence has been recorded and local newspapers have also reported incidents of prostitution. Thirdly, health problems have been a major concern for the management agencies. Finally, there has been a serious impact on the local forest resources, as evidenced by the following comment: 'When we came here first this was a big forest and grazing place. There is nothing now.'[37]

Repatriation negotiations

High-level delegations from UNHCR, the EU, the US and many other countries have visited Bhutan, Nepal and India to negotiate the return of these refugees. Refugee organizations have raised their plight in UN Commissions and Sub-commissions, and a range of other international bodies. In most of these meetings the RGB expressed its firm commitment to find solutions within the framework of the bilateral talks with HMG.

Although a six-member Joint Ministerial Level Committee (JMLC) of Bhutan and Nepal has been meeting since July 1993 – and has now reached 15 rounds – repatriation issues have become more complicated. A solution remains as elusive as it was in 1993. The process of identification of refugees was conducted on the basis of four 'mutually agreed' categories: i) bona fide Bhutanese if they have been evicted forcibly; ii) Bhutanese who emigrated; iii) non-Bhutanese people; and iv) Bhutanese who have committed criminal acts. This process started in March 2001 and it has been tedious, slow and frustrating.

There are widespread allegations that many refugees are not genuine Bhutanese citizens. RGB also alleges that many of them are Nepalese who moved into refugee camps for economic reasons. In fact, Bhutan has stressed this point at every JMLC meeting. When the hordes of people with children on their backs and in their arms arrived in very harrowing circumstances, the first reaction of the Nepalese authorities was, as expected, a purely humanitarian act of providing them with food, medicines and shelter, and they were accepted in various camps. Nobody knew that their political and nationality status had to be verified and confirmed as refugees under the defined clauses of the 1951 UN Convention and 1967 Protocol. Nepalese authorities at the district level revealed that they did not know about these procedures.[38] UNHCR admits that the large majority of those who arrived at the Nepali border prior to June 1993 and who claimed to be refugees were recognized *prima facie* and admitted to camps, without undergoing individual refugee status determination. Since June 1993 this status determination has been carried out by the NUCRA and RCU in collaboration with UNHCR.[39]

The form to be completed by the refugees at the camps during the verification process was proposed by the Bhutanese side and approved by the JMLC in the third round of talks. Very few refugees would be able to withstand this screening. This is what was clearly evident from the last verification process. For instance, out of the 10 different sections in this form to be completed by the refugees, section 8 deals with the forcibly evicted category. Refugees belonging to this category have to answer five questions, namely: a) the date of eviction; b) the authority by whom the eviction was carried out (civilian official, military, police or others); c) whether they have any proof of eviction; d) whether any appeal was made to a higher authority and if so, to whom? If not, why not?; and e) any other relevant details. How can illiterate refugees fleeing such life-threatening situations give exact accounts in this manner? The soundness and purposefulness of such categorization were not accepted by international observers or refugee representatives, yet it was made operational. Furthermore, the Joint Verification Team (JVT) did not have representation from the refugees or even from the representatives of the villages from which they had fled.[40]

As widely expected, the first verification exercise conducted by the JVT that verified 12,183 refugees in Khudnabari camp (started in March 2001, results declared in June 2003) found only 2.4% (293 persons – 74 families) to be bona fide Bhutanese citizens who had been forcibly evicted from Bhutan and had the right to return with full citizenship. Of the rest, 8,595 (70.5%) were Bhutanese who had left Bhutan voluntarily, thereby losing their citizenship in accordance with Bhutanese laws, 2,498 (24.2%) were non-Bhutanese and 347 (2.8%–85 families) were refugees who had committed criminal acts. It is interesting to note that the team found people from the 'anti-national' and 'criminal' categories in many cases from within a single family, including children: 'Devi Poudel, age 8 of Hut 9, Sector A of Khudnabari is placed in fourth category ... 18-month-old Kiran Gautam is also placed in the same category. Nima Dorjee Tamang is classified as non-Bhutanese and his own brother Lakpa Dorji Tamang placed in voluntary migrant category'.[41]

Of those verified, 94% appealed the results. Despite the assurance given by the Nepalese government, no appeal process with international legal standards was put in place. More seriously, besides the very unscientific manner in which these verification exercises were conducted, the way the Bhutanese members of the JVT unilaterally announced the harsh conditions for prospective returnees badly upset the refugees. Under the repatriation arrangement, category 1 refugees (citizens wrongly evicted) could return to Bhutan but would get neither rehabilitation support nor their confiscated or usurped properties back. Similarly, category 2 refugees (who had left voluntarily) were to be placed in refugee camps in Bhutan with an employment provision of one person per family. They could regain their citizenship after eight years provided they pass an examination to prove loyalty to *tsa-wa-sum*. This implied even less freedom than they had enjoyed while living in exile.[42] Naturally such blatant injustice led to protest among the refugees and in reaction they threw stones. In the name of the security threat to the Bhutanese JVT members, this exercise has been indefinitely postponed since 23 December 2003.

There has been widespread protest against the very process of verification. The Bhutanese Refugee Women's Forum demanded that 'the refugees should be categorised only as Bhutanese and non-Bhutanese',[43] and also sought the involvement of UNHCR in the status determination and repatriation process. The United States has pleaded for the 're-examination of the verification report' and mentioned that 'the US government is also concerned about the absence of guarantees provided to Bhutanese refugees returning to their homeland'.[44]

The JMLC's performance has been lacklustre and bureaucratic. The Home Minister of Nepal recognized the irregularities and injustice of the process and demanded the setting up of an independent panel to recategorize the refugees. Bhutan does not agree to this proposal. With no real international pressure and India's seemingly neutral behaviour (Nepal has invariably taken it as a pro-Bhutan policy), Bhutan is apparently buying time to see the controversy die a natural death.

For Nepal, which has been thrown into unprecedented political instability, the Bhutanese refugee issue has always been a low priority. Every change in government in Nepal has resulted in a change in its position on the problem, said a cryptic editorial in *Kuensel*.[45] The Nepali method of negotiation is incoherent and unclear and at each stage of negotiation its position has been compromised.[46] There is neither a group to monitor the refugee situation in the Foreign and Home Ministries, nor a mechanism to review the state's policy on the issue. Nepal failed to involve refugees and UNHCR in the negotiations for repatriation and also wrongly perceived that India would intervene. These mistakes proved quite costly to Nepal and to the refugees.[47] Today the issue is essentially only discussed between Bhutan and Nepal whereas it should have been between Bhutan and the Bhutanese refugees.

Recent developments

There have been five major recent developments. Firstly, UNHCR declared that it would start phasing out its assistance to the refugees in the camps by the end of December 2005. Secondly, the army action against the Indian militant camps in Bhutan has on the one hand further distanced India from any participation in the refugee return process, and, on the other, mobilized the anti-monarchy forces more vigorously. Thirdly, Bhutan has blatantly shown its unwillingness to accept the refugees back by withdrawing from the negotiation process, by initiating the settlement of northern Bhutanese in the lands and villages of Lhotsampa refugees in southern Bhutan, and by again denouncing the refugees as terrorists. Fourthly, there are definite indications that Bhutan had to undertake reforms in its political system in the face of changing global opinion and constant pressure. And finally, some members of the CORE Group including the US and Canada have offered to resettle over 60,000 refugees. UNHCR and the Nepal government completed a systematic re-enumeration exercise in the camps in May 2007 that confirms the population of registered Bhutanese refugees.

India's role

Interestingly, India's representative on the Executive Committee of UNHCR (ExCom) made a very direct attack on the UNHCR's proactive role in the crisis. India, while blaming UNHCR for creating the current situation, stated that the 'misguided approach adopted by the UNHCR through the funding of the camps, has led to the creation of a vested interest in their perpetuation'. He further remarked that 'the fact that the inmates have a better standard of living than the local population in surrounding areas itself serves as a magnet which creates its own problems'.[48] This highly subjective statement by India on better living conditions in the camps was a mere repetition of Bhutan's decade-old stand. Bhutan has always wanted to quickly demolish camps to avoid constant international embarrassment. In contrast, anyone visiting a camp site would realize how devastating it has been for the refugees to remain in the camps. The social indicators are horrendously dismal. Most of the refugees had homes back in Bhutan.

Though Nepal has been demanding India's intervention in finding a lasting solution, the latter has consistently remained aloof, saying it is a bilateral issue between Bhutan and Nepal. India also hosts a number of Bhutanese refugees but they are not in refugee camps. There is ample evidence to prove India's consistent apathy towards the refugee problem: the complete absence of Indian relief participation in refugee camps; the arrest and imprisonment of refugees going back to Thimphu via India under the banners of the Appeal Movement Coordinating Council in 1996, the Bhutan National Democratic Party in 2005, and finally the Bhutanese Movement Steering Committee in 2007; and the arrest and detention of Rongthong Kuenley Dorji, Chairperson of the United Front for Democracy in Bhutan and Druk National Congress in Delhi in April

1997. Why is the government of India worried about having a democratic regime in Bhutan? Does democracy in Bhutan bring insecurity to India?

This is unlike the way India treated the Tibetans in 1960s, the East Pakistani (now Bangladesh) refugees in 1971–72, the Sri Lankan Tamil refugees over the last two decades, and the Chakmas from the Chittagong Hill Tracts of Bangladesh in the 1980s and 1990s. The critical and strategic difference is that, at least in the three cases above, the affected state governments (West Bengal, Tamil Nadu and Tripura) had readily accepted the refugees, unlike the case of the Bhutanese refugees.

Bhutan and Nepal do not have a common border, so the first country of asylum is India. The refugees did attempt to settle down in the bordering Indian districts of Jalpaiguri and Darjeeling, but they were physically removed and chased away by the security forces of West Bengal. This blatant violation of international norms – including the fundamental principle of non-*refoulement* – and parochialism by the ruling elites in West Bengal are widely discussed by the refugees.

India has remained indifferent and undeterred by the appeals made by refugee leaders, the international community and organizations including the European Union. It apparently has a much larger stake and interest in Bhutan than it has in making the case for the repatriation of refugees. India's strategic interest in Bhutan, which serves as a buffer between China and India, is of paramount consideration. Bhutan falls within India's 'security umbrella'. In addition, India has a solid stake in the hydroelectric power resources of Bhutan, which has served as one of the cheapest and most reliable sources of power supply in recent decades.

Final uprooting

There has been a very conscious policy to resettle people from the north and eastern zones to southern Bhutan on the land belonging to the Lhotshampas. The tenor of proceedings in the 71st Session of the National Assembly in 1992 clearly showed this when it stated: 'if landless people from other *dzongkhags* were re-settled on the vacated land in the south, it would generate a greater sense of security among the local people and neutralize any plan by the emigrants to return and claim the land they had sold and abandoned'.[49] In fact, as the South Asia Forum for Human Rights (SAFHR) has observed: 'The first resettlement programme was carried out in the Samdrup Jongkhar district in the Bhangtar sub-division of the Bakuli block, with 58 ex-Royal Bhutan Army families. From 1998 the RGB has undertaken a massive land resettlement programme in six southern districts. This is reported by both the refugees with whom we have been interacting in the camp sites and also independent human rights organisations and other NGOs ... During the 76th National Assembly session held in August 1998, the Bhutanese government announced that around 1,027 households from the north and east of Bhutan had been rehabilitated in southern Bhutan'.[50] Some of the lands have been given to army and police officers or their relatives.[51] There is also a strong coercive method used in this process.

A government announcement mentioned that 'landless people from other *dzongkhag* who got a land allotment in Tsirang *dzongkhag* under the resettlement programme have failed to report despite repeated requests from *dzongkhag*. Therefore, Tsirang *dzongkhag* administration once again requests them to report immediately as the cultivation season has already set in. Non-compliance shall be viewed very seriously and *dzongkhag* administration shall not be held responsible if any complication arises in future on the matter'.[52]

Political reforms

The end of the 1990s saw some significant changes in the Bhutanese political system. The refugee issue exposed problems within the Bhutanese political and socioeconomic structure in an unprecedented manner. There has been exhaustive discussion of the Bhutanese citizenship laws, the discriminatory population census, the imposition of traditional etiquette including dress codes, and severe and silent human rights violations at both the regional and international level. More critically, a large number of international human rights organizations and donor agencies started imposing severe conditions on Bhutan with regard to the treatment of the Lhotshampas. An increasingly visible chasm between the ruling class, from Ngalongs in the west (invariably referred to as Drukpas) and other ethnic groups in Bhutan like the Sharchops in the east, the Kheng, Brogpa, Doya and Tota, and globalization-led information deluge have further facilitated these political changes.[53]

The King's *kasho* (royal edict) of 1998 announced three-fold political changes: i) all cabinet ministers should henceforth be elected by the National Assembly; ii) the Assembly should decide on the role and responsibilities of the cabinet; and iii) the *Tshogdu Chhenmo* (National Assembly) should adopt a mechanism to move a vote of confidence in his Majesty the Druk Gyalpo. Tek Nath Rizal, a leader of Bhutan's democracy movement and a 'Prisoner of Conscience', was released along with 200 other prisoners (including 39 political prisoners) on 17 December 1999.[54] The first 'democratic' constitution of Bhutan has been framed. King Jigme Singye Wangchuck has now abdicated the throne and handed it over to his son Jigme Khesar Namgyel Wangchuck.[55] After the two-party nationwide election was over in March 2008, a new Druk Phuensum Tshogpa party led by Jigme Thinley formed the government. Two private organizations are for the first time allowed to publish newspapers like the *Bhutan Observer* and *Bhutan Times*.[56]

Security ramifications

This protracted refugee situation has generated four major security dynamics. Firstly, the security of the camps itself has been a major concern. Well before the agreement was reached between the Seven Party Alliance and the Maoist forces in Nepal in April 2006, the widespread conflict between the Maoist and the government forces had seriously affected humanitarian access to the refugee camps. The security situation in and around the camps was exacerbated further by the withdrawal of the police in 2003. The UNHCR then encouraged the Community Watch Teams of refugees. In addition to the disruption of the east–west Mahendra highway several times,[57] recently two refugees were killed at Beldangi camp in a police attempt to control a massive showdown between those who favoured third country resettlement and those who did not.[58]

Secondly, the lurking fear that peace marchers might clash with the security forces of Bhutan and India, potentially undermining what little progress that has been made in the negotiations, has further gained ground after the revelation in the National Assembly that Bhutan has officially started training its own brand of militia troops to tighten security in the south. The Army Chief of Bhutan, Goongloen Lam Dorji, reported in the 73rd National Assembly that between 1991 and 1995 a total of 38,230 volunteers had come forward to serve in the militia. Since then, 9,895 men and 48 officers from the civil service had been trained. He also reported that the total expenditure incurred for the training and deployment of the militia amounted to Nu. 280.3 million. Quoting him, *Kuensel* mentions, 'while the expenses were considerable, I am happy to report to the National Assembly that we have been able to cover it with assistance from the government of India and by utilizing the army welfare fund of more than Nu. 300 million held by the Finance Ministry and we have thus been able to meet all the expenditure without applying any burden on the national exchequer or the planned budget'.[59] The attempt by thousands of refugees to cross the Nepal–India border in order to march to Bhutan in May–June 2007 turned violent as the marching refugees clashed with Indian police.

Thirdly, India's hesitant response could also be partly because of the presence of Indian militants in Bhutan, which the latter could have used to foment other problems if the former had showed any inclination to press for the return of refugees. However, the army campaign in December 2003 against the militants hiding in the jungles of south-eastern Bhutan for the previous 12 years brought forward an array of sensitive issues.

These militants mainly belonged to the United Liberation Front of Assam (ULFA, fighting for a separate nation), the Kamtapur Liberation Organization (KLO) and the National Democratic Front of Bodoland (NDFB, fighting for a separate statehood outside Assam). It is also widely believed that the Bhutanese government encouraged them to enter Bhutan. A senior retired Indian security official wrote:

> When the BdSF [Bodo Security Force] and later ULFA asked the Bhutan government for permission to set up camps in the remote jungles close to the Indian border, they gladly permitted it. The ULFA and the BdSF became the unofficial border guards of the Bhutan government against the Nepalese reinfiltration. This was confirmed to me by several top-level sources from Bhutan ... The ULFA and probably the NDFB invested a lot of their funds collected from extortion in Bhutan. The businessmen of Bhutan benefited from this and naturally lobbied for them with their government ... The strength of ULFA cadres in Bhutan at any one time was nearly 1000. By this time the link with the Bhutan government was fully established. The ULFA were using the diplomatic bags of the Bhutan government for sending money to their contacts abroad. The ULFA managed to go to Tibet and possibly China from Bhutan ... In fact an official from the Bhutan government accompanied the ULFA leader to Tibet to meet Chinese officials for purchase of arms ... The second consignment of arms was to be delivered on the Tibet border at Tremo La in Chumbi valley. A senior ULFA cadre went to this remote outpost to take delivery of the consignment accompanied by a senior official from the Bhutan Foreign Ministry. At the last moment the Bhutan government asked their officer not to accept the consignment as the Indian Embassy in Bhutan got wind of the deal.[60]

Finally, the radical elements have noticeably increased in the camps, and refugee political leaders have no control over them. As UNHCR has stated, 'Refugee parents have expressed concern surrounding the nature of such youth activities'.[61] The Bhutan Communist Party (Marxist-Leninist-Maoist) is already said to be mobilizing refugees to launch an armed struggle to topple the government of Bhutan. Given the blatant anti-monarchy slant of the Maoists, it may prove to be a problem for the Bhutanese monarchy. The recent bomb blasts in Bhutan during election time and the arrest of Bhutanese refugees in Siliguri (both in April 2008) after a fatal blast in a residential locality indicate that militancy is already gaining ground.

The road ahead

Bhutan, and to a large extent India, never expected that the refugees would remain in the camps with so much perseverance. The recent widespread protest generated by the UNHCR's proposal of third country resettlement does show that the Bhutanese refugees are determined to remain in the camps. Refugee camps remain widely politicized. The recent developments in Nepal have substantially changed the situation and may ultimately bring about a strategic shift in the attitudes of various actors involved, including the governments in India, Bhutan and Nepal, as well as UNHCR, thereby hastening the process of achieving a durable solution. Some of the refugee leaders openly told the author that 'Bhutan is making another set of Lhotsampas from among the remaining 25% of the Nepali-speaking Bhutanese population ready to throw them out of the country'.[62] The following scenarios and policy suggestions can be made.

Bilateral process: finalization a must

Recently, major political parties have come together to put pressure on all the authorities concerned.[63] A series of mass movements have been launched. This joint front, the Bhutanese Movement Steering Committee, launched in February 2006 and led by Rizal, exhorts total repatriation and declares 1985 as the cut-off year for deciding the citizenship status of Lhotsampas. The Bhutan Press Union (BPU), a united forum of Bhutanese journalists in exile set up in 2002, is vociferously campaigning for repatriation and advocating democracy in Bhutan.[64] Young students have become vocal and their expressions radical. Their mouthpiece, entitled 'Vidyarthi Pratirodh', calls upon the Bhutanese refugees to launch a massive protest movement in the camps, Kathmandu, New Delhi and other parts of India, and inside Bhutan.[65] Attempts to cross the India–Nepal border are likely to be frequent and the mode of crossing could also change drastically. These types of mobilization and formations are new. All these could change the course of the movement of the refugees towards their repatriation.

Therefore, it is essential for the Nepal–Bhutan bilateral process to undertake a final round of negotiations to find a durable solution in a timely manner. If finalization of a negotiated solution within a specific timeframe is not agreeable between these two parties, this protracted bilateral process should be terminated.

SAARC-level commission for determination and repatriation

Given the very strong opposition to the existing categorizations of refugees, the alternative is to categorize them into Bhutanese and non-Bhutanese and work in a timely manner to determine their exact status, on a family basis rather than an individual one. A much more convincing effort would therefore be to appoint an independent South Asian Association for Regional Cooperation (SAARC)-level Commission to examine all the issues of identification, determination and repatriation in a timely manner. The commission's findings should be accepted by all interested parties. UNHCR should be involved in the entire verification process and also in the dignified reintegration of those repatriated to Bhutan.

The inevitability of India's pivotal role

India's intransigence in this humanitarian crisis has actually emboldened Bhutan to prolong the crisis. This is very aptly indicated by the absence of even a word about this protracted crisis in the 85th Session and the portrayal of the entire refugee community in the camps as terrorists in the 86th Session of the National Assembly. Foreign Minister Khandu Wangchuk said that 'the camps are infiltrated by Maoist elements, and radical parties are formed with the declared objective [of] carrying out armed struggle to overthrow the government of Bhutan. Allowing the highly politicized camp people into Bhutan would mean importing ready-made radical political parties and terrorists to duplicate the violence, terror and instability the Maoists have unleashed in Nepal'.[66]

In the same way the refugee arrival in the early 1990s created a stir among the local authorities in West Bengal, the organized attempts to cross the India–Nepal border by the refugees could similarly become a serious issue in a very sensitive locality. In a meeting held between the Foreign Minister of India and the Chief Minister of West Bengal in the aftermath of the latest attempt to cross the border by the refugees in June 2007, India for the first time called it an 'international problem' rather than a 'bilateral problem'.[67]

However, there is also an undercurrent of opinion that if the (Maoist) Communist Party of Nepal remains at the helm of affairs after the recently held Constituent Assembly election, it could completely change the parameters of Nepal–India and Nepal–Bhutan relations. In such a situation the government of India may intervene and ask Bhutan to provide a durable solution to this problem. The fact that the Maoists have strong support among Indian left-wing parties would further complicate the situation.

Third country resettlement: a partial yet precursory approach to a durable solution

The offer of third country resettlement has been a welcome relief to the hapless refugees in the camps. Though there have been several instances of physical threats and intimidation against those who have expressed their support for this move, the majority in the camps see it as a major step towards a durable solution. Many pragmatic refugees have started imagining and concentrating on a 60:40 resettlement and repatriation solution. Those in favour of resettlement, largely led by an advocacy platform called the Bhutanese Refugee Durable Solution Coordination Committee, indicate that they could actually be a more robust support group for the democratization of Bhutan as a diaspora. Some of them are considering a 'government in exile'. However, in the absence of adequate information flows about the direction, nature and content of resettlement offers and options, the refugees remain confused and prone to rumours that could lead to serious disturbances within the camps. A second group of the refugees has already gone abroad for resettlement. Though all the countries that have offered resettlement options are signatories to the 1951 Convention and 1967 Protocol, there is still very little information flow about the destination countries, the unit (individual, family or camp) of resettlement and the actual conditions of resettlement. There are three clear ways forward in this regard. (i) an extension of voluntary choice to the refugees for resettlement, with a formal and prior bestowing of Bhutanese citizenship. This would mean refugees willing to opt for resettlement would also have Bhutanese citizenship so that they could go back to Bhutan whenever they desired. (ii) The resettlement process should be carried out in a transparent manner within a specific timeframe and without any major hold-ups in the sending process. (iii) A broad assurance to the refugees not opting for resettlement about their well-being in the camps until they are repatriated to Bhutan, and transparent conditions for their repatriation. This means a negotiated, durable settlement of the entire Bhutanese refugee issue before the resettlement process is actually set in motion.[68]

The King and the royal prerogative

The young King of Bhutan could be persuaded to use his royal prerogative to provide a one-time amnesty and dignified settlement of all the refugees as a basis for national reconciliation. The King could issue a *kasho* to this effect as a previous king did in the past. For instance, as per Section 4 (c) of the 1958 National Law, 'if any person has been deprived of his/her Bhutanese nationality or has renounced Bhutanese nationality, forfeited his/her Bhutanese nationality, the person cannot become a Bhutanese national unless His Majesty grants approval to do so'. This implies that the King has the ultimate prerogative to give a reprieve to at least 75–85% of the refugees – those belonging to the first, second and fourth refugee categories.

In the past, despite severe opposition and strong reservations from the members of the National Assembly, it was the King who ultimately pressed upon the members of National Assembly the urgent need to flush out the militants. Again it was the King who initiated the entire process of introducing a new constitution in Bhutan, deciding to hand over the throne to his son in 2007 and to hold an election in 2008.

In the new constitution of 2007, also under the larger umbrella of royal privileges, the King may employ, at his discretion, 'citizenship and land', both of which he may grant [Article 2(16)(b)]; 'criminal justice', including the granting of amnesty and the commutation of sentences [Article 2(16)(c)]; 'legislative power' to command laws through legislature, free from governmental hindrance [Article 2(16)(d)]; and 'residuary powers' over all matters which are not provided for under this constitution or other laws [Article 2(16)(e)]. The new King has the challenging responsibility of intervening to end the inhuman suffering of his own people, in the larger national interest.

Notes

1. Lhotsampas is the official Bhutanese term used to represent the peoples in the southern belts of Bhutan, mainly from districts like Chirang, Sarbhang, Chukha, Dagana and Samdrup Jongkhar. Lhotsampas are mostly Nepali-speaking Bhutanese. The total population of Bhutan has always been a major issue of conjecture. The following data are provided by UNHCR, Kathmandu, 2006. The camp-wide spread of the refugee population is as follows: Beldangi I (18,148), Beldangi II (22,285), Beldangi II Ext (11,461), Goldhap (9,360), Khudunabari (13,408), Timai (10,253), all in Jhapa and Sanischare in Morang (20,781); 49% of the refugees are women. There are over 270 registered refugees and over 15,000 non-registered refugees residing outside camps in Nepal. In addition, there are over 30,000 Lhotshampa refugees living in asylum in India, not in designated refugee camps but scattered in cities, towns and villages across the country.
2. D. N. S. Dhakal and Christopher Strawn, *Bhutan: A Movement in Exile*, Jaipur: Nirala Prakashan, 1994.
3. Tessa Piper, 'The Exodus of Ethnic Nepalis from Southern Bhutan', WriteNet, UK, 1995; D. B. Thronson, *Cultural Cleansing: A Distinct National Identity and the Refugees from Southern Bhutan*, Kathmandu: INHURED International, August 1993.
4. Leo E. Rose, *The Politics of Bhutan*, London: Cornell University Press, 1978; Sunanda K. Datta-Ray, *Smash and Grab: Annexation of Sikkim*, New Delhi: Vikas, 1984; B. S. Das, *The Sikkim Saga*, New Delhi: Vikas, 1983; A. C. Sinha, *Politics of Sikkim*, Faridabad: Thomson Press (India) Limited, 1975; and Mahendra P. Lama, *Sikkim Human Development Report 2001*, New Delhi: Social Science Press, 2001.
5. Manorama Kohli, *From Dependency to Interdependence: A Study of Indo-Bhutan Relations*, New Delhi: Vikas, 1993. Also see Rajesh S Kharat, *Foreign Policy of Bhutan*, New Delhi: Manak, 2005.
6. The Bhutan Citizenship Act, 1977, as quoted by D. N. S. Dhakal and Christopher Strawn, *Bhutan: A Movement in Exile*.
7. Ibid.
8. Bhutan Marriage Act, 1980 (Marriage with a non-Bhutanese).
9. During my 65-minute audience with His Majesty Jigme Singhye Wangchuk in his office in Thimphu on 7 July 1993 he mentioned that the Bhutanese authorities lately realized that the citizenship certificate which was officially distributed could easily be, and was in

fact, largely faked by many people as it had not been printed at a security press (it was printed at Calcutta's Croxton Press).
10. Kanak Mani Dixi, 'Bhutan: The Dragon Bites its Tail', *Himal* 5, no. 4, 1992.
11. The Bhutan Citizenship Act, 1985.
12. International Institute for Human Rights, Environment and Development, *Bhutan: An Iron Path to Democracy*, Kathmandu: INHURED, 1992.
13. National Assembly Resolution, 20th Session, Autumn, 1964.
14. This was disclosed to the author by many expelled members of the National Assembly and Royal Advisory Council who are living in camps and other places in Nepal.
15. As quoted by Tessa Piper from Michael Hutt, 'Refugees from Shangri-La', in *Index on Censorship* 22, no. 4, April 1993, p. 10.
16. Interviews conducted with the refugees at the camp at Beldangi in January 1998.
17. Jigme Y. Thinley, 'Bhutan, A Kingdom Besieged' in *Bhutan: A Traditional Order and the Forces of Change, Three Views from Bhutan*, Ministry of Foreign Affairs, RGB Government of Bhutan, 1993, p. 24 (Thinley was the then Home Secretary of Bhutan and is currently the Prime Minister).
18. Ibid., pp. 24–27.
19. This petition, signed by Tek Nath Rizal and B. P. Bhandari on 9 April 1988, mentioned many other discriminations and confusions created by the ongoing Census.
20. Bhutan now earns more than US$100 million annually by selling electricity to India from projects located in southern Bhutan including Chukha, Kurichu and Tala. It plans to enhance the installed capacity of hydropower to 3000 MW by 2017, as against the current capacity of over 1000 MW.
21. Mahendra P. Lama, 'Bhutanese Refugees in Nepal and Indian Responses I and II', *The Rising Nepal*, Kathmandu, May 1996.
22. Mahendra P. Lama, 'Managing Refugees in South Asia: Protection, Aid, State Behaviour and Regional Approach', Occasional Paper 4, Refugee and Migratory Movements Research Unit, Department of International Relations, Dhaka University, 2000.
23. One of the first major actions by the RGB after the refugee problems became serious was to ask all the farmers in southern Bhutan to compulsorily market all their produce through the State Trading Corporation.
24. The Home Minister mentioned this during the 71st Session of the National Assembly held from 16 October to 6 November 1992. Secretariat, National Assembly of Bhutan, Tashichho-Dzong, Thimphu.
25. 'India Is Not Tilting towards Us', interview with Dawa Tsering, Foreign Minister of Bhutan, *Outlook*, Delhi, 20 March 1996.
26. Mahendra P. Lama, 'Separatism and Armed Conflicts in North-East India' in K. M. de Silva, ed., *Conflict and Violence in South Asia*, Kandy, Sri Lanka: International Centre for Ethnic Studies, 2000, pp. 333–377.
27. *Kuensel*, Bhutan's official and only newspaper, 12 July 1993.
28. Royal Government of Bhutan, 'Driglam Namsha – *Traditional Etiquette*' (mimeo), September 1992.
29. *Kho* and *kira* are Bhutanese ethnic (national) dresses worn by men and women respectively. These garments are full dresses suitable for a relatively cooler climate.
30. Michael Hutt, *Unbecoming Citizens: Culture, Nationhood, and the Flight of Refugees from Bhutan*, New Delhi: Oxford University Press, 2003, pp. 183–187.
31. Amnesty International, *Bhutan: Forcible Exile*, AI Index: ASA 14/04/94, London, August 1994.
32. Ibid.
33. Reported by refugee Tsirimjip Lepcha to *Outlook*, New Delhi, 13 August 1997, p. 48.
34. Thinley, 'Bhutan, A Kingdom Besieged', p. 27.

35. Statement issued by UNHCR Representative Abraham Abraham, quoted by *Himalayan Times*, Kathmandu, 4 June 2006.
36. Reported by a refugee, Sukumaya, to Nepalnews.com, 18 November 2005.
37. Dhanmaya Basnet, who runs a small canteen at Sanischare camp (Morang), mentioned this to the author on 8 July 2006 at the camp site.
38. Interviews with Nepalese district officials who handled these refugees, initially conducted in 1993 and 1998 by the author.
39. 'Bhutanese Refugees and Asylum Seekers in Nepal, May 1997', briefing note for Camp Sadako participants, Geneva: UNHCR, Centre for Documentation and Research, 1997.
40. This pro forma was discussed and approved by the third Joint Ministerial Committee meeting held in April 1994 and continues to be considered a classified document.
41. *BPU Bulletin*, Press Secretariat of Bhutan Press Union, Vol. 1, Jhapa, Nepal, April 2006.
42. UNHCR, *Situation of Bhutanese Refugees in Nepal*, Kathmandu: UNHCR, 2006.
43. Bhutanese Women and Youth Empowerment Program, Birtamode, Jhapa and Lamidara Gewog, Chirang Dzongkhag, 2006.
44. Statement by the US Envoy in Nepal.
45. *Kuensel*, Thimphu, Bhutan, 13 April 1996.
46. Lok Raj Baral, 'Bhutanese Refugees in Nepal: Insecurity for Whom?' in S. D. Muni and Lok Raj Baral, eds., *Refugee and Regional Security in South Asia*, Delhi: Konark, 1995, p. 167.
47. K. P. Sharma Oli, 'Bhutani Saranarthi Samasya and Tyasko Samadhan', presentation made in the national seminar on 'Different Dimensions of Bhutanese Refugee Problems: Implications and Lasting Solutions', organized by the Institute of Foreign Affairs, Kathmandu, 21 May 2007, p. 4. Oli was until very recently Deputy Prime Minister and Foreign Minster of Nepal.
48. UNHCR, 54th Session of the Executive Committee (ExCom), Geneva, October 2003.
49. Proceedings and Resolutions adopted during the 71st Session of the National Assembly held from 16 October to 6 November 1992, distributed by Secretariat, National Assembly of Bhutan, Tashichho-Dzong, Thimphu, p. 61.
50. *SAFHR'S E–Briefs: Bhutanese Refugee Situation*, South Asia Forum for Human Rights, Vol. 1, Issue 1, Kathmandu, February 2000.
51. A study conducted by the Habitat International Coalition in 2001.
52. *Dzongkhag* Administration, Tsirang, DAT/ADM-/98-99, *Kuensel*, Thimphu, 20 March 1999.
53. Mahendra P. Lama, 'Experiments in Democracy in Bhutan' (cover story), *Frontline*, Chennai, 2–16 January 2004.
54. The High Court of Bhutan sentenced Tek Nath Rizal, a prominent Lhotshampa leader, to life imprisonment on 16 November 1993, for violating the National Security Act of 1992.
55. He abdicated the throne in favour of his son, making him the Fifth King on 9 December 2006 after 34 years of ruling the country of 20 *dzongkhags* (districts), *Bhutan Times*, 14 December 2006.
56. See www.bhutanewsonline.com.
57. Reported on www.nepalnews.com, 2 August 2006.
58. Two refugees, Narapati Dhungel, a minor, and Purna Bahadur Tamang were killed and several others were injured in the police firing. www.nepalnews.com, 27 May 2007, and also press release issued by UNHCR office in Kathmandu, 28 May 2007.
59. *Kuensel*, Thimphu, 26 August 1995.
60. E. N. Rammohan, *Insurgent Frontiers: Essays from the Troubled Northeast*, New Delhi: India Research Press, 2005, pp. 60–62. Rammohan retired as the Director General of the Border Security Force in 2000 and was deeply involved in the crusade against the insurgency in the north-east region of India.

61. UNHCR, *Situation of Bhutanese Refugees in Nepal*.
62. As told to the author by a number of senior and young Bhutanese refugee activists during his visit to the camps in July 2006.
63. These parties include the Bhutan People's Party, a faction of the Druk National Congress and another faction of the Bhutan Gorkha National Liberation Front.
64. 'Call for the End of Darkness', *BPU Bulletin*, Press Secretariat of Bhutan Press Union, Vol. 1, Jhapa, Nepal, April 2006.
65. See 'Vidyarthi Pratirodh', a Nepali monthly exclusively brought out by Bhutanese refugee students, July 2006.
66. *Kathmandu Post*, Kathmandu, 29 December 2006.
67. *Telegraph*, North Bengal Edition, Siliguri, 10 June 2007.
68. These observations are based on my recent discussions with a broad range of the refugee population, both within and outside the camps. They include heads of families, journalists, students, social workers, teachers, political leaders, women activists and also individuals of the Bhutanese refugee diaspora who have already been given asylum in countries like the US and the UK.

14

Burmese refugees in South and Southeast Asia: A comparative regional analysis[1]

Gil Loescher and James Milner

For nearly 60 years, the Burmese regime in Myanmar (Burma) has remained in power through a dual campaign of preventing democratic change and waging war against the country's numerous ethnic nationality parties and minority groups. Indeed, the human rights and political situation there is one of the most intractable in the world. As a direct consequence of decades of political repression, conflict, poor governance, arbitrary personal power and the underdevelopment of remote border areas populated by ethnic minorities, huge numbers of people have been forcibly displaced.

Unlike the other case studies examined in this book, this chapter analyses the prolonged exile of Burmese refugees from both a host state and regional perspective. The ongoing conflict in Myanmar has created at least four separate but related protracted refugee situations. For the past several decades, there have been large and protracted Burmese refugee populations in Thailand, Bangladesh, India and Malaysia. However, little understanding of the regional dynamics and connections between these refugee situations exists. At the same time, the prolonged presence of these refugee populations has come to have an impact on bilateral and regional relations. Given the particular regional and geo-strategic location of these refugee populations – on the axis between South and Southeast Asia and at the centre of regional competition between India and China – this chapter argues that situating the related protracted Burmese refugee situations within a broader comparative and regional context will prove more useful in the formulation of a comprehensive solution.

Protracted refugee situations: Political, human rights and security implications,
Loescher, Milner, Newman and Troeller (eds),
United Nations University Press, 2008, ISBN 978-92-808-1158-2

This chapter argues that the protracted presence of Burmese refugees in South and Southeast Asia has not only had an impact on individual host states in the region, but has also had an impact on relations between states. While more research needs to be done in the region on the interconnection between these protracted refugee situations, a comparative analysis of the political and strategic implications of refugee flows from Myanmar for host states and regional relations suggests that such linkages and regional dynamics exist. Based on preliminary fieldwork in the region and interviews with stakeholders engaged in negotiation with the regime in Rangoon, this chapter argues that in the long run a regional response, both to the situation in Myanmar and to the associated refugee populations, will likely be more successful than the current international response that principally relies on US and European trade sanctions against Myanmar.

This chapter focuses on the situations in Thailand and Bangladesh. The goal is to consider not only the causes, consequences and patterns of the individual refugee situations, but also their interaction and the impact of refugee movements on regional relations. This chapter has four main sections. The first provides an overview of the root causes of conflict in Myanmar and the patterns of displacement and traces the significant refugee flows to the neighbouring states of Thailand and Bangladesh. Section two outlines the political and strategic impact of refugees, both for individual host states and at a regional level. In light of these concerns, section three examines how these two main host states have responded. Based on this comparative analysis, section four considers what lessons can be learned to find solutions, both to specific short-term challenges and to the refugee situations themselves.

Principal causes of refugee flows and internal displacement

There have been two principal causes of forced displacement.[2] The first, and the cause which attracts the greatest amount of international attention, is a result of the suppression of the pro-democracy movement, led by the Nobel laureate Aung San Suu Kyi. Her party, the National League for Democracy (NLD), was overwhelmingly elected to power in 1990. The Burmese army refused to honour the outcome and forcibly and illegitimately held on to power. In the wake of these events, the Burmese military launched an intense nationwide campaign to crush civil protest and to exterminate support for Aung San Suu Kyi and the NLD. In fear for their lives, thousands of Burmese students and political activists fled to neighbouring countries. In 2007 widespread demonstrations against the military's political repression and economic mismanagement broke out throughout Myanmar. The army responded again with brutal force, imprisoning and torturing large numbers of political activists, including Buddhist clergy who had led some of the demonstrations.

The second cause of refugee movements from Myanmar is a result of conflict between the military regime and ethnic minority groups, such as the Karen, Karenni, Mon and Shan among others, who live in the eastern borderlands of Myanmar. The military has attempted to unify the country under a single territorial sovereignty and a strong central government. This has resulted in armed conflict against minority groups who are fighting for political autonomy in previously semi-autonomous border regions along the eastern border with Thailand. This struggle has produced by far the largest number of refugees.

Most displacement has occurred as a consequence of these protracted conflicts and counter-insurgency operations. The military strategy employed in the border areas seeks to undermine ethnic minority political and military organizations by targeting their civilian support base. Continuous armed conflict has directly undermined human and food security throughout Myanmar and has impoverished large parts of the civilian population. Thus huge numbers of people have been displaced as a consequence of military occupation, social control and/or state-sponsored development activities. The Burmese army forcibly confiscates land and relocates civilians to new government-controlled villages as part of their counter-insurgency strategies and in an effort to obtain free labour and other resources. Large infrastructure projects such as the construction of dams, roads, bridges and airports and the extraction of natural resources such as timber and minerals have required massive forced recruitment of labour. The International Labour Organization has repeatedly criticized the Burmese army for using large numbers of ethnic minority people as forced labour for military and development purposes.

In the wake of military repression and government economic policies in the eastern borderlands, at least half a million Burmese are currently internally displaced and without significant international assistance. Liable to various taxes and extensive forced labour, many civilians are unable to support themselves. Food insecurity, lack of education and basic health services, and the outbreak of major health crises (malaria, cholera, HIV/AIDS) have resulted in a major humanitarian crisis in Myanmar.

The impact of refugees on host states in the region

According to Refugees International, an estimated 3 million people have been uprooted by insurgency, counter-insurgency, repression and economic mismanagement by the military junta in Rangoon and have fled to neighbouring countries.³ This section provides an overview of the impact of these refugee movements on states in the region, before subsequent sections consider the response of host states and elements of a possible solution.

Thailand[4]

The best-known protracted Burmese refugee situation is those groups who have taken refuge in camps or settlements just across the border in Thailand. In addition to the more than half a million internally displaced persons in eastern Myanmar, tens of thousands of refugees have fled to Thailand since the mid-1980s. By 2008, there were over 150,000 Karen, Karenni and other national minority people in nine refugee camps strung along the 2,100 kilometre border.[5] In addition, there are now probably at least 300,000 refugees outside camps in Thailand, including 250,000 Shan refugees, plus at least 1.5 million Burmese migrants in Thai provincial towns and cities as well as over half a million internally displaced persons inside eastern Myanmar.

Thailand has been a major receiving country for refugees from neighbouring countries over the past five decades. More than 1 million Vietnamese, Lao, Hmong and Khmer refugees sought refuge in Thailand during and after the conflicts in Indo-China, by far the largest refugee burden of any Southeast Asian state. The resolution of these refugee problems was ultimately tied up with Cold War rivalries and regional politics. During the 1980s, particularly in Cambodia, external patrons, such as China and the United States, sustained the continuing resistance to Vietnamese rule in Phnom Penh through military aid and political support. The West also generously financed international humanitarian relief programmes to various client refugee warrior groups encamped along Thailand's eastern border. Protracted refugee situations developed, lasting decades in some places. Indeed, it took until 2004 to resettle the last Lao Hmong refugees from camps and settlements in Thailand.

The first major flows of refugees fleeing human rights abuses in Myanmar to Thailand occurred in 1984. Then, in 1988, the military regime known as the State Law and Order Restoration Council (SLORC)[6] seized power in Myanmar and cracked down on widespread political demonstrations, causing yet more outflows of politically active people. Following the overwhelming victory of the NLD in the 1990 national elections, SLORC declared the election void. Aung San Suu Kyi was placed under house arrest and thousands of her supporters fled to Thailand. Most of these politically active dissidents, called 'students' by Thai authorities, took up residence in Bangkok and other Thai cities. Initially, some of the 'students' were forcibly repatriated to Myanmar, but by the early 1990s the Thai government recognized that many had a valid fear of persecution. While Thailand is not a signatory to the 1951 Refugee Convention, does not have national refugee status determination procedures, and refused to recognize the 'students' as refugees, it did permit UNHCR to register them and to provide assistance.

During this period, a far greater number of ethnic minority people fled *tatmadaw*[7] offensives and forced labour and relocation programmes aimed at pacifying and controlling the border regions. Hundreds of thousands of Karen, Karenni, Mon and Shan poured across the border to Thailand, where they have been confined to camps for the past 20 years. Unlike its treatment of 'students', the Thai government terms the Myanmar ethnic minority groups 'temporarily displaced people' and until the late 1990s permitted UNHCR only limited access to them.

From the time it first set up camps in 1984 until the mid-1990s, the Thai military provided covert support to the Karen National Union (KNU) and other ethnic national parties, including the Karenni National Progressive Party (KNPP). The Thai military permitted insurgent groups to administer 'liberated zones' along the border, where they served as a buffer between the fighting in Myanmar and the western border of Thailand. Inside Thailand the refugee camps provided a civilian support base for the insurgent armies, and a source of recruits and safe haven for the armed groups. Because the refugee communities fled into exile together with their political parties and some of their resistance forces, there existed close links between the KNU and KNPP and their civilian supporters. Thus, the Thai army used refugee settlements and camps to support the resistance struggles and to contain Myanmar.

After the fall of insurgent bases at the border to Burmese forces in 1997, it was no longer possible to maintain a buffer zone between Thailand and Myanmar. The Thai military withdrew their support to their former clients in Myanmar in favour of a policy of 'constructive engagement' and building economic and trading ties with the government in Rangoon. At the same time, the Association of Southeast Asian Nations (ASEAN) began to pursue a policy of drawing Myanmar within its sphere of influence as a counter to the growing power of China in the region. Myanmar became a member of ASEAN in 1997. In the following years, Myanmar pursued a policy of opening up economic and military relations with China and Thailand and other ASEAN states, which enabled the army to continue to receive substantial arms imports and to further consolidate and extend its power within the country.[8]

These developments had a disastrous impact on the livelihoods of Burmese refugees in Thailand. In an effort to shore up its defences along the border, Thailand consolidated the 25 small and difficult to defend camps/village settlements – which refugees had inhabited on a mostly self-reliant basis – into 9 fortified camps. The Thai military placed security around the camps and enforced severe restrictions on the more than 100,000 refugees living there. Refugees could no longer move freely between camps or beyond the camp perimeters and were not allowed to work locally on Thai farms or as day labourers. Refugees became entirely dependent on international aid and, in effect, were 'warehoused' until conditions permitted their return to Myanmar.

With the election of Thaksin Shinawatra as the Thai Prime Minister in January 2001, Thailand further increased its efforts to exploit economic opportunities with Myanmar. Over the next several years, Thailand became increasingly reliant on imports of Burmese natural gas from Myanmar and became Myanmar's major trading partner. The Thaksin government pursued a dual policy, initiating an economic cooperation strategy to generate development in the border region while increasing pressure on the KNU and other ethnic nationalist parties to negotiate peace with Rangoon and enter into ceasefire agreements with the military. At the same time, Thailand, along with its ASEAN allies, encouraged the State Peace and Development Council (SPDC) to pursue political reform and to initiate a process that would lead to the passage of a new National Convention.

As illustrated by this overview, the long-term presence of Burmese refugees has had significant implications both for Thai domestic security as well as for regional cooperation and security. The refugee problem has at times been a source of inter-state conflict, straining bilateral relations between Thailand and Myanmar and even led to serious border clashes in 1995 and early 2001. Until recent years, the political activism of Burmese dissidents in Thailand particularly irritated the Myanmar authorities. Burmese exiles not only held frequent pro-democracy demonstrations in Bangkok and Thai border towns, but also participated in more extreme forms of activism. For example, on two occasions in 1989 and 1990, Burmese political dissidents hijacked commercial planes. A group of Karen student activists seized Myanmar's embassy in Bangkok in October 1999, causing a major bilateral crisis and the closure of the border for several months. In January 2000, Karen guerrillas took hundreds of hostages in a Thai provincial hospital, raising Thai public and government concerns about the domestic national security threats posed by the presence of Burmese activists and insurgents on Thai soil.

Moreover, Thailand and Myanmar have had numerous disputes over the demarcation of their long, mountainous border. The border is porous and difficult to police, making it easy for refugees, migrant workers and drug smugglers to cross. Among the most serious and direct security concerns for Thailand are the movement of insurgents and ethnic armed opposition groups in and out of camps, forcible military recruitment, involving not only adults but also child soldiers, diversion of food and medicines for military purposes, and the harbouring of insurgents in camps. The existence of refugee warriors in Thai camps has in the past strained bilateral relations between Thailand and Myanmar, leading to attacks on refugee camps and serious border clashes. Perhaps the most serious cross-border attacks occurred in 1995, when Myanmar's military and its proxies, such as the Democratic Karen Buddhist Army (DKBN), extended the conflict inside Myanmar to the refugee camps in Thailand. These attacks not only aimed to destroy potential sources of supply and bases for insurgent forces but also provided forced labour for their military operations. These incursions violated Thai sovereignty and strained bilateral relations.

The protracted Burmese refugee situation has also at times been a drain on local resources in an already poor region of Thailand and a source of social tensions.[9] Thai communities frequently complain that refugees and illegal migrants compete for local jobs and for natural resources, particularly during periods of economic downturn such as the late 1990s. Thai labour unions complain that factory owners and businessmen hire illegal workers and refugees because they are cheaper than Thai workers. Consequently, following the Asian economic crisis in 1997, the Thai government came under public pressure, especially from labour unions, to deport large numbers of illegal workers and refugees. Refugees and migrant workers are also frequent scapegoats for social problems in Thailand.[10] Displaced people are viewed as being responsible for health problems along the border and for the increase in transborder crime. Women in the border refugee camps, including children as young as 10, are exploited by the sex trade in Thailand.

Government authorities have also frequently accused refugees of perpetuating the illegal drug trade, one of Thailand's primary domestic security concerns. Myanmar is the world's second-largest producer after Afghanistan of illicit opium and heroin, and also exports large quantities of amphetamine-type stimulants. These drugs are an important source of income for insurgent movements and diverse criminal activities. Instability in Myanmar, combined with corrupt management and the illicit narcotics trade, generates associated illegal activities including small arms trafficking, money laundering, smuggling of forced sex workers, and illegal migration schemes across Thailand and other ASEAN states. These developments have had a negative impact on Thai state security, rendering the Thai state vulnerable in that they erode both state sovereignty and state capacity at various levels.

In response to these direct and indirect security concerns and as part of an effort to improve relations with Myanmar, the Thai authorities began to impose tighter border controls and restrictions on politically active refugees from Myanmar, beginning in 2002. Some activists were forcibly returned to Myanmar. At the same time, Thailand initiated a policy of relocating all refugees, including student activists, from urban areas to border camps, shutting down UNHCR programmes for Myanmar students and clamping down on political demonstrations. As discussed later in the chapter, however, this restrictive approach to Burmese refugees in general began to change in 2005 in response to the purge of military leaders in Myanmar and lack of progress towards political reform there.

Bangladesh

During the past several decades, Bangladesh has hosted the largest numbers of Burmese refugees in Asia. There has been a long history of interaction and cross-border migration of the peoples of the Arakan/southern Bangladesh region[11] and the plight of the Muslims of Arakan (the Rohingya) is among the most tense and difficult of all the ethnic problems in Myanmar. When Burma gained independence in 1948, a group of Muslims in Arakan rebelled, demanding their own independence. The armed revolt continued until 1954 and was followed by repeated attempts at secession, which laid the foundations for the current anti-Muslim perceptions and distrust that continue today. In 1974, Arakan was officially renamed Rakhine state. The Rohingya comprise a Muslim minority population of some 725,000 inhabiting three townships in northern Rakhine state adjacent to Bangladesh. Years of political and armed turmoil have exacerbated tensions between the majority Rakhine population, who are Buddhists, and the Rohingya Muslims. But the immediate cause of the exodus of refugees to Bangladesh in recent years has been the policies of exclusion, discrimination and repression against the Rohingya community, by the Burmese military regime.

In 1978 and again in 1991, successive crackdowns by the Burmese military against the Rohingya Muslim minority in northern Rakhine state triggered massive outflows of refugees to Bangladesh.[12] The immediate cause of the 1978 exodus was the launch of a series of massive military campaigns by the Burmese military targeted against the Muslim populations in the north. Employing brutal tactics ostensibly to check identity cards, the army reportedly murdered, tortured and forcibly relocated local villagers and destroyed Muslim mosques in an attempt to flush out insurgent forces and their sympathizers. Over 200,000 Muslims fled across the border into Bangladesh.

While Bangladesh and UNHCR provided immediate assistance to the refugees, government authorities announced that only temporary asylum would be provided and that all the refugees had to be returned as soon as possible. Myanmar also had an interest in seeing that the Rohingya return. Rangoon officials believed that the adverse publicity arising from the exodus of the Rohingya would damage their international legitimacy and their hope for greater economic assistance from the international community. The UN informed the government in Rangoon that 'the high daily cost to the international community of keeping refugees could well jeopardize contributions to the development projects of both countries – the implication being that Burma would suffer most'.[13] Consequently Bangladesh and Myanmar quickly reached a repatriation agreement, despite the fact that the conditions within Myanmar which had led to the flight of the Rohingya had not changed. UNHCR did not have a role in the agreement and there were also no provisions in the agreement guaranteeing the voluntary nature of the return. Essentially this was a bilateral repatriation arrangement which stemmed not from the interests of the refugees but from a convergence of the two states' national interests.[14]

Not surprisingly, the refugees fiercely resisted repatriation under these conditions. In response, Bangladeshi troops stepped up pressure on the refugees to return, primarily through intimidation tactics and restricting food supplies to the refugees. Conditions in the camps were appalling and by the end of 1978 some 10,000 refugees had died of epidemic illnesses and malnutrition.[15] Nevertheless, UNHCR, believing that conditions were unlikely to improve in Bangladesh, sanctioned the repatriation. UNHCR felt it had little leverage as neither country was a signatory to the international refugee instruments and there was little interest on the part of major donor countries in preventing the repatriation of Rohingya. Faced with deprivation and death in the camps, the Rohingya decided to return home, and by the end of 1979 nearly 190,000 had gone home.

During the 1980s, tensions in northern Rakhine state remained high. Rangoon continued its policies of deliberate exclusion and discrimination against the Rohingya Muslims. Shortly after the repatriation of 1978–79, the central government promulgated the Citizenship Law of 1982, which excluded and discriminated against a number of ethnic and religious minorities, particularly Rohingya Muslims. Rangoon claimed that, despite their well-established presence in the country, the majority of the Rohingya Muslims were in fact illegal Bengali immigrants who had previously crossed into Myanmar as part of a general expansion of the Bengali population in this region of Asia and should therefore return to Bangladesh.[16] As non-citizens and stateless persons, the Rohingya faced discrimination in the form of severe restrictions on their freedom of movement, access to employment and education. Under these adverse conditions, many Muslims continued to leave Myanmar, complaining of official harassment and persecution.

In 1991, the army introduced a border development programme in northern Rakhine state. During the campaign, numerous Muslim villages were destroyed; Muslim-owned land was confiscated and handed over to Burmese or Rakhine Buddhists moved into the area by the Burmese military; villagers were forcibly conscripted to work as unpaid labourers on development projects; and there were frequent reports of destruction of religious shrines and buildings and of killings, beatings and rape.[17] Consequently, some 260,000 Rohingya Muslims, about 30% of the population in northern Rakhine state, fled across the Naaf River into Bangladesh during 1991–1992.

The refugees arrived in a desperate condition beginning in mid-1991. Initially, Bangladesh mounted basic relief programmes and set up improvised and rudimentary camps, as it had done in the immediate aftermath of the 1978 refugee influx. UNHCR was asked to coordinate the international response and appealed for aid.

From the beginning of the crisis, however, Bangladesh made it clear that only temporary asylum would be provided and that the Rohingya had to be repatriated as soon as possible. In the autumn of 1992, on the basis of a memorandum of understanding with the government of Myanmar, Bangladesh began to deport thousands of refugees and denied UNHCR access to the camps where the refugees were accommodated. In response to UNHCR's public protests about the forcible repatriation, the government of Bangladesh signed an agreement with the Office in October 1992, permitting UNHCR to verify the voluntary nature of the repatriation. Despite the agreement, UNHCR was initially denied a monitoring role and therefore it withdrew from the programme.

When Bangladeshi authorities engineered a mass repatriation of more than 10,000 Rohingya in early 1993, there was an international protest which forced Bangladesh to call a halt to the programme. Eventually, Bangladesh signed a memorandum of understanding with UNHCR that permitted the Office access to the camps and the right to interview the refugees independently in order to determine whether their decisions to repatriate were indeed voluntary. In return, UNHCR agreed to promote repatriation once an international presence in Myanmar was established that could verify the existence of reasonable conditions for the returnees. In November 1993, UNHCR concluded a separate agreement with Myanmar which permitted the refugees the right to return to their places of origin and allowed UNHCR access to all returnees in Rakhine state in order to monitor their safe reintegration. Human rights NGOs criticized these agreements on the grounds that they failed to state explicitly that the refugees could repatriate only after the human rights and security situation in northern Rakhine state had changed substantially.

In 1994, UNHCR began to actively promote the repatriation of Rohingya and, by the end of 1995, all but 20,000 of the Rohingya in the camps had returned to Myanmar. Although human rights conditions in both Myanmar and northern Rakhine state had not changed substantially, UNHCR determined that, given the appalling conditions in the camps in Bangladesh, the refugees could be better served by returning as soon as possible while their return and reintegration could be monitored.[18] NGOs severely criticized UNHCR for promoting a repatriation programme for which conditions in the home country were unsafe for returnees and in which the principles of voluntary repatriation could not be adhered to.[19] Not only had the situation in Myanmar not changed, but NGOs claimed that UNHCR's capacity to monitor the returnees and to provide them with long-term security was severely limited because of the significant geographical and logistical constraints faced by its field staff.[20]

During the past decade, Rohingya have continued to be denied Burmese citizenship and are rendered stateless under the 1982 Citizenship Law. The human rights situation has not improved much since the early 1990s.[21] They continue to be discriminated against on the basis of their ethnicity and religion. Even though there has been a reduction in forced labour after UNHCR and the World Food Programme (WFP) took over responsibility for building local roads, the Rohingya are routinely subject to compulsory labour on other projects, arbitrary physical abuse, extortion, illegal taxation and constant humiliations. Their freedom of movement is severely limited and Rohingya are not free to travel from village to village without official permission, thus restricting their access to jobs, markets, health and education facilities. The ability of the Rohingya Muslims to practise their religion is also limited. In an effort to control the growth of the Rohingya population, Rohingya are required to obtain marriage licences, which are made difficult to obtain and expensive by local officials. As noted by Chris Lewa, these policies are designed to make life impossible for Rohingya and to encourage their departure to Bangladesh.[22]

Although there has been no recurrence of mass exodus of Rohingya refugees to Bangladesh during the past 10 years, there has been a steady stream of new arrivals to the Cox's Bazar region near the border with Myanmar. New arrivals are not permitted access to the camps. While the presence of newcomers is generally tolerated, the 100,000–200,000 Rohingya living illegally outside of camps in the Teknaf-Cox's Bazar region of Bangladesh must survive in extremely precarious situations as undocumented migrants without any protection or humanitarian assistance from UNHCR or the local authorities. There also exists considerable secondary migration from Bangladesh to Malaysia, Pakistan and Saudi Arabia for the Rohingya who can afford to pay the fees of human traffickers. In Bangladesh, some 27,000 Rohingya refugees, mostly the residual caseload from the 1990–1991 influx, remain in two camps, Kutupalong and Nayapara, located near the border with Myanmar.

For the past decade and a half, the conditions in the camps have been among the worst in Asia.[23] The overcrowded shelters housing the refugees had not been repaired for years and were poorly ventilated. Access to clean water and sanitation facilities was extremely limited. Bangladeshi officials who ran the camps prohibited international agencies from making improvements in the camp facilities. The withdrawal of Médecins Sans Frontières (MSF)-Holland at the end of 2003 and of CONCERN at the end of 2004 from the camps accentuated these problems and led to numerous protection concerns and protests by UNHCR and some donor governments.

The overwhelming majority of Rohingya in Bangladesh today, however, are not confined to camps but live in limbo among local populations. There is little information about the undocumented population of some 100,000–200,000 Rohingya and their links with the much smaller camp-based population. Although the irregular migrants provide cheap farm labour to the local economy, they compete with the local population for scarce resources in one of the most impoverished regions of Bangladesh. Locals complain of more expensive prices for essential goods, a decline in wages and environmental pressures as a consequence of the influx of Rohingya. The newcomers are unregistered and undocumented and have no access to any of the rights enjoyed by the local Bangladeshi community despite the fact that the Rohingya Muslims are closely related to the local Bangladeshi people and have a common language and religious background. Because the authorities are fearful of creating a pull factor they prohibit the provision of any national or international assistance or protection to the newcomers.

Without formal documentation or the legal right to work, migrants are vulnerable to harassment and exploitation. Therefore, many of the migrants are forced into begging and accused of theft and even murder, giving them a bad reputation among local people who view them as 'local gypsies'. Moreover, the fact that the Rohingya refugees have frequently held violent and well-publicized protests against poor living conditions and the government's repeated attempts to forcibly repatriate them have caused the public to have an extremely negative perception of the Rohingya. According to C. R. Abrar, the media have taken an 'unsympathetic view of the refugee presence in Bangladesh, portraying them to be an aggressive and disorderly bunch of people'.[24] Rohingya migrants regularly face arrests, deportations or evictions.

The failure to find a solution to this long-neglected protracted refugee situation and to the large-scale presence of undocumented Rohingya constitutes a source of significant tension and a major security concern in Bangladesh and for the region. Since a series of bombing incidents across the nation in August 2005, Bangladeshi security forces have not only targeted underground militant groups as security threats but also foreign nationals staying illegally in the country, including some migrants and refugees. Most significantly, however, the neglect of extremely poor conditions within the camps and within the undocumented Rohingya communities in the region has caused a recent alarming rise in instability and insecurity. Cross-border trafficking networks of drugs, arms and people have increased in the area, making the region fertile ground for an increase in criminal activities and for the expansion in Islamic fundamentalism.

Beginning in late 2006, a new exodus of Rohingya from Myanmar and Bangladesh spilled out into neighbouring Southeast Asian states. From October 2006 to January 2008, an estimated 6,000 boat people departed from both countries for Malaysia. This resulted in an unexpected boat people crisis in southern Thailand when large numbers of Rohingya were captured by Thai authorities on their way between Bangladesh and Malaysia and were subsequently deported to Myanmar.[25] These developments underscored the lack of regional mechanisms in place to deal with cross-border flows other than perceiving the crisis as a people-trafficking and smuggling issue with security implications.[26]

Regional dynamics

The arrival and prolonged presence of Burmese refugees in Thailand and Bangladesh have, at various times, caused tensions in bilateral relations between Myanmar and its neighbours. Myanmar's neighbours increasingly worry about Myanmar's destabilizing exports: HIV/AIDs, trafficking of drugs and narcotics and, in the case of Thailand and Bangladesh, uncontrolled refugee influxes which have at times become a regional political issue.

Until very recently, ASEAN leaders adhered strictly to the policy of non-interference in the domestic affairs of Myanmar despite the security problems it routinely exported across its borders. ASEAN placed primary emphasis on a policy of 'constructive engagement' and strongly opposed the imposition of economic, political or military sanctions, arguing that a policy of exclusion was not likely to achieve any positive change in Myanmar.

Despite these tentative efforts to encourage political reform in Myanmar, the Burmese military leadership steadfastly resisted all attempts to negotiate an agreement that would permit a political transition that was not strictly managed or guided by themselves. Sensing that the insurgencies in eastern Myanmar were drawing to a close, the SPDC perceived that it had the upper hand in determining the political future of the country. The military negotiated a series of ceasefire agreements with most of the armed ethnic movements and continued to push the exhausted remnants of the Karen and Karenni forces out of their remaining strongholds. At the same time, the Prime Minister and Military Intelligence Chief, General Khin Nyunt, announced in August 2003 the resumption of a National Convention to draft a new constitution, part of a seven-stage 'road map to democracy'.[27] However, the consultation process was labelled a sham by most outside observers and dragged on to mid-2007. Most of the 1,000-plus delegates to the convention were selected by the government and the two main opposition parties, the NLD and the United Nationalities Alliance (a coalition of ethnic nationality parties), refused to join the convention.

The overthrow of Khin Nyunt in October 2004 by hard-line elements within the military brought political dialogue on democratic change in Myanmar to a complete standstill. The Burmese government placed new restrictions on aid programmes for humanitarian purposes in eastern Myanmar. The contraction of humanitarian space prompted several international NGOs to suspend their programmes in conflict-affected areas and to become more dependent on community-based organizations and local capacities. Hope for political change in Myanmar receded even further in 2005 and 2006 when the Ministry of National Planning and Economic Development issued new restrictive guidelines for UN agencies and NGOs. International mediation efforts failed to break the impasse. The SPDC reconvened the National Convention for the 11th and final time in July 2007. The military offered a draft constitution which ignored the aspirations of the ethnic nationalities to govern their own affairs and maintained controlling power for the military in government affairs. In early 2008, the SPDC declared that it would hold a constitutional referendum in May 2008 and multi-party elections in 2010 from which the NLD would be banned from participating. At the same time, there continued to be no improvement in the human rights situation in the border areas of Myanmar.

In recent years, the lack of improvement in the human rights situation in Myanmar and the ongoing outflow of refugees have begun to play a role in regional relations, especially in the debate over the nature of the relationship between Myanmar and ASEAN. While ASEAN continues to advocate for 'constructive engagement' with the regime in Rangoon, a number of voices in the region began to express concerns about the limitations of this approach. In July 2006, for example, the Chair of the ASEAN Inter-Parliamentary Myanmar Caucus, Zaid Ibrahim, argued that 'the security of all of Asia is threatened by the trans-border flow of refugees, human trafficking, drugs, HIV' from Myanmar, and made clear that relations between the regional body and Myanmar will be frustrated so long as the problems of Myanmar spill over into its neighbours.[28]

Refugees are perhaps the most dramatic manifestation of the spill-over of the crisis, and regional actors now cite the prolonged presence of refugees and their impact on neighbouring states as the primary way in which the situation in Myanmar has a broader regional impact. The frustration with Myanmar contributed to a shift in ASEAN's approach to the state beginning in July 2005. At their annual summit in Vietnam, ASEAN foreign ministers took the unusual step of publicly expressing their frustration with Myanmar's lack of progress towards political reform, leading to Myanmar being forced to give up its rotational chair of ASEAN. In the build-up to the July 2006 ASEAN Foreign Ministers' Summit in Kuala Lumpur, Malaysia's Foreign Minister, Syed Hamid Albar, made increasingly critical statements announcing that ASEAN had reached the stage where it was not possible to defend Myanmar if it did not cooperate with ASEAN or deliver tangible progress on economic and political reforms, including the process of meaningful reconciliation with the NLD and ethnic minority groups.[29] Despite these warnings, no progress was forthcoming and during the so-called 'saffron revolution' in September 2007 the Burmese military again brutally cracked down on protesters. ASEAN foreign ministers expressed their 'revulsion' at the killings in Rangoon and sternly demanded the Burmese military stop using violence against the demonstrators.[30] Separately, Singapore Prime Minister Lee Hsien Loong, in his capacity as chairman of ASEAN, made public a letter to General Than Shwe, stating that the Burmese military's actions in Myanmar 'have serious implications, not only for Myanmar but also for ASEAN and the whole region' and he 'strongly urged' Shwe to work with the UN Special Advisor on Myanmar, Ibrahim Gambari, 'to try to find a way forward'.[31]

These are clearly signs of increasing regional discomfort with continued repression in Myanmar and indicate how the prolonged political crisis there and the associated long-term exile of refugees have had an impact on regional relations. The growing scale of these concerns may, therefore, eventually lead to a shift in the regional approach towards Myanmar, which may, in turn, prove significant in trying to formulate a comprehensive solution to the Burmese protracted refugee situation, as outlined below.

Host state response

In the absence of a solution to the crisis in Myanmar, coupled with the perceived impact of refugees on host states, as outlined above, Thailand and Bangladesh have started to improve their hosting policies towards Burmese refugees. Thailand has recently indicated its willingness to adopt a more solutions-oriented approach to Burmese refugees, while Bangladesh is starting to permit some improvements in camp conditions for the Rohingya. This section briefly considers the response of Thailand and Bangladesh to the prolonged presence of Burmese refugees.

Thailand

For more than 20 years, Thailand has been opposed to providing any opportunities for Burmese refugees to integrate locally.[32] The government and public increasingly viewed refugees as a security burden, as a strain on state capacity and as a threat to social cohesion. They resented the actions of student activists; they perceived refugees and illegal migrants as competitors for jobs, resources and limited social services; and they viewed the Burmese as being the cause of increased crime and health risks.

With the change of regime in Myanmar in 2004, however, Thailand became increasingly concerned with the increasingly protracted nature of the Burmese refugee situation. Consequently, there has recently been a positive shift in Thai government policy towards the protracted refugee situation on its border with Myanmar. In 2005, the Thai government conceded privately that, given the continued resistance of the SPDC to political reform, Burmese refugees were unlikely to be sent back to Myanmar in the near future. After refusing for decades to permit third countries to resettle Burmese refugees, Thailand agreed for the first time in 2005 to the overseas resettlement of Burmese refugees from the camps.[33] By September 2006, the US began to resettle Burmese refugees after legal obstacles to resettling them were finally removed from US legislation.[34] From 2005 until the end of December 2007, over 63,000 refugees had been referred by UNHCR for resettlement consideration; some 27,000 had been accepted by resettlement countries and 22,000 had departed Thailand.[35] In total, 11 countries were resettling refugees from Thailand.

While resettlement is a welcome development and contributes to a solutions-oriented approach to the protracted refugee situation, there is widespread concern among NGOs regarding the way the programme is handled. Former combatants remain ineligible for resettlement to the US, which results in some families being split up. Several other resettlement countries accept only the most trained and educated from the camps, resulting in a drain of vital, skilled staff workers from the camps. NGOs complain that this will have a major impact on camp management, community services and assistance projects supported by NGOs.[36]

In recent years, Thai authorities have demonstrated a new openness towards thinking about alternative ways of dealing with the protracted refugee situation along the border. In particular, they have taken tentative steps towards improving the livelihood and solutions prospects for Burmese refugees. The Thais were influenced in their thinking by a 2005 UNHCR–NGO joint letter to the Thai government pointing out that refugees were still living in the same conditions as those who first arrived over 20 years ago. They were confined to basic bamboo and thatch camps with limited education and skills training opportunities and almost no income-generating or employment possibilities. It was suggested that long-term confinement of refugees was detrimental not only to the refugees but also to future stability in Thailand. NGOs and UNHCR argued that, if refugees were given more skills training, further education and income generation opportunities, this would prepare them well for whatever solution awaited them in the future, whether that was in a third country, back in Myanmar or during their stay in Thailand. They argued further that refugees were a resource that could contribute positively to the Thai economy during their exile.

In December 2005, NGOs and UNHCR capitalized on these initiatives, producing a comprehensive plan for 2006 at a workshop in Chiang Mai. In response, the Thai government subsequently approved extended skills training projects designed to produce household income and improve livelihoods and employment opportunities. Thai authorities also agreed to support education in the camps by setting up learning centres with a focus on teaching the Thai language. Results from the authors' preliminary fieldwork in Thailand in 2006 indicated a willingness on the part of the government to implement such a new approach, but political uncertainty in Thailand both before and after the overthrow of the Thaksin government, coupled with a lack of donor engagement, delayed action for most of 2006. In 2007, Thailand finally issued identity cards to some 85,000 refugees in the camps, which are an important prerequisite for exploring self-sufficiency opportunities for refugees both inside and outside the camps. At the same time, the Thai authorities began to permit limited skills training and education in the camps.

Bangladesh

Unlike refugees in camps along the Thai–Burmese border, Rohingya refugees in the Cox's Bazar area have not been permitted any role in administering and governing themselves by forming elected refugee management committees to oversee health, education, sanitation, food distribution and security in the camps. Rather, camps are administered by government-appointed local Bangladeshis and refugee camp volunteers, called 'mahjees', who for the past several decades have exerted control through intimidation, physical abuse and bribery. Women, in particular, are at risk of sexual and gender-based violence. There are reports of trafficking of women and children. Because UNHCR and international NGO staff have not been permitted to remain in the camps after dark when much of the violence and abuse of refugees takes place, the refugee population is provided with no real international protection.

Rohingya refugees are also denied other basic rights. Freedom of movement and freedom for refugees to engage in income-generating activities are extremely restricted. For decades children have been denied basic formal education beyond primary school and there are no opportunities for vocational or skills training. In creating and maintaining such a harsh and inhumane camp environment, the authorities aim both to encourage camp-based refugees to return home and to discourage Rohingya from fleeing Myanmar.

The search for a solution to the camp-based Rohingya refugees has been deadlocked for years. The repatriation process has virtually come to a standstill. The majority of refugees continue to oppose repatriation as long as conditions in northern Rakhine state remain unchanged. Moreover, Myanmar refuses to accept any of the residual refugee caseload in the camps. At the same time, the size of the refugee population in Bangladesh remains nearly constant as the number of children being born in the camps annually exceeds the number of refugees repatriated each year.

To break this deadlock, UNHCR began to adopt a more proactive approach, beginning in the late 1990s, and proposed a number of new programmes to make the refugees self-sufficient and to end their dependency on international aid, thereby enabling them to become contributing participants in the host community and also in Myanmar when they eventually returned home. In 2003, UNHCR proposed a policy of 'temporary self-reliance' which would entail dismantling the camps and allowing refugees to engage in income-generating activities. However, Bangladesh steadfastly refused to consider any proposal that called for refugees to move out of camps, arguing that it was an over-populated country and could not afford such a scheme and that this reversal of temporary local integration policy on its part would likely trigger a fresh flow of Rohingya from neighbouring Myanmar. As a result of this stalemate, the donor community repeatedly expressed its reluctance to continue to fund the Rohingya relief operation for an indefinite period without any durable solution in sight. Cutbacks in donor funding to UNHCR and to local service programmes for the refugees in turn made the Bangladeshi government even more resistant to consider local integration for the Rohingya.

For the past few years, UNHCR and the international community have lobbied strongly for the improvement of conditions in the camps and have found the government more responsive than in the past. In 2006, the government finally agreed to permit UNHCR to build new shelters in the camps and construction began in 2007–2008. At the same time, the government began to allow other UN agencies and NGOs to provide much-needed services to the refugees. Consequently, UNICEF, the United Nations Population Fund, MSF-Holland and Handicap International began new programmes in the camps.

Donor governments have also recently become more interested in working towards a solution for the long-staying Rohingya. There are ongoing discussions between the UN country office and diplomatic missions about the need to get Bangladesh to work with the international community towards a solutions-based approach and the need for donor states to take the lead in this initiative and to commit themselves to a solution. Finally, for the first time ever, the government permitted the initiation of a small resettlement programme from the camps. Beginning in 2006, Rohingya have been resettled to Canada, New Zealand and the UK.[37]

Towards a comprehensive solution

The inability of Thailand and Bangladesh to resolve their protracted Burmese refugee situations independently points to the benefit of a comprehensive solution at a regional level. However, such an approach does not currently appear to be immediately feasible, given the highly limited prospects of sustainable repatriation in the foreseeable future. In fact, efforts to find a solution to the protracted refugee situation in the region have been stymied by the political and military impasse in Myanmar. The military continues to use force to quell opposition to its rule and engages in systematic human rights abuses. The Burmese junta perceives refugees and the displaced as part of the insurgent forces and refuses to discuss any possible solutions apart from the total defeat of the opposition. Thus, the military favours repatriation only after it has secured ceasefires on its own terms and/or after it has secured complete control of the ethnic nationality forces and their border territories. In the case of the Rohingya, the prospects for a solution are particularly dire. The root causes of the continuous outpouring of Rohingya result from the Burmese government's policies of exclusion against an entire minority to the point of denying them citizenship rights. Therefore it is very unlikely that the SPDC government in Myanmar would cooperate in a comprehensive solution.[38] Moreover, the states hosting Rohingya have legitimate concerns about creating pull factors.

For their part, Burmese refugees throughout the region fleeing war and persecution refuse to return home until their physical safety can be assured. It is also the case that communities in Myanmar's border regions are among the most impoverished in the world and will not be able to support and reintegrate large influxes of returnees without substantial new international economic assistance. The impasse between the regime in Myanmar and its opponents, both in the pro-democracy movement and among ethnic minority communities, has played a predominant role not only in frustrating repatriation but also in perpetuating the conflict.

In the long term, a greater presence of humanitarian and development actors will be necessary to overcome the current crisis and contribute to the future stabilization of the situation in Myanmar. Humanitarian NGOs and development actors will be needed to respond to the current economic and human security crisis in Myanmar, particularly in the border regions.[39] During recent years, several Thai-based NGOs have provided cross-border assistance to civilians displaced by armed conflict. Consequently, there has been a significant growth in community-based organizations inside Myanmar that have initiated a number of extensive health and education programmes. The important assistance and protection work of local civil society actors will be crucial in rebuilding a future post-conflict Myanmar. By ensuring humanitarian aid for vulnerable groups, by planning for economic development and by supporting local institutions and agencies, the international development community can create a positive environment for political and social change.[40] While the engagement of these humanitarian and development actors within Myanmar will be important, their presence is no substitute for the engagement of political actors and a resolution of the conflict within Myanmar.

Addressing the political situation in Myanmar

Interest in resolving the crisis in Myanmar is not new, and many individuals and groups have been committed to the long and slow process of trying to effect change there for decades. The military crackdown in August–September 2007 and the continued political impasse in Myanmar have, however, recently led to greater international and regional concern about the situation there. The international community has become increasingly vocal about the prolonged conflict in Myanmar and the link between these human rights abuses, displacement and regional security.

Myanmar has been an important issue within the United Nations for nearly a decade and a half.[41] In 1994, the United Nations General

Assembly requested the Secretary-General to use his good offices to facilitate a dialogue between the NLD, the various organizations representing minority groups in Myanmar and the regime in Rangoon. Between 2000 and 2004, the Secretary-General's Special Representative for Myanmar, Razali Ismail, paid more than 12 visits to the country in an effort to build confidence between the various parties. However, tentative moves towards greater international engagement led to a power struggle within the SPDC in 2004, leading to the removal of Khin Nyunt and the extension of the house arrest of Aung San Suu Kyi. After the purging of the internationally minded elements of the SPDC during the period October 2004–May 2006, neither the UN Secretary-General's Special Representative nor the UN Human Rights Rapporteur were allowed to enter Myanmar.

Frustration with the unwillingness of the SPDC, led by General Than Shwe, to cooperate with efforts to resolve the conflict resulted in increased international pressure on Myanmar. In late November 2005, in a letter to the President of the UN Security Council, the US expressed concern about the situation in Myanmar as a threat to international peace and security, citing the large number of refugees fleeing the country, narcotics trafficking and the country's human rights policy. Ibrahim Gambari, the then UN Under-Secretary General for Political Affairs, subsequently briefed the Council in December 2005 on the political, social and economic situation in Myanmar. In the following debate, there was consensus that there were serious cross-border activities that needed to be addressed by international action. While the issue of Myanmar was not placed on the agenda of the Security Council, it was clear from the debate that member states had begun to see the situation there as being one of regional and international concern. But high-level visits to Myanmar by the UN and ASEAN during 2006 and 2007 failed to end the stalemate there. The junta continued to keep Aung San Suu Kyi under house arrest and during 2006 escalated its military activities along the Thai border. The widespread suppression of civil dissent in Myanmar from September 2007 on brought further limited external pressure on the military regime to negotiate with the political opposition but few outside observers expected immediate change.

The fact remains that while Myanmar has been an important topic in recent years there has been no real international consensus on how to effect change there. The United States favours direct pressure on the regime, mainly through sanctions on trade, investment and other financial dealings, in addition to placing strict conditions on the provision of humanitarian assistance. In contrast, the European Union favours the use of targeted sanctions, humanitarian assistance and quiet diplomacy to bring about change. Regionally, ASEAN continues to favour constructive engagement, and views the issue primarily as a regional one. For their part, China and India, the two most significant powers in the region, see influence in Myanmar as a point of competition. Both countries are concerned about the potentially destabilizing impact of the ongoing conflict in Myanmar and the impact of cross-border movements and smuggling, but both countries are also keen to maintain friendly relations with the SPDC to facilitate economic relations and access to Myanmar's abundant natural resources. All regional and international efforts to urge the SPDC to stop attacking civilians and protect its people have failed. The non-binding UN Security Council resolution introduced by the US in January 2007, which included a call to the government in Myanmar to cease attacks on the country's minorities, was vetoed by China and Russia. As of early 2008, China, Russia and most ASEAN states continued to oppose sanctions against Myanmar. Until the members of the Security Council and ASEAN develop a united approach to Myanmar, Asia's worst protracted refugee situation will persist.

As a resolution to the crisis in Myanmar is not possible without the sustained involvement of these stakeholders, it is important to begin by understanding what interests can foster greater cooperation between these actors. Apart from addressing the spill-over of the Burmese crisis into its neighbours' territories, including the impact of refugees, there are significant geo-political and economic reasons for resolving the stalemate in Myanmar. As noted in the introduction, Myanmar lies at the crossroads between South and Southeast Asia and between India and China. Both regions, and especially these two regional powers, have made economic cooperation and development a primary objective in their external relations. Given its location, Myanmar will be key to facilitating exchange and growth in the region. As argued by David Arnott, however, 'a rather serious impediment to this undertaking is that Burma, on account of her political and human rights misbehaviour, is not allowed to receive funding from the world financial institutions such as the World Bank, the IMF and the Asian Development Bank'.[42]

The need to access such resources, especially to support infrastructure development, regional trade and economic growth, potentially provides a significant motivation to pressure the regime in Rangoon to resolve the two main sources of domestic conflict. For its part, China has an interest in promoting economic growth in its south-west border provinces, traditionally an underdeveloped region with chronic security and narcotics problems. Such concerns, especially relating to the flow of narcotics, crime and HIV/AIDS, prompted China as early as the UN Security Council in December 2005 to begin to express its frustration with the lack of change in Myanmar.[43] As Myanmar's principal patron, China's role in a process of change will be crucial. However, Beijing's geo-strategic and considerable economic interests in Myanmar have prevented it from abandoning its regional ally.

Given competition between China and India, the engagement of the government in New Delhi will also be important. This position may also be affected by concerns about security, development and the impact of refugees from Myanmar. Specifically, the situation in India's north-east, characterized by insecurity and underdevelopment, has long been a concern for the government in New Delhi, which is also keen to capitalize on the potential economic benefits of securing a bridge through Myanmar to markets in Southeast Asia. Together, India and ASEAN have sought to reduce China's influence over Myanmar by drawing it into regional organizations and trade groupings.

In fact, ASEAN could potentially prove to be the key actor in initiating a process that leads to change in Myanmar and a resolution of the refugee situation. ASEAN's increased assertiveness towards Myanmar in recent years, as outlined above, reflects its growing frustration with the protracted crisis in Myanmar, its concerns about the impact of Myanmar on ASEAN's international credibility as an effective regional organization, and its inability to deal with the spill-over of the crisis in Myanmar into neighbouring states. ASEAN members also stand to benefit significantly from the same economic incentives that would benefit India and China should the crisis in Myanmar be resolved.

ASEAN can also play a central role in drawing in the two major regional powers. Given competition between China and India in the region, a third party with regional credibility and important links to the two regional powers is needed to initiate and guide a process leading to change in Myanmar. ASEAN could not only provide the forum to initiate such action, but also place additional pressure on India and China. ASEAN has had formal links with both powers for some time, given their status, along with the US, the EU and the UN system, as ASEAN 'dialogue partners'. Building on these links, ASEAN could provide the focus for a regional process that fosters consensus on the part of key stakeholders and a sense of regional ownership over the process and ensure the necessary regional political buy-in that an externally driven process would not necessarily achieve.

While tying Chinese, Indian and ASEAN trade to economic and political liberalization in Myanmar might be the external pressure to which the junta might respond, the US and Europe could offer Myanmar certain incentives, such as the gradual easing of sanctions, if the junta met certain benchmarks, such as embarking on real political dialogue with the country's minorities. A united international response combining both specific incentives and disincentives will likely yield more promising results than a continuation of the present policy.

Addressing the refugee situation

The international community cannot keep Burmese refugees in neighbouring countries in a perpetual state of limbo while awaiting a political solution to the conflict in Myanmar. A regional response to the prolonged presence of Burmese refugees in South and Southeast Asia will likely be the most effective. Immediate steps need to be taken to address the refugee situation in the region. These steps relate to both the stabilization of the current refugee situations and the adoption of an approach that could lead to a shift to a longer-term solutions-oriented approach.

In the short term, it is important to address the gaps in the current response to the protection and assistance of Burmese refugees in the region. As outlined above, there is significant disparity between the policies of the main host countries towards Burmese refugees. There is very little evidence that asylum policies are part of bilateral and regional dialogue between host countries, and positive examples of solutions-oriented approaches are consequently not shared. Instead, a regional dialogue among key stakeholders – including host governments, donor governments and UNHCR – could identify both gaps in current programmes that need to be filled and common methods for shifting towards a more solutions-oriented approach. Key benefits of such a regional approach would be the engendering of greater confidence among otherwise reluctant host states, the marshalling of greater resources, and the learning of best practices from within the region and elsewhere.

The early lessons from Thailand's shift towards a more solutions-oriented approach could provide a useful example for other host states in the region and a basis for stabilization of the Burmese refugee situations throughout the region. Through a process encouraged by UNHCR and NGOs, the Thai government has recently issued identity cards to refugees in the camps and has started to overcome its long-standing reluctance to engage in discussion on issues like freedom of movement for refugees, educational opportunities and the introduction of employment opportunities for refugees outside the camps. This shift came about largely as a result of not only Thailand's desire to better prepare refugees for an eventual solution but also recognition of the limitations of the current encampment policies. Given Thailand's long experience with the hosting of refugees, and the range of impacts refugee movements have had on Thai security, Thailand's willingness to discuss such a shift is a very important development. As Thailand is a regional, political and economic leader as well as a front-line state in its relations with Myanmar, it may be able to use its regional status to encourage other host states in the region to adopt a similar policy.

Donor governments would have a vital role to play in ensuring the success of such a process. As seen in the case of Thailand, addressing the gaps of current programmes and developing new initiatives as part of a more solutions-oriented approach need to be supported with additional resources. Additional resettlement opportunities, and greater coordination of the strategic use of resettlement to effect change in host country policies, would also make an important contribution. More fundamentally, however, donor missions in the region need to be engaged with the process and demonstrate political support for a change in approach. In Dhaka, for example, the donor community, through its individual missions and through the UN Country Team, has started to become more directly and collectively engaged in discussions about the direction of Bangladeshi policy towards the protracted Rohingya population. A framework for durable solutions is being devised with the support of the government and the UN Country Team. Similarly, in other host countries, sustained and meaningful donor and UN Country Team engagement in discussions on the future of asylum policies will likely engender greater confidence on the part of host states.

UNHCR would also have a crucial role to play in a process of regional stabilization and a shift towards a more solutions-oriented approach. As seen in Thailand, UNHCR, working closely with NGO partners, was able to play a significant leadership role in persuading the Thai authorities to reconsider their position on encampment. Likewise, in Malaysia, UNHCR has recently established a regional protection hub to negotiate a more dependable protection environment for refugees throughout the region. At a country level, UNHCR can also do more to share examples of locally developed initiatives to address specific problems in protracted refugee situations in other regions. Sharing these examples of 'what has worked' elsewhere can provide a useful basis for considering practical responses to problems faced by host states in South and Southeast Asia.

UNHCR could also play an important role in helping states in the region develop a common legal foundation that would contribute to an effective response to their refugee concerns. None of the countries that host Burmese refugees has developed domestic legal systems to deal with refugees or acceded to the UN Refugee Convention. At the national levels, refugees are subject to the same laws as illegal aliens. Consequently, refugees are treated in an ad hoc manner, subjected to arbitrary and discriminatory measures and denied basic rights. Because refugees are perceived as a security and economic burden, host states have been reticent about providing asylum or agreeing to new legal obligations for the care and protection of refugees on their territories.

In South and Southeast Asia, refugees are perceived to be national and regional security concerns rather than a humanitarian and human rights issue. Host states and countries of origin mainly deal with the Burmese protracted refugee situation as a bilateral political issue. Pragmatic rather than human concerns generally dominate these discussions. The priority for host states is the rapid return of refugees to the refugee-producing states. To avoid the continued politicization of the refugee issue in these regions and to lay an essential foundation for a solutions-oriented approach, it is essential that an agreement among regional stakeholders be reached on the application of legal standards on the treatment of refugees, including repatriation.

Without such standards, foreign policy, national security and domestic political considerations will continue to prevail over protection principles, making future repatriation unsustainable and putting refugees at risk. While the political interests of states need to be addressed as part of a solution, a common legal framework would provide greater dependability for refugees and states alike. The development of regional and national legal frameworks, in addition to accession to the UN Refugee Convention, would better reconcile the concerns of states and refugees, primarily by giving states a more dependable and transparent mechanism for addressing their legitimate security concerns.

UNHCR should also be prepared, once political and economic stability is restored in Myanmar, to initiate region-wide discussions focused on finding solutions among the major host countries and the new government in Myanmar. As the primary organization in all host states mandated to find solutions to refugee situations, UNHCR has a unique role to play in this regard. In fact, this was UNHCR's most significant contribution to the process leading to the Comprehensive Plan of Action for Indo-Chinese refugees (CPA) in the late 1980s, as described in the chapter by Alexander Betts. In a similar way, UNHCR could act as a catalyst to bring together the full range of stakeholders needed to devise and implement a regional approach to the Burmese refugee situation in South and Southeast Asia.

The historical precedent of the CPA points to the benefits of moving beyond a country-by-country approach and adopting a regional approach to complex refugee situations. As outlined above, current programming for Burmese refugees is divided into distinct country operations, with limited interaction between these programmes. In contrast, the CPA worked from a regional perspective and was consequently able to effectively engage with host states across Southeast Asia. Given the regional impact of the Burmese refugee situation, the vulnerability of most host states, and the links between refugees and problems ranging from regional stability, drug trafficking, migration and economic development in border regions, the benefits of a regional approach become more apparent. Just as making linkages between the issue of refugees and areas of broader concern contributed to the success of the CPA, so too could the recognition of these linkages contribute to a more effective response to the Burmese refugee problem today.

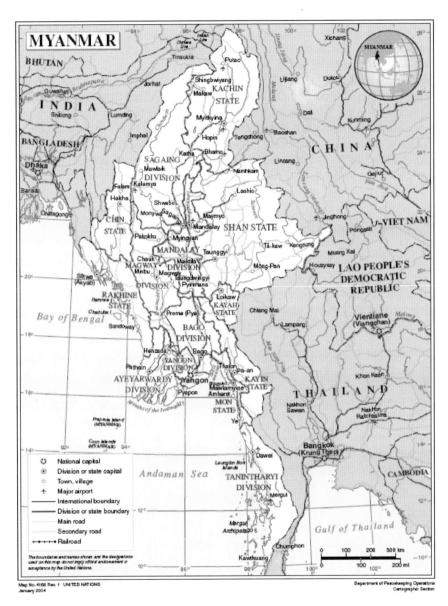

Figure 14.1 Map of Myanmar

The lessons of the CPA also illustrate that sustained engagement in resolving a common refugee problem can further engender the confidence and solidarity required for states in the region to collectively address the cause of refugee movements. In this way, a regional process focused on responding to the Burmese protracted refugee situation could foster the confidence necessary for an ASEAN-led response to the situation in Myanmar itself, as outlined above. Central to such an approach, and a resolution to the conflict, is a solution to the three causes of refugee movements from Myanmar: the conflict between the regime and ethnic minorities, the denial of citizenship rights to the Rohingya, and the suppression of the pro-democracy movement. All these causes need to be more fully addressed before the Burmese protracted refugee situation can be fully resolved.

Conclusion

Just as the presence of refugees has had a negative impact on bilateral and regional relations, combined efforts to find solutions to the refugee problem can have a positive impact on regional cooperation. As outlined above, in addition to urgent humanitarian needs there are sound economic and political reasons for resolving the situation in Myanmar. The impact of refugees, representing a significant human rights and a security problem, is a further motivation. Individual host states have been challenged by the scale of these problems for several decades. There are signs that it is time to take a more engaged and united regional approach that will likely contribute not only to short-term responses to specific challenges posed by the prolonged presence of Burmese refugees in the region, but also to the longer-term objective of designing and implementing a comprehensive solution for the conflict in Myanmar and for the associated refugee populations.

Notes

1. A fuller version of this chapter, which includes an analysis of all four major host countries (Thailand, Bangladesh, Malaysia and India) for Burmese refugees, can be accessed under 'Publications' at: http://www.prsproject.org.
2. Ashley South, 'Burma: The Changing Nature of Displacement Crises', Refugee Studies Centre, Oxford University, Working Paper no. 39, February 2007.
3. Refugees International, 'Burma: Military Offensive Displacing Thousands of Civilians', Bulletin, Washington, DC, 16 May, 2007.
4. This section draws on the authors' fieldwork in Thailand in January 2006.

5. Thailand Burma Border Consortium (hereafter 'TBBC'), *Burmese Border Refugee Sites with Population Figures*, Bangkok: TBBC, January 2008.
6. A newly reconstituted military junta, the State Peace and Development Council (SPDC), replaced SLORC in November 1997.
7. *Tatmadaw* is the Burmese term for the military.
8. David Arnott, 'China-Burma Relations', in David Arnott et al., *Challenges to Democratization in Burma: Perspectives on Multilateral and Bilateral Responses*, International IDEA, 14 December 2001, pp. 37–56.
9. Carl Grundy-Warr, *The Silence and Violence of Forced Migration in Southeast Asia: Overcoming the Walls of Sovereign Indifference*, paper for the International Association for the Study of Forced Migration, Chiang Mai, Thailand, January 2003.
10. Hazel Lang, *Fear and Sanctuary: Burmese Refugees in Thailand*, Ithaca, NY: Cornell University Southeast Asia Program Publications, 2002. See also: Human Rights Watch, *Unwanted and Unprotected: Burmese Refugees in Thailand*, New York: Human Rights Watch, 1998.
11. For background, see Martin Smith, *Ethnic Groups in Burma: Development, Democracy and Human Rights*, London: Anti-Slavery International, 1994; and Martin Smith, 'The Muslim Rohingyas of Burma', Conference of Burma Centrum Nederland, 11 December 1995.
12. Ibid., and C. R. Abrar, 'Repatriation of Rohingya Refugees', unpublished paper, 1996.
13. United Nations Quarterly Economic Review of Thailand, Burma, 4th Quarter (1978), p. 11, cited in Tony Reid, 'Repatriation of Arakanese Muslims from Bangladesh to Burma, 1978–79. "Arranged" Reversal of the Flow of the Ethnic Minority', unpublished paper, p. 15.
14. Ibid., p. 22.
15. Cato Aall, 'Disastrous International Relief Failure: A Report on Burmese Refugees in Bangladesh from May to December 1978', *Disasters*, issues 3 and 4, 1979, p. 429.
16. For a discussion of this disputed claim, see Smith, *Ethnic Groups in Burma*, pp. 56–57.
17. See the reports of Amnesty International, Asia Watch and other human rights organizations during this period.
18. Loescher's interviews with Milton Moreno, Asia Bureau and Irene Khan, UNHCR, Geneva, March 2000.
19. Human Rights Watch, *Burma: The Rohingya Muslims Ending a Cycle of Violence?*, New York: Human Rights Watch, September 1996; and US Committee for Refugees, *The Return of Rohingya Refugees to Burma: Voluntary Repatriation or Refoulement?* Washington, DC: US Committee for Refugees, March 1995.
20. Northern Rakhine state is a region difficult to access, with few paved roads and transportation links. UNHCR officials' visits to returnees are closely monitored by the authorities thus making it difficult, if not impossible, to get an accurate account of the difficulties returnees faced. See Loescher's interviews with newly arrived Rohingya refugees near Teknaf-Cox's Bazar, February 1998, and David Petrasek, 'Through Rose-Coloured Glasses: UNHCR's Role in Monitoring the Safety of the Rohingya Refugees Returning to Burma', unpublished paper, 1999.
21. Refugees International, 'The Rohingya: Discrimination in Burma and Denial of Rights in Bangladesh', *Refugees International Bulletin*, 21 July 2006.
22. Chris Lewa, 'The Situation of Burmese Refugees in Bangladesh', Bangkok: Forum Asia, November 2003.
23. Loescher visited the camps in 1998 and considered conditions there among the worst he had ever seen in refugee camps he had visited in Africa, Asia and Central America.
24. Abrar, 'Repatriation of Rohingya Refugees', p. 9.
25. TBBC Programme Report, January–June 2007, pp. 6–7.

26. See Chris Lewa, 'Asia's New Boat People,' *Forced Migration Review*, issue 30, April 2008, pp. 40–41.
27. Ashley South, *Mon Nationalism and Civil War in Burma: The Golden Sheldrake*, London: Routledge Curzon, 2003.
28. See Zaid Ibrahim, 'ASEAN Can Do More for Change in Myanmar', statement from the ASEAN Inter-Parliamentary Myanmar Caucus, Kuala Lumpur, 31 July 2006.
28. Human Rights Watch, *Out of Sight, Out of Mind: Thai Policy Toward Burmese Refugees*, New York: Human Rights Watch, 2004.
29. BBC World News, 23 March 2006.
30. Reuters, 'ASEAN Assails Myanmar Crackdown', *International Herald Tribune*, 28 September 2007.
31. 'Burma Generals Keep Envoy Waiting', *International Herald Tribune*, 2 October 2007.
32. Ibid.
33. Most refugees selected for resettlement go to the US. Similarly, Canada, Australia, the UK, Sweden, Finland, Norway, the Netherlands, Denmark and New Zealand are involved.
34. A provision in the USA Patriot Act denies entry to anyone who has provided material support to a terrorist or armed rebel group. A large number of the Burmese refugees (particularly the Karen and Karenni but also the Chin) were excluded from resettlement to the US on the grounds that they had formerly paid taxes, provided food and offered support to insurgent armies that controlled their communities. After much lobbying in Washington, DC, Congress passed a waiver to the statute which would allow the resettlement of these refugees.
35. Susan Banki and Hazel Lang, 'Difficult to Remain: The Impact of Mass Resettlement', *Forced Migration Review*, issue 30, April 2008, pp. 42–44.
36. Correspondence with Jack Dunford and Sally Thompson, TBBC, Bangkok, Thailand.
37. See Pia Prytz Phiri, 'Rohingyas and Refugee Status in Bangladesh,' *Forced Migration Review*, issue 30, April 2008, pp. 34–35.
38. We are indebted to Chris Lewa for making this point to us. Indeed it is unlikely that either the present or any future Burmese government will be willing to grant the Rohingya citizenship rights.
39. International Crisis Group, *Myanmar: The Role of Civil Society*, Report no. 27, Brussels, 6 December 2001.
40. International Crisis Group, *Myanmar: Sanctions, Engagement or Another Way Forward?*, Asia Report no. 78, Brussels, 26 April 2004.
41. This section is based on discussions between the authors and UN representatives in New York in May and December 2006. See also Jurgen Haacke, *Myanmar's Foreign Policy: Domestic Influences and International Implications*, Adelphi Paper 381, London: International Institute for Strategic Studies, 2006.
42. David Arnott, 'China-Burma Relations', p. 81. See also Haacke, *Myanmar's Foreign Policy*.
43. Based on meetings with senior UN officials, New York, May and December 2006.

15

Afghan refugees in Iran and Pakistan

Ewen Macleod

The Afghan refugee situation has evolved significantly since its genesis at the beginning of the 1980s. From a solutions perspective it is arguable whether it can now be understood or effectively addressed through refugee policy frameworks and humanitarian arrangements alone. Indeed, further dependence on the latter may only contribute to deepening the intractability of many of the complex political, economic and social issues now confronting policymakers and practitioners. Evidence and experience since the end of the Soviet occupation suggest that the pursuit of classic refugee solutions and approaches is compromised by the range and scale of post-conflict challenges inside Afghanistan, by contemporary population movements, by poverty and exclusion, and by past and present policies and practices. Without greater political convergence on an achievable and pragmatic set of solutions, progress will remain in doubt. But neither the current policy environment in Iran and Pakistan nor the situation inside Afghanistan is favourable. In the meantime, continuing work on developing new approaches that goes beyond the standard refugee paradigm is essential to the future prospects for finding solutions.

Background and origins of the situation

By the mid-1980s approximately 6 million Afghan refugees had fled in almost equal numbers to Iran and Pakistan. Despite two political accords negotiated under international auspices in 1988 and 2001, and two huge

Protracted refugee situations: Political, human rights and security implications,
Loescher, Milner, Newman and Troeller (eds),
United Nations University Press, 2008, ISBN 978-92-808-1158-2

repatriation movements in 1992–1993 and in 2002–2005, some 3.5 million Afghans still remain in Iran and Pakistan, 27 years after the crisis first began.[1]

The immediate cause for refugee flight was the Soviet occupation of Afghanistan, its brutal support for an isolated and unpopular communist regime, and the rapid escalation of the conflict from a local dispute to an international crisis. Many factors have combined to underpin this protracted situation – international and regional geo-politics, poverty and conflict, the geographical, economic and social structure of Afghanistan, and the varied effects of prolonged displacement and exile. The core of the problem has been and remains the weakness and incapacity of the Afghan state to withstand external influence and to deliver sustainable progress and development for its population.

Afghanistan's history has been shaped by its harsh, arid climate and limited natural resources, by its landlocked location as an ancient transit point for trade and commerce, and by its position on the geo-political front-line of imperial rivalry. In the modern era it has been the last major theatre of Cold War confrontation, the object of regional manipulation and the victim of the darker face of globalization in the form of terrorism and narcotics.

Contemporary Afghanistan has an estimated total Gross Domestic Product (GDP) of approximately US$5 billion (some 45% of which is estimated to derive from the narcotics sector) and a population of approximately 23 million (not including Afghans outside the country), resulting in one of the lowest per capita incomes in the world. The country's socio-economic and other diversified indices place it well below the average poverty levels in the developing world. But even before the 1979 Soviet occupation and the subsequent civil conflicts, it was among the world's poorest countries. It had a tiny formal economy and the majority of the population were involved in subsistence agriculture. Cultivable land area was around 11%, with heavy dependence on spring snow melt for water resources and irrigation. Attempts at land reform were largely motivated by ideology and poorly prepared through lack of technical and judicial capacity. Few formal title deeds for land have ever been issued in Afghanistan, resulting in often violent claims and counter claims.

The state's limited compact with its citizenry was based on and intermediated through the patronage bestowed on tribal leaders by the king, his court and the government. The government was heavily dependent on foreign budget support to cover even basic operating costs. Cold War rivalry sustained these transfers, which largely relieved the authorities of the need to develop a fiscally viable basis for running the state. Political change occurred with only marginal effects on the population at large. This largely benign public management incompetence remained until the late 1970s when a coup by local communist factions in 1978 was followed by ill-conceived attempts to impose new reform policies on a deeply conservative, rural population.

The strength of resistance to these policies provoked a political crisis that was temporarily stabilized by the Soviet intervention. What now appears to have been a muddled expedition with unclear objectives to quell local instability on its southern border provided the *casus belli* for the last significant confrontation of the Cold War. But it also set in motion a train of consequences that continue to impact on the politics, economies and societies of Afghanistan and its regional neighbours almost 20 years after the withdrawal of the last Soviet troops.

Since 1979 Afghanistan has lurched from socialist central planning to the most extreme experiment in fundamentalist Islamist government yet attempted. This volatility was arrested when the Taliban were overthrown by a coalition of international and local Afghan forces. The subsequent Bonn Agreement of December 2001 was not a classic peace agreement based on compromises painstakingly negotiated among the political protagonists of the conflict. Neither did it contain comprehensive and detailed provisions for stability, reconciliation, reconstruction and development. Its contents were dictated mainly by the imperative of filling the vacuum occasioned by the sudden overthrow of the Taliban and resuming humanitarian assistance to a traumatized and hungry population. The Agreement essentially set out a timetable of events designed to re-establish political processes and government institutions. It was anticipated that peace and reconciliation would be built through its actual implementation and strong (but discretionary) donor support for security and reconstruction. However, the same critical political dilemmas that provoked political crisis and collapse in the 1970s – an appropriate structure of government, reform and development in a deeply traditional, conservative society, the role of religion in the state – have still to be worked through and carry important implications for conflict resolution, the state-building agenda and solutions for the refugee situation.

Demography, displacement and population mobility

Since 1979, when the first large-scale arrivals from Afghanistan were recorded in Iran and Pakistan, the populations of all three countries have grown significantly. Although accurate figures for Afghanistan's present population are hard to obtain as no census has been carried out since the 1970s, a figure of around 26 million (including Afghans outside the country) is widely used. This represents a near doubling of the pre-Soviet occupation population of around 14 million. There are no official estimates but most projections of population growth are in the order of 3%.

In Iran, the population has risen rapidly to around 68 million since the Islamic Revolution of 1979. However, birth rates have fallen to under 2% since the mid-1980s following the introduction of family planning policies. Pakistan's population has also risen substantially, to approximately 150 million, and is projected to reach 200 million by 2015. A recent registration exercise carried out in Iran indicated the presence of 920,000 Afghans with refugee documentation. There are also an unknown number of undocumented Afghans in Iran. The registered Afghans account for 1.4% of the overall population. In March 2005 the results of a joint government of Pakistan and UNHCR census of all Afghans in Pakistan were released and indicated a total of just over 3 million persons, or 2% of the overall population. In February 2007 some 2.15 million Afghans were registered in a joint registration exercise. They were issued with Proof of Registration cards permitting them to remain in Pakistan until the end of 2009.

Among the many striking features of both Afghan populations in Iran and Pakistan are the numbers that have been born and grown up in exile (almost 50%) and the significant proportion who have spent more than 20 years outside their country (approximately 80%). As will be explored in more depth below, these figures indicate the extent of the quantitative and qualitative challenges facing policymakers.

The 6 million Afghans in Iran and Pakistan at the height of the Soviet occupation accounted for roughly 40% of the known Afghan population of the time. Such has been the visibility, size and longevity of the Afghan refugee situation that it has become the default framework within which the presence of all Afghans outside their country are situated and analysed.

Historically, there have been three major causes of population movements from Afghanistan – political conflict and violence, natural disasters, and economic migration. These factors have all contributed to varying degrees to the current composition of the Afghan populations outside their country. As will be argued below, this has greatly complicated both the resolution of the refugee question and the design of the more contemporary management arrangements for population mobility that are now required. This perception underpinned UNHCR's position with respect to the census and registration of the Afghan population in Pakistan. The agreements with the government of Pakistan referred specifically to Afghan citizens in Pakistan rather than Afghan refugees.

There have been four distinct movements of Afghan refugees into the neighbouring countries of Pakistan and Iran, each driven by specific aspects of successive conflicts. The first and largest displacement occurred consequent to the communist-led coup and the subsequent Soviet occupation between 1979 and 1989; the second during the conflict between the Najibullah government and the mujahideen (1989–1992); the third during the inter-factional fighting and the rise of the Taliban movement (1994–2001); and the fourth at the time of the coalition intervention and the overthrow of the Taliban in the autumn of 2001.

Afghanistan is prone to natural disasters. Their effects are exacerbated by a harsh climate, poor natural resource management, underdevelopment and weak public services. In 1971, widespread crop failure following severe drought caused many inhabitants of the affected provinces of northern and north-western Afghanistan to migrate to Iran. The same pattern was evident following the onset of protracted drought in 1998. Its impact was more widespread and profound due to the aggregate effects of the long conflict, and triggered internal as well as external movements to both Iran and Pakistan.

There are many historical instances of Afghan migration for economic and commercial purposes. In the modern era it was initially characterized by the annual and seasonal movement of limited numbers of nomadic pastoralists and traders. By the early 1970s, the seasonal migration of Afghan male labourers to Iran and the Gulf states in search of employment was more commonplace. It accelerated greatly after the oil price rises of 1973 fuelled a construction boom. Since the end of the Soviet occupation, and especially since 2001, there has been a rapid growth in both formal and informal trade through Afghanistan into the neighbouring countries. This has encouraged cross-border movements and linkages among Afghan retailing networks on both sides of its frontiers.

Over the last decade the nature of population movements to and from Afghanistan has become more complex. In addition to conflict and drought, many Afghans now cross the border in both directions to look for seasonal employment, to trade, to access services and to maintain social and family connections. There is emerging evidence to support the view that these networks have become an important component in the livelihood systems of many Afghan families. With a major increase in population, stasis in the area of land under cultivation and increased pressure on limited employment opportunities in rural areas, both internal and external migration is rising. The volume of bilateral trade (both official and unofficial) with both Iran and (especially) Pakistan has grown rapidly. The reverse phenomenon of Afghans temporarily returning to their homeland (especially to provinces close to the border) for seasonal employment, for trade or to maintain active social relations has also been evident for some time.

Contemporary cross-border movements to and from Afghanistan to Pakistan are substantial and largely unregulated, even at official crossing points. Many more persons arrive by unofficial crossing points undetected and unrecorded. Previous immigration concessions granted for temporary movements to local people straddling the borders are no longer adequate to address the depth and range of cross-border travel. Stricter border controls, and the need for updated documentation, have made illegal and multiple movements to and from Iran a more costly and risky enterprise. Concurrently, the issuance of passports to Afghans and the granting of different forms of visas by Iranian diplomatic missions have risen considerably. The growth in the officially regulated movement of Afghans to Iran is a positive development. Yet its volume is also a concern to Iranian policymakers since it is clear that it is motivated largely by economic migration.

The refugee experience – key characteristics and social change

An assessment of Afghan refugee experiences has to take into account the different phases of the displacement cycle – initial flight, stabilization and consolidation, return and reintegration or continued asylum. It has also to include some political contextualization to understand how changes in the international, regional and domestic circumstances impact on the situation. A more fine-grained study would be needed to examine the original reasons for flight, prior social, family and individual capital, skill and asset levels before and during exile, the risks and challenges inherent in return to different locations, and the institutional and policy actions of governments, donors and assistance providers. This chapter attempts to capture only the most salient features of the Afghan refugee experience and to highlight some insights relevant to the overall focus of the study.

It is one of the many complexities of the Afghan refugee situation that the displacement cycle has been repeated several times, something that substantial numbers of Afghans have experienced. In that respect, and in a number of others, the description of the situation as 'protracted' does not do proper justice to its evolution or inherent dynamics. Moreover, the composition of the Afghan presence in Iran and Pakistan has evolved during the last two decades. It contains strata from four displacement periods, ethnic minorities from all regions, and an overall mix of refugees, asylum seekers and different categories of economic migrants. These groups have varying attitudes to return and make their decisions in what they judge to be their own interests. There is a direct correlation between the length of stay in exile and the willingness to return on a permanent basis.

The majority of Afghans who arrived in Iran and Pakistan during the 1980s came from rural areas, especially from adjacent provinces. They were mostly farmers and share croppers, though a significant proportion in Pakistan were nomads (*kuchi*) whose way of life had become untenable during the war. During the later period of the civil war, there were more arrivals from cities, including civil servants and middle-class professionals. Latterly, the proportion of refugees from among ethnic minorities rose prominently as a consequence of the Taliban's occupation of the centre and north of Afghanistan.

As far as was possible, tribal coherence, practices and customs were maintained in the refugee camps. But the arming of the mujahideen, and their affiliation with the political factions, facilitated the emergence of new types of power holder. Prior to the Soviet occupation, religious fundamentalism was not a significant political force in Afghanistan's formal public life. *Jihadi* politics provided the path by which previously marginal figures acquired immense power and influence within the refugee community. The emergence of these largely fundamentalist factions also facilitated the growing power of networks of commanders affiliated to them. That legacy, in the form of a breakdown of traditional tribal authority, is still potent inside Afghanistan today, especially in the Pushtun belt along the border with Pakistan.

In Pakistan, the Afghan population remains predominantly Pushtun, though there are also substantial numbers of Hazaras (particularly in Baluchistan), Baluchi, Tajiks, Turkomens and Uzbeks. The physical settlement of the Afghans in Pakistan has diversified and developed over nearly 30 years. Although conditions vary, the majority of Afghans now live in permanent dwellings, most of which have been self-established. It is therefore something of a misnomer to talk of Afghans in camps.

Until the 2005 census, a precise breakdown of the location and numbers of Afghan communities in urban contexts in Pakistan had never been established. It is believed that many Afghan refugees first moved to the cities during the 1980s when local employment opportunities in the North West Frontier Province (NWFP) and Baluchistan were oversubscribed or after seasonal agricultural work ended. Many male breadwinners continue to look for employment in cities, often leaving their dependants in the refugee villages.

The degree of social interaction between Afghan refugee and local Pakistani village communities has not been documented in detail, although this gap in knowledge has recently been partially addressed. Available data do suggest that Afghans tend to be consistently poorer than their host communities, have high levels of debt and are often in poor health.

Inter-communal relations have been predominantly harmonious in NWFP, where there are close cultural and ethnic affinities. In Baluchistan, however, historical concern with the local ethnic balance has made the presence of the Afghans more sensitive. This is particularly the case when it has impacted on economic opportunities or placed pressure on the environment and the delivery of social services.

At the same time, exposure and access to more urbanized environments, different labour markets, social services (health and education), electricity, transport and communications networks have enabled them to experience a standard of living few knew previously in Afghanistan. Moreover, many Afghans have been able to start businesses and to develop marketing and retailing outlets. New forms of social organization and civil society have also emerged within the Afghan communities in response to their situation.

In Iran, a registration exercise carried out in 2005 noted that approximately 1 million Afghans were living as families and had been in the country for a long time. (There were thought to be several hundred thousand Afghan migrant workers present in the country before the start of the Soviet occupation). It indicated that there were equal proportions (roughly one-third each) of Tajiks and Hazaras, with the balance being made up of the other ethnic groups. Only a fraction of Afghans (less than 2.5%) have been housed in refugee villages, mostly in the east of the country.

For a long period, Afghans were generously permitted free access to Iran's social services (health and education) and to work in designated sectors. Until recently there has been little formal documentation or analysis of the employment profile and the impact of Afghans on the Iranian economy at national or local level or in individual sectors. During the 1980s, Afghans filled a significant gap in the workforce during the war against Iraq. Despite stricter labour legislation, there is considerable anecdotal evidence to show that Iranian employers continue to appreciate Afghan workers as a source of cheap and reliable labour in important sectors (agriculture, construction and services).

Social interaction between Afghan refugee and Iranian communities has only recently been assessed. As in Pakistan, there have been few recorded instances of serious communal strife. There are many documented cases of marriages between Afghan men and Iranian women. Recent changes to Iranian legislation offer the opportunity for the children resulting from these unions to apply for Iranian citizenship. Overall, it is fair to surmise that exposure to the quality of life in Iran has had a significant impact on generations of (especially) young Afghans. There are, for example, some indications that literacy and numeracy levels among young Afghan refugees in Iran are notably higher than those of their parents. This followed from Iran's generous policy of granting Afghans free access to health and education through the 1980s and much of the 1990s.

For convenience, the following reflections on the overall Afghan refugee experiences are divided into two periods – from the beginning of the Soviet occupation to the fall of the Najibullah government (1979–1992), and from the onset of the civil war (1993) to the present. The subsequent paragraphs on some key qualitative issues apply to the whole period.

Soviet occupation and withdrawal

The first major refugee arrivals in Pakistan and Iran occurred following military and bombing campaigns conducted by the Soviet and Afghan armies in the border provinces in the east, south and west of the country. Most large-scale refugee arrivals occurred during the period 1980–1984. Initially, most Afghans within these emergency influxes were housed in tents or ad hoc temporary accommodation along the respective frontiers. Many families subsequently moved into or built more permanent accommodation or migrated to cities. Those who have arrived subsequently have tended (by and large) to settle in cities, partly because of accumulated knowledge and partly because of ethnic and kinship ties. In Iran over 97% of Afghans live in cities or towns and only 3% are in camps. In Pakistan, approximately 50% of Afghans now live in cities and towns. What are still referred to as camps and more accurately resemble small towns, peri-urban settlements and villages.

Initially, both Tehran and Islamabad publicly welcomed the refugees with strong expressions of Islamic solidarity and sympathy. The hospitality of the ordinary people of Iran and Pakistan towards refugees may be judged extremely positively. Despite the huge numbers, there are virtually no known instances of significant inter-communal violence. An important part of this harmony can be attributed to religious and ethnic affiliation. There were also strong linguistic and cultural affiliations between the refugees and their host communities. In fact, within the Afghan refugee communities themselves, there is greater evidence of political, ethnic and social violence and tensions. This was more visible during the 1980s when recruitment and competition for power, influence and resources among the mujahideen factions were at their height.

Iranian public advocacy for the refugees remained largely humanitarian and rhetorical in character. In later years there was an increase in political and cross-border resistance activity. The government has never permitted a significant international assistance presence on its territory, a policy that has been maintained steadfastly to this day. In Pakistan, the Afghan refugee presence not only became an international *cause célèbre*, it also presented an opportunity to local policymakers to further their own geo-strategic objectives. Indeed, refugees became an important component of the resistance to the Soviet occupation, with many camps effectively providing both sanctuary and logistical back up.

The government of Pakistan was active in establishing seven political factions whose composition, outlook and orientations tended to reflect its own objectives in Afghanistan and the region. The factions provided convenient vehicles through which to recruit, organize, arm and supply the resistance. The establishment of camps was also conducted along political and factional lines and permitted the factions' leaders control over the delivery and distribution of food and other kinds of assistance. This patronage negatively influenced the accurate and impartial allocation and distribution of aid. It also influenced the content and practice of aid inputs and delivery. Many religious schools madrassahs for Afghan refugee children were established during this period.

During the 1980s, the international and national humanitarian response was significant. Pakistan opened its doors to a broad range of assistance agencies. For most of the 1980s, aid programmes worth hundreds of millions of dollars annually operated in NWFP and Baluchistan. Immediate essential needs (shelter, food, health and basic education) were met. Iran's more cautious foreign policy limited international presence and support. However, it contributed generously from its own public resources, allowed free access to its social services (health and education) and permitted access to its labour market (though mostly in menial, low-value employment).

In overall terms, the period of the Soviet occupation may be characterized as one where physical safety, basic stability and the essential needs of the Afghan refugee population were achieved relatively quickly. The international and regional political constellation favoured a benign protection climate but disguised a number of more malign trends. Indeed, the origins of later problems that have sustained the conflict may be traced directly to this period – the militarization of camps, the growth of extremism and intolerance, the rising influence of a new and unaccountable group of power holders (mujahideen), the manipulation of political and civil rights, and restrictions on aid access and delivery.

The civil war period and its aftermath (1993–present)

When the Soviet-backed government of President Najibullah fell in 1992, the atmosphere and attitudes towards Afghan refugees changed, especially at the political level. Both Tehran and Islamabad expressed the view that it was time for the Afghans to return home, given that the original cause for flight had now been removed. A range of more restrictive measures was introduced and assistance programmes were wound down in expectation of mass return. Initially, there were large repatriation movements, but they tailed off significantly as rivalry among the Afghan factions vying for power turned into widespread violence in the capital and elsewhere.

Many Afghans connected with the former regime had begun to flee to Iran and Pakistan and further afield soon after the victory of the mujahideen. They were followed by others escaping the different states of the conflict. Their reception in the neighbouring countries was markedly different from that of their predecessors. Moreover, in the case of Pakistan, this coincided with both a closer relationship with some of the protagonists inside Afghanistan and a significant downturn in both international interest and funding. Yesterday's courageous fighter against communist oppression had become today's violent extremist or illegal migrant. At the same time, the emergence of other major humanitarian challenges in the Great Lakes region of Africa and the former Yugoslavia rapidly claimed a higher proportion of international humanitarian resources.

The apparent neglect of the Afghan refugee situation by the international community certainly contributed to an erosion of the quality of protection and asylum in Iran and Pakistan during this period. Afghans now leaving their shattered country were periodically described as economic migrants. On several occasions borders were closed, albeit for brief periods. Public comment on the negative influence of the Afghan presence became both more shrill and regular. Afghan refugees were linked to drugs, arms and economic distress. The accusation that the world had abandoned the asylum countries to shoulder the burden of the Afghan presence alone was voiced and periodically cited as justification for more restrictive polices and actions.

In 2001, the Iranian authorities carried out a new registration exercise of Afghans and stopped issuing documentation to new arrivals. This growing antagonism reached a climax in 2001 when the Pakistani authorities refused to provide acceptable sites or to permit access and assistance to Afghans fleeing from the Taliban campaign in northern Afghanistan. This was resolved only after several months of difficult negotiations had produced an agreement under which UNHCR agreed to screen the population in question to determine whether they were indeed fleeing from war and persecution.

The Taliban ascendancy ushered in a period of relative stability and security in Afghanistan. It also coincided with a major drought, a large and fast-growing problem of internal displacement, and the further deterioration of an already degraded and war-distorted economy. These factors combined to generate new arrivals, particularly to Pakistan, one of only three states to grant official recognition to the Taliban. There were markedly different responses from Iran and Pakistan. For a while, the level of pressures on Afghans appeared to relent as Tehran's relations with the Taliban leadership went from bad to worse. This did not, however, stop a serious rise in deportations of Afghans judged by the Iranian authorities to be illegal migrants.

In Pakistan, there were renewed (though unsuccessful) calls to relocate Afghans to internally displaced persons (IDP) camps inside Afghanistan and to prevent further arrivals adding to economic and social pressures. The extremist policies of the Taliban did spark a certain revival of international interest in the plight of Afghans both inside and outside the country. This may also have been influenced by the fact that by 2000 Afghans had become the fastest-growing group of asylum seekers in Europe. However, in material terms the volume of assistance to the asylum countries did not increase notably, although there was greater support and engagement on human rights issues.

The period since the removal of the Taliban (2001 to the present) has seen some of the largest repatriation operations in modern history. Despite these movements, there has not been a particularly positive evolution in policies towards Afghans in Iran and Pakistan. There has been an increase in pressure for repatriation, with an attendant rise in insecurity among the remaining populations. The Iranian authorities have imposed many restrictions (both old and new) on Afghans. In recent years, deportations have averaged almost 100,000 persons annually. In Pakistan, the government has closed a substantial number of refugee camps and announced the closure of others, citing security as the principal justification. The focus on camps followed the realization that the majority of the returns during 2002–2004 were from urban locations.

The overall circumstances of Afghan refugees in the 16 years since the fall of the Najibullah government have been characterized by increasing uncertainty. The pressures on them to return following the fall of the communist government have clearly grown. Much may be attributed to the belief that it is economic and social conditions in the asylum countries rather than classic refugee and protection criteria that prevent their return home. The pressures are also reinforced by a tendency to attribute a disproportionate share of the blame for contemporary economic and social problems to the refugees.

The majority of the Afghans still in Iran and Pakistan have now been in exile for two decades or more. Yet their status and prospects for remaining under improved terms of stay in Iran and Pakistan remain unclear. The decline in international humanitarian support in their situation has certainly affected local political attitudes in the asylum countries, especially in Pakistan. However inaccurate and unjust, this has also lent weight to the perception that Afghans are a burden that their hosts should no longer bear.

Policy responses

Following the collapse of the communist-backed administration of President Najibullah in 1992 and the installation of the interim mujahideen government, policymakers largely assumed that the Afghan refugee populations would repatriate within a few years. Indeed, almost 2 million Afghans did return, the larger part from Pakistan. But these repatriation movements, although huge, represented only one-third of the estimated refugee numbers in Iran and Pakistan. By 1994, political instability, civil conflict and deteriorating economic circumstances inside Afghanistan halted national recovery and reconstruction. Not only did the situation stall repatriation, it also generated new population displacement and movements inside and outside the country. These trends accelerated following the ascendancy of the Taliban movement between 1996 and 2001. By 2000, applications for asylum from Afghans had been received in almost 70 countries around the world and accounted for the largest number of asylum seekers in Europe.

Policy responses to the Afghan refugee situation have been shaped principally by international, regional and national political and security interests. From the period of the Soviet occupation through to the response to the events of 11 September 2001, the engagement of states has oscillated between strong humanitarian commitment (during periods when Afghanistan was perceived to be of vital geo-strategic importance) and benign neglect. It has also been characterized by shifts in the policies of regional states in relation to their national interests in Afghanistan. These have embraced classic refugee definitions and principles during periods of international engagement (and aid largesse) and less accommodating descriptions ('illegal economic migrants') when this has been absent. More recently, the government of Pakistan, seemingly in response to accusations of its alleged continued support of the Taliban, linked the presence of Afghans on its territory (particularly those remaining in camps) with terrorism and the insurgency.

This definitional promiscuity has tended to obscure the range of complex political, economic and social changes within the Afghan populations in exile that have occurred and which make clear-cut definitions increasingly hard to apply. These same changes have also had an impact on the viability of the approach to solutions doggedly pursued by most stakeholders. Ever since the signature of the Geneva Accords in April

1988, large-scale repatriation has been the preferred, if elusive, prescription for resolving the Afghan refugee situation. The combination of policy shifts and the dominance of repatriation as the default solution has also prompted particular difficulties for UNHCR as it sought to maintain coherence over its protection and legal responsibilities towards a refugee population itself undergoing important transformations.

At the institutional level, Iran is a signatory to the 1951 Convention and the 1967 Protocol, whereas Pakistan is not. Neither country has developed a body of domestic refugee law. Policies have therefore been influenced essentially by political considerations and delivered primarily through administrative instructions. Nevertheless, to their considerable credit, for most of the last 29 years the two governments have respected international refugee and humanitarian principles and have largely reflected these in the arrangements established and applied to Afghans.

As indicated above, humanitarian aid responses to the Afghan refugee arrivals of the early 1980s were substantial and consistent for almost a decade. But they were mostly founded on the premise that only short-term, essential humanitarian assistance was required since official policy held that one day the refugees would go home. These responses were also reflected in the establishment of official refugee departments that were expected to disappear after a relatively short period.

By the time the Afghan refugee displacement had entered its second decade, there was awareness of the need to graduate from relief assistance to more predictable longer-term programmes. Some progress was made – albeit on a modest scale – especially in critical sectors like education, health, water and sanitation. But with declining international interest in Afghanistan for most of the 1990s, there was insufficient political will to push for a change in programming arrangements. There was also a persistent policy stance within the asylum countries which held that more development-oriented investments would discourage rather than enable repatriation. In consequence, shrinking humanitarian aid budgets were obliged to offer social welfare functions at a level wholly incompatible with needs. Opportunities for equipping Afghans with the skills and knowledge they might have needed to better re-establish themselves in their homeland (and thus enhancing solutions) were lost. This is reflected today by the fact that the majority of the Afghan populations in Pakistan and Iran remain poorly educated and are predominantly engaged in menial day-labouring tasks. Furthermore, the same organizational arrangements, heavily oriented to relief functions, remain largely in place after more than quarter of a century.

With respect to the political arrangements for solutions, Annex IV of the 1988 Geneva Accords provided for a bilateral agreement between Afghanistan and Pakistan on the voluntary repatriation of refugees. It was not until the early 1990s that the first Tripartite Agreements governing the legal and operational aspects of voluntary repatriation between UNHCR, Afghanistan, and Pakistan and Iran respectively were signed. These arrangements subsequently lapsed during the Taliban period when repatriation declined to negligible levels.

The Bonn Agreement of December 2001 contained no references to the refugee and displacement challenge. But large-scale voluntary repatriation assisted by UNHCR resumed in March 2002 and the Tripartite Agreements (and their supporting commissions) have been re-established. To date over 5 million Afghans have returned since March 2002, internal displacement has largely stabilized (although solutions for many IDPs have still to be found), and secondary movements to countries outside the region have declined dramatically.

The high return figures of the period 2002–2004 mirrored the optimism and the political progress achieved under the Bonn Agreement. They also revived the belief in repatriation as the primary solution to the Afghan presence in the neighbouring countries. Yet an estimated 3 million Afghans still remain in Iran (920,000) and Pakistan (2.1 million). In 2006 there was a sharp decline in assisted voluntary repatriation, particularly from Iran. This downturn coincided with a rise in violence in the southern and south-eastern provinces and a loss of the initial confidence in the prospects for reconstruction. It may also be attributable in large measure to the fact that the majority of Afghans remaining in Iran and Pakistan have been there for more than two decades, 50% of whom have been born in exile.

The policy responses in Iran and Pakistan to these trends have been instructive. Ever since the fall of the Najibullah regime in 1992, the Iranian authorities have consistently asserted that Afghans should return to their homeland because the original reasons for their flight as refugees no longer exist. Despite subsequent periods of instability in Afghanistan, they have maintained this position. They introduced registration exercises in 2001, 2003 and 2005 which were oriented to return rather than protection in that the actual documents issued are literally registration slips for repatriation (*amayesh*) with time-limited, if renewable, validity.

Despite these renewed pressures and a series of restrictive policies against the Afghan refugee population in Iran, no significant progress in elevating repatriation figures has been achieved since 2005. More recently, the Iranian government indicated that it would put in place a comprehensive response to the challenge of population movements by expelling all undocumented Afghans by the end of March 2008 and introducing a mechanism to assess the continued need for protection under the international refugee regime of the remaining 920,000 registered Afghans. It may be assumed that the calculation underpinning this latter strategy is that the majority of those processed may not qualify for continued protection as refugees. But the Iranian authorities have stated their willingness to offer work and residence permits to registered Afghan refugees provided they return first to Afghanistan with their families. It is the first official acknowledgement of the links between the Afghan presence in Iran, the needs of the Iranian economy and the labour migration pressures in Afghanistan.

In Pakistan, the government has recently announced that all non-registered Afghans discovered without passports or valid visas will be treated as illegal migrants and deported. It has further announced that a three-year repatriation plan has been agreed by the cabinet under which it foresees the return of all the remaining registered Afghans (2.15 million) in Pakistan. To date, the government of Pakistan has not deviated from its stated goal of returning all Afghans despite the obvious tensions between this ambition, the declining return figures and the principles of voluntary and gradual return.

A critical element of the solutions equation is the capacity of Afghanistan and the reconstruction programme to absorb those who have repatriated to date and to create conditions conducive to return in future. All the portents suggest that this will be a long and arduous task. The inclusion of provisions for supporting refugee return and reintegration in documents such as the Afghanistan Compact (agreed in London in January 2006) and the Afghanistan National Development Strategy do show greater political willingness to embrace the refugee and displacement issue. But so far they have not led to significant changes in resource allocations or management arrangements. To date, UNHCR and its counterpart, the Ministry of Refugees and Repatriation, have largely been left alone to raise funds for programmes addressing immediate needs – shelter, water and sanitation. Though many returning Afghans have certainly benefited from national development programmes, there is currently no formal linkage between resource allocations and the patterns and presence of returning refugees. This may change in future. The high visibility and press coverage of the large-scale deportation of undocumented Afghans from Iran, and the assertions by Pakistan that it will repatriate all 2.15 million registered Afghans before the end of 2009, have pushed the refugee issue up the national agenda.

In anticipation of the downward trends in repatriation, of the changing nature of population movements within the region, and of its own limitations in addressing such complex challenges, UNHCR launched a new policy initiative in 2003–4. Its principal aims were to highlight the fact that reliance on the refugee and humanitarian framework alone would not resolve the situation and that a broader range of stakeholders and other approaches and resources would be required. UNHCR also supported a series of research studies on cross-border movements, transnational networks, the reintegration process and needs assessments, to emphasize the economic and social drivers of contemporary population movements. Its purpose was to show how retaining the refugee paradigm as the sole basis for policy approaches was increasingly being undermined by complex socioeconomic realities.

In its analysis of return prospects, UNHCR drew attention to the strong economic dimension of refugee decision-making and the range of political, social and cultural challenges for repatriation. It argued that the developmental and migratory dimensions of the Afghan refugee situation needed to be addressed in a more holistic approach to solutions. Whilst acknowledging implicitly that not all Afghans were in need of international protection as refugees, it contended that, due to the continuing fragility of the political, security and reconstruction process in Afghanistan, there would continue to be an important refugee and protection dimension to the situation. It also encouraged the governments of the region to develop policies and arrangements to manage population movements as normal socioeconomic processes rather than as refugee flows.

With a view to exploring how a new consensus on solutions for Afghanistan might be forged, UNHCR supported a series of consultations with the governments of Afghanistan, Iran and Pakistan, major donors and international organizations during 2004–2005. A major objective was to emphasize UNHCR's wish to broaden stakeholder engagement with the problematic, particularly those aspects which as a humanitarian and protection agency it had neither the technical nor the financial capacity to address. Rather than embrace the opportunities that a more diversified approach would produce, the response of the asylum countries to this comprehensive approach was to call for renewed emphasis on repatriation and greater donor support for Afghanistan's reconstruction.

At the beginning of 2007, the poor regional and bilateral political and security climate, and the deteriorating circumstances in Afghanistan, again cast a shadow over the prospects for future solutions of the Afghan refugee situation. Furthermore, the refugee problem – as an object of political tensions between states – has re-emerged to obscure how the problems of refugees – as an issue of political, economic and social complexity – can be tackled. The modest progress made in advancing a more differentiated and adapted approach to this challenging human problem has again been overtaken by political rhetoric and recourse to simple mathematical solutions in the form of multi-year repatriation plans. More worryingly, it has been accompanied by a further erosion of the refugee protection and asylum climate. There is as yet little clarity as to how population movements will be managed in future or what will be the future status of the old refugee population. All policy development appears to be predicated unsteadily on the success of repatriation programmes that are under pressure to relax the critical principle of voluntariness.

As with other development and reconstruction challenges in Afghanistan, tackling the refugee and displacement problematic will depend on Afghanistan's political and security circumstances and the ability of the Afghan authorities to overcome familiar governance and public management shortcomings. The country's ability to absorb and retain a further 3 million Afghan returnees and to reverse an out-migration trend that has grown appreciably in recent years was uncertain even before the resurgence of violence in 2005. The level of investments required to arrest the previous cycle of return and departure witnessed since the beginning of the 1990s cannot easily be quantified. But it certainly goes far beyond the level of financing witnessed to date. Without substantial commitments for a decade or more, the prospects for the remaining populations to repatriate sustainably and voluntarily, or to find a more secure and predictable status, appear uncertain. Moreover, as long as political expediency continues to trump pragmatism and realism in the planning and management of solutions and approaches to population movements, the complexities surrounding the Afghan refugee problematic will continue to proliferate as a third generation of displacement draws closer.

Note

1. This chapter is written by a long-time practitioner and draws upon his recent experience with the Afghan Unit at UNHCR both in Geneva and in the region. The data in this chapter relating to Afghanistan and countries hosting refugees from Afghanistan can be found in the UNHCR Statistical Yearbooks (published periodically by UNHCR, see www.unhcr.org/statistics.html); the Global Human Development Reports (published annually by the UN Development Programme, see http://hdr.undp.org/en/reports/); and National Human Development Reports (published periodically by the UN Development Programme in cooperation with national authorities, see http://hdr.undp.org/en/reports/).

Part III
Policy conclusions and recommendations

Part III

16

A framework for responding to protracted refugee situations[1]

Gil Loescher and James Milner

As the earlier chapters of this volume make clear, protracted refugee situations pose a significant challenge to refugees, the agencies that care for them and a wide range of other actors. While important work has already been published on addressing certain aspects of these challenges, especially the challenge of refugee livelihoods,[2] the objective of this chapter is to present the elements of a framework for resolving protracted refugee situations. This chapter argues that solutions will best be achieved if they are pursued within a broader political and strategic context, and linked to conflict management, peacebuilding and development activities. This chapter argues that two elements of a response are required to find solutions for protracted refugee situations: first, it is necessary to address the current challenges posed by protracted refugee situations; second, it is necessary to develop comprehensive solutions. To do this, a shift is required, from the current 'care and maintenance' mindset, focused on managing protracted refugee situations, to a more 'solutions-oriented' approach, involving more than just humanitarian actors.

A common characteristic of many protracted refugee situations today is the emphasis placed on 'care and maintenance': providing basic, life-saving assistance, long after the emergency phase of the operation has ended. As outlined in previous chapters, there are many reasons for this phenomenon, including the typically restrictive policy responses of host states and the low and dwindling levels of external and donor engagement. The combined effect of restrictive asylum policies, especially those that make refugees dependent on international aid, and declining donor

Protracted refugee situations: Political, human rights and security implications,
Loescher, Milner, Newman and Troeller (eds),
United Nations University Press, 2008, ISBN 978-92-808-1158-2

support for these assistance programmes makes the delivery of even a basic care and maintenance programme a significant challenge.

An effective response to protracted refugee situations must overcome this dynamic. The first objective should be to shift the emphasis of the programme from a care and maintenance approach to a more 'solutions-oriented' approach, where the focus of the programme points toward the resolution of the situation itself. Initiating such a shift will depend on the ability to more fully engage and address the interests and concerns of a range of state actors, especially host and donor states. At the same time, such a shift will require not only a change in approach on the part of actors currently engaged in the refugee programme, but also the direct involvement of a number of other actors from the peace and security and development communities. Without the support of these broader actors, UNHCR can only be expected to adopt what Jamal terms in this volume 'ad hoc, modest and segmented approaches'. As such, it is only with the support of a broader set of actors that solutions can be truly comprehensive.

This chapter outlines how such a shift may be initiated. It begins by examining a number of specific examples of innovative responses to specific challenges in protracted refugee situations that have involved the engagement of diverse actors, and argues that these examples of 'what has worked' in individual situations needs to be more broadly and systematically shared. Reponses to specific challenges, however, are no substitute for a comprehensive solution to the refugee situation itself. The chapter will continue by drawing on lessons of past and contemporary efforts to find solutions to argue for the need for real engagement of peace and security, development *and* humanitarian actors for solutions to be truly comprehensive. Building from these lessons and drawing from earlier chapters in this volume, the chapter will then present a 'solutions framework', outlining how these three sets of actors can cooperate in the short, medium and long term to develop and implement comprehensive solutions, before considering how the recent creation of the UN Peacebuilding Commission provides a possible context for the implementation of such an approach.

Central to this approach is a recognition of the important link between the resolution of protracted refugee situations and successful peacebuilding efforts. While there has been considerable innovation in the literature on peacebuilding in recent years,[3] there has been limited consideration of the linkages between refugee populations and peacebuilding in the country of origin.[4] Often, the successful return and reintegration of refugees into their home country is taken as an indicator of the success of peacebuilding efforts. It must also be recognized, however, that a failure to address refugee-related challenges may undermine peacebuilding.

Refugee-populated areas in neighbouring states may harbour elements that seek to undermine peacebuilding in the region, and refugee populations may be drawn into a campaign of destabilization. Likewise, the concerns of host countries and the limitations on their willingness to host refugees must be taken into account. If the concerns of host states relating to the potentially negative impact of the prolonged presence of refugees on their territory are not addressed, host states may pursue early and coerced repatriation, placing fragile institutions in the country of origin under significant strain and further undermining peacebuilding efforts. Recognizing these links further illustrates the need for a comprehensive and collaborative response to the challenge of protracted refugee situations.[5]

Addressing current challenges

As outlined in chapter 2, and as clearly illustrated in the case studies of this volume, prolonged exile poses a number of specific challenges for refugees, the agencies that are charged with their protection and assistance, and the countries of asylum and origin. These challenges include increased levels of crime and insecurity in refugee-populated areas, high levels of domestic and gender-based violence, dependence on dwindling international assistance and a lack of alternative livelihoods, and a range of direct and indirect security concerns. In many ways, these problems can frustrate efforts to shift from a care and maintenance to a solutions-oriented approach, and must consequently be addressed in the short term as part of a comprehensive solution.

A number of programmes in different regions have developed innovative ways of addressing some of these concerns, often involving the sustained engagement of a broader range of actors beyond humanitarian agencies alone. While formulated and implemented in specific operational contexts, and while the 'successes' of these programmes are themselves the source of some debate, it is important to briefly consider the nature and benefits of some of these initiatives.

Examples from Kenya

As detailed in the chapter by Kagwanja and Juma, Kenya has hosted over 135,000 Somali refugees since 1992. The overwhelming majority of these refugees live in three camps near the town of Dadaab, in the north-east province of Kenya. During the 1990s, these camps were renowned as the most violent refugee camps in the world, where rape, murder and armed robbery were almost daily occurrences. Violence was endemic not only in the camps but also in the areas surrounding the camps, as bands of *shiftas*, or bandits, attacked convoys of humanitarian relief, aid workers, and refugees collecting firewood outside the camps.

A series of interventions was introduced by UNHCR in the late 1990s to address these concerns, including a mobile court system to try those suspected of criminal offences, additional support to the Kenyan police to substantially increase their presence in and around the camps, and the firewood project, designed to provide refugees with 30% of their firewood needs, with the objective of reducing the exposure of refugee women to sexual violence by reducing the amount of time they would be required to spend in the insecure areas around the camps. The six years following the introduction of these initiatives witnessed a dramatic decline not only in the number of reported cases of rape in the Dadaab camps, but also in murder and armed robbery. In 1998, there were over 300 reported cases of violent crime in the Dadaab camps, of which 104 were cases of rape. By 2003, that number had fallen dramatically: to 36 reported cases of violent crime, of which 15 were cases of rape.

Humanitarian workers in Dadaab believe that this change has largely been the combined result of the range of innovations introduced since 1998.[6] For example, the firewood project has had a number of direct and indirect benefits for refugee security. Specifically, it has created jobs for the local population and has encouraged young men who would otherwise pursue banditry as a means of livelihood to participate in the more lucrative trade in firewood. The firewood project has thereby contributed to the mitigation of a number of concerns. First, it has reduced the strain on the scarce environmental resource of firewood in and around Dadaab by ensuring that the firewood is collected in a managed way across a wider area. Second, it has ensured an income for the local population, thereby reducing grievances that may arise between refugees and Kenya. Third, by providing a context within which the refugees and the local population can cooperate in a large-scale, mutually beneficial project, better understanding has been developed between the two groups, which serves as an important basis for future conflict resolution at a local level. Although a costly programme, the firewood project is one example of a development-related project that has played a significant role in addressing the security implications of the protracted presence of Somali refugees in northern Kenya.

Examples from Tanzania

Special Programmes for Refugee Affected Areas (SPRAAs) are a second example of possible development-related interventions. The positive effects of SPRAAs have been most striking in Kibondo, a district in western Tanzania that has hosted over 100,000 mostly Burundian refugees since 1993.⁷ The protracted presence of refugees in western Tanzania has resulted in a rise in local grievances against the refugees and a common belief that their presence has resulted in a rise in banditry, crime, disease and environmental degradation, in addition to placing a significant strain on the local infrastructure and public services. As a result, relations between refugees and local authorities have deteriorated and a sense of insecurity now prevails.

In an attempt to reverse this trend, humanitarian agencies have undertaken a wide range of programmes to directly benefit the local population and counter the negative effects of the presence of such a large refugee population. Programmes have focused on the rehabilitation of roads used by aid convoys, the improvement of water supply to local communities, the development of local communication infrastructure, the building of local schools and health centres and the planting of trees.

A report by the Centre for the Study of Forced Migration at the University of Dar es Salaam comprehensively reviewed the allegations made by Tanzanian officials, both locally and nationally, that the presence of refugees in districts like Kibondo is a burden to the host state and a source of insecurity.⁸ In assessing the cost of hosting refugees against the benefits that have accrued to the local population, both directly through the SPRAAs and indirectly through the creation of employment and larger markets, the report concludes that the hosting of refugees has been a benefit to Tanzania. Indeed, local community and business leaders at the local level recognize the efforts that have been made to ensure that the presence of refugees benefits local development, and they have worked closely with UNHCR in the formulation of SPRAAs.

Activities such as SPRAAs and the firewood project contribute to an improved security and protection environment by reducing competition between refugees and the local population over scarce resources and by reducing local grievances towards refugees. At the same time, however, it has been argued that such programmes, and more generally the presence of refugees and refugee programmes, could, if effectively managed, significantly contribute to longer-term local and national development.⁹ There is, therefore, a double benefit in the short to medium term: development-related projects targeting refugee-populated areas can foster an environment of greater security and protection for refugees and the local population, while also contributing to broader national development objectives.

Examples from Guinea

Similar cooperation is possible with peace and security actors, as illustrated by the case of Guinea.¹⁰ As a result of continued insecurity in both Liberia and Sierra Leone, the refugee population in Guinea climbed to approximately 450,000 in 1999. Then, in mid-2000, the regional conflict spread to Guinea. Between September 2000 and April 2001, some 100,000 Guineans were displaced as a result of a string of some 30 cross-border incursions.[11]

These incursions also had particular implications for the refugee population. First, tens of thousands were themselves displaced by the fighting. The majority of refugee settlements in the region were destroyed, along with the refugees' livelihood. In the midst of the conflict, refugees were subjected to harassment, forced recruitment (both as combatants and as porters to ferry looted goods back into Sierra Leone), physical and sexual abuse, arbitrary detention, and direct attacks by all sides of the conflict.[12] Finally, the killing of the UNHCR Head of Office in Macenta in September 2000 resulted in the evacuation of all UNHCR staff from field offices and the suspension of all UNHCR activities outside of Conakry, leaving some 400,000 refugees without assistance for months.

As the violence subsided in early 2001, UNHCR developed a three-pronged strategy to restore stability to the refugee population and to address the protection needs of the refugees. This strategy included a massive exercise to relocate refugees to new camps, the return of many refugees to Sierra Leone, and efforts to resettle the most vulnerable refugees to third countries.

As part of this strategy, a new Guinean policing body, the Mixed Security Brigade (Brigade Mixte de Sécurité, BMS), was formed to provide security for humanitarian personnel, to promote law and order in the camps, and to ensure that the new camps did not become militarized. Building on the success of the 'security package' approach developed in Tanzania and Kenya, UNHCR hoped that the equipping and training of security personnel specifically responsible for the camps would ensure greater security within them.

Following a number of violent incidents and allegations of abuse of refugees by the new force, new initiatives were undertaken to provide more effective operational training to effectively police the camp populations. In particular, the Canadian government agreed to deploy two Royal Canadian Mounted Police (RCMP) officers to southern Guinea to train the BMS in basic policing and human rights principles and to ensure effective coordination between UNHCR, the BMS and the Guinean authorities. A mid-term review of the programme in July 2003 concluded that the deployment had achieved 'mixed results'.[13] There was concern over the lack of previous training of the BMS and the fact that the RCMP programme had to start with the most basic principles of policing. There was also a concern that the policy of rotating BMS officers out of the camps and back into regular duties meant that the benefits of the training were not retained in the camps. These concerns notwithstanding, fieldwork in 2004 found that the contribution of the Canadian deployment had raised the standards of camp security to a level unrecognizable from 2001.[14] Most significant was the improvement in relations between the BMS and refugees. In this way, the deployment made an important contribution to addressing a specific concern through the cooperation of security and humanitarian actors.

The case of Guinea, more generally, highlights the importance of addressing the direct security implications of protracted refugee situations.[15] Since the late 1990s, for example, UNHCR has developed a number of operational responses to deal with the problem of militarization of refugee camps. It introduced a 'ladder of options' to prepare and respond to these situations and introduced humanitarian security officers and military advisers from the Department of Peacekeeping Operations (DPKO) in selected African camps. UNHCR's first efforts to operationalize its new policy response to armed elements in refugee camps were its attempts to implement the 'security package' in western Tanzania and northern Kenya and the relocation exercise in Guinea. While these actions helped create greater security for some of the refugee camps and communities in Tanzania, Kenya and Guinea, they have not uniformly led to greater security in the wider refugee-populated areas.[16] A similar effort by UNHCR and the DPKO in the Democratic Republic of the Congo in mid-2001 successfully separated armed elements from refugees.[17]

From these experiences, it is evident that the future success of the ladder of options depends on the practical partnerships and 'security packages' that UNHCR is able to form with the DPKO and governments and regional organizations. While discussions between DPKO and UNHCR have set the groundwork for future cooperation between the two offices, serious differences of approach and political and resource constraints remain. On the one hand, UNHCR and other humanitarian aid organizations fear that too close an association with the military compromises their impartiality and neutrality, and, on the other, governments are reluctant to authorize military forces for such functions. Protection for refugees in militarized situations also depends critically on the willingness and ability of host states and countries of refugee origin to observe international humanitarian norms regarding the treatment of refugees and non-combatants.

Towards comprehensive solutions

Such targeted interventions do not, however, constitute a solution to protracted refugee situations. These short-term interventions can only help manage the situation until a resolution can be found. In the long term, the implications of protracted refugee situations can be fully addressed only through the formulation and implementation of comprehensive solutions. Such a response would employ the full range of possible solutions for refugees – repatriation and reintegration, local integration in the host country and resettlement to a third country. As outlined in the chapter by Jamal, however, more needs to be done to reinforce the individual solutions and build on the complementary nature of the three durable solutions.

Comprehensive solutions to protracted refugee situations based on the three durable solutions are not new, as detailed in the chapter by Betts. Such an approach was central to resolving the situation of displaced people remaining in Europe long after the Second World War, that of millions of Indo-Chinese refugees, and the Central American refugee situation in the 1980s. By approaching the particular character of each refugee situation and by considering the needs, concerns and capacities of the countries of first asylum, the country of origin, donor and resettlement countries, along with the needs of refugees themselves, the international community has successfully resolved the plight of numerous refugee populations in the past 50 years. More generally, Betts argues that successful comprehensive plans of action should be:

- Comprehensive: drawing on the entire range of durable solutions.
- Cooperative: based on inter-state burden-sharing. Countries of asylum cannot solve protracted refugee situations on their own. There needs to be engagement by both regional actors and the international community.
- Collaborative: involving a broad range of UN agencies and NGOs. UNHCR and humanitarian agencies alone cannot find a solution. There is a need for peace and security and development actors to play a sustained role.

Recent interest in solutions to protracted refugee situations

Notwithstanding these lessons of history, there was little research and policy interest in protracted refugee situations during the 1990s. Until the very recent exceptions outlined in the Introduction, researchers and policymakers have largely ignored the problem of protracted refugee situations for decades. A few key studies by practitioners addressed this issue in the 1970s and 1980s.[18] More recently, the Evaluation and Policy Analysis Unit at UNHCR undertook a series of important studies on the issue.[19] While these studies provide important new insights into protracted refugee situations in Africa and elsewhere, the primary focus of the research has been on addressing the daily security concerns of refugees and on refugee livelihoods. While these issues are clearly important dimensions of the problem of protracted refugee situations, the lessons of history suggest that such initiatives undertaken in isolation from the wider political and strategic context of protracted refugee situations will not lead to solutions for refugees. Instead, the examples from this volume clearly indicate that solutions must also address the interests of a range of other actors, and engage with the links between local and regional security and protracted refugee situations.

The rising significance of protracted refugee situations has, however, been given a higher profile within intergovernmental settings in recent years. In December 2001, there was an African Ministerial Meeting on protracted refugee situations[20] and the issue has been considered at recent UNHCR Executive Committee sessions[21] as well as within the framework of the UNHCR Global Consultations on Refugee Protection.[22] While there was some preliminary discussion on comprehensive solutions for the most prominent protracted refugee situations,[23] discussions focused largely on issues of livelihood and burden-sharing and not on either the links between regional security and chronic refugee situations or the security problems refugees pose for host countries in regions of refugee origin.

Policy discussions in recent years have also tended to concentrate on the need to develop the refugees' potential to engage in economically productive activities, to foster refugees as 'agents of development', and to promote community-based assistance, including aid to host communities, as a pillar of UNHCR's future programmes. While recent research has highlighted how the long-term presence of refugees can contribute to the development of infrastructure and state-building,[24] there appears to be little recognition in these discussions of the history of UNHCR's earlier and often unsuccessful efforts to promote self-reliance in Africa's rural refugee settlements.[25] The current policy proposals and solutions advanced by UNHCR and others need to be examined critically and addressed within a historical perspective so as not to simply repeat past policy failures.[26]

More recently, UNHCR's thinking in the area has started to consider a broader range of political aspects of both the causes of, and the preconditions for, protracted refugee situations. In a paper from June 2004, UNHCR recognized that, 'as the causes of persistent refugee situations are political, solutions must be sought in that arena' and that it was important for the organization to 'understand the political forces and opportunities' underpinning responses to protracted refugee situations.[27] In September 2004, UNHCR rightfully introduced the need to develop a step-by-step framework approach to resolving long-standing refugee situations[28] that recognizes not only the differences between the protracted refugee situations of today and those of the past, but also the diversity of contemporary situations.

While UNHCR has highlighted the important political dimensions of protracted refugee situations, it has only recently come to recognize that it cannot address these dimensions on its own. Also, as clearly highlighted in the chapter by Crisp and Slaughter, there is an increased awareness of how UNHCR, acting independently, may be contributing to some of the challenges of protracted refugee situations. While it is essential that agencies involved in protecting refugees are sensitive to host governments' concerns regarding chronic refugee populations, actions by humanitarian agencies, such as UNHCR, without the support of peace and security and development actors will not lead to truly comprehensive solutions. So long as discussions on protracted refugee situations remain exclusively within the humanitarian community, and do not engage the broader peace and security and development communities, they will be limited in their impact.

Framework for a truly comprehensive response

Despite the need for a multifaceted approach to protracted refugee situations, the overall response of policymakers remains compartmentalized, with security, development and humanitarian issues mostly being discussed in different fora, each with their own theoretical frameworks, institutional arrangements and independent policy approaches. There remains little or no strategic integration of approaches and little effective coordination in the field. Neither the UN nor donor governments have adequately integrated the resolution of recurring regional refugee problems with the promotion of economic and political development, conflict resolution and sustainable peace and security, as outlined in the chapters by Mattner and by Morris and Stedman. Meaningful comprehensive solutions for protracted refugee situations must overcome these divisions and adopt a new approach that incorporates recent policy initiatives by a wide range of actors.

Such an approach needs to be rooted in an understanding of the relationship between forced migration and security since the end of the Cold War and in an understanding of the security concerns of third world states. Just as Morris and Stedman highlight in their chapter on our changing understanding of the dynamics of protracted conflict in recent years, it is important to understand how the nature of protracted refugee situations in the developing world has changed. During the Cold War, these situations were addressed as part of the interest of the superpowers, primarily the United States. In recent years, declining donor engagement in many refugee situations coupled with the new sense of vulnerability of many host states has led to a new political and strategic environment within which solutions must be crafted. In this sense, it is important to emphasize that the task is not simply to replicate past solutions but to fashion new solutions, drawing on the lessons of the past but appropriate to the new environment.

For solutions to be truly comprehensive, and therefore effective, they must involve coordinated engagement from a range of peace and security, development and humanitarian actors. Within the multilateral context, it is important to begin by identifying the full range of actors implicated before specifying the role they should each play. First, from the **peace and security** sector, sustained engagement is necessary not only from the UN Security Council (UNSC), the Office of the UN Secretary-General (UNSG) and the Department of Peacekeeping Operations (DPKO), but also from regional and sub-regional organizations, such as the African Union (AU), the Economic Community of West African States (ECOWAS), the South Asian Association for Regional Cooperation (SAARC), the Association of Southeast Asian Nations (ASEAN), and foreign and defence ministries in national capitals. Second, **development** actors, from the UN Development Programme (UNDP), the World Bank and international development NGOs to national development agencies, such as the Department for International Development (DFID) and the US Agency for International Development (USAID), would have an important role to play at all stages of a comprehensive solution. Finally, **humanitarian** actors, including UNHCR, the UN Office for the Coordination of Humanitarian Affairs (UNOCHA), and the full spectrum of international humanitarian NGOs, need to bring their particular skills and experience to bear. As argued by Ferris in her chapter, human rights actors also have a potentially important role to play.

Working with the host states and country of origin, these three sets of actors will each have individual but related responsibilities in the formulation and implementation of comprehensive solutions for protracted refugee situations. Central to the success of such an approach will be the ability to effectively identify the causes of the impasses in the particular situations and gain the necessary political support and resources from both donors and region actors to resolve these obstacles. The following framework has been developed to outline how such coordinated action can be devised to respond to the types of protracted refugee situations detailed in the case studies of this volume: where refugees are warehoused in isolated and insecure camps, where host states have enacted restrictive asylum policies, where donor engagement is diminishing, and where there has been no resolution to the underlying cause of flight in the country of origin. The objective of the framework is to outline how to move the situation from one of impasse to one where comprehensive plans of action, involving the three durable solutions, can be implemented. While few of the activities included in this framework are new, the importance is the linking and sequencing of the engagement of the three clusters of actors.

The framework outlines how the three sets of actors should engage in a related set of short-, medium- and long-term activities, which would combine to form a comprehensive plan of action (CPA). The first phase of any CPA should be an analysis phase, concentrating on the identification of the sources of impasse in the current situation. Peace and security actors could not only concentrate on the nature of the ongoing conflict in the country of origin but also consider the impact of protracted refugee situations on the host state and the region. Development actors could not only assess the conditions in the country of origin to identify rehabilitation priorities ahead of return, but also assess the positive and negative impacts of the prolonged presence of refugees in countries of asylum in the region. Finally, the humanitarian agencies could conduct a comprehensive survey of the refugee population to determine the applicability of the various durable solutions to various sub-groups within the population. At the same time, humanitarian actors could identify gaps in the current protection environment of refugees in the region that need to be addressed pending the implementation of a comprehensive solution.

This analysis phase should be followed by a stabilization phase. This phase would bring the three sets of actors together to develop a coordinated action plan based on the findings of their analysis. During this phase, benchmarks should be established to determine at what point the comprehensive plan of action moves through the three phases of the solution. This action plan could then serve as the basis for engaging major donors to ensure that the necessary support has been secured to launch the short-term objectives of the CPA.

The objective of the short-term phase of the CPA should be the stabilization of the current situation and the establishment of dialogue between key stakeholders. For the peace and security actors, this would involve a range of confidence-building measures with host states in the region, the country of origin and major donor states, leading to a durable ceasefire. For development actors, this would involve the implementation of targeted development assistance projects to help respond to the perceived burdens of the host state, while also implementing programmes that build on the positive contribution that refugees make to host areas and develop possibilities for future local integration. During this phase, development actors should also undertake targeted activities in the country of origin to address immediate needs and rebuild essential services. Finally, humanitarian actors should be engaged in the stabilization of the protection environment for both IDPs in the country of origin and refu-

Table 16.1 A framework for formulating and implementing comprehensive solutions for PRS

	Peace and security (UNSC, UNSG, DPKO, regional organizations, foreign and defence ministries)	Development (UNDP, World Bank, NGOs, national development agencies)	Humanitarian (UNHCR, UNICEF, NGOs, UNOCHA)
Analysis phase	Examine ongoing political conflict in country of origin and identify obstacles to resolution of conflict and refugee issue Analyse impact of protracted refugee situation on host state and regional security	Assess the positive and negative impacts of the protracted presence of refugees in host states, and assess targeted development needs Assess reconstruction needs in country of origin	Conduct comprehensive survey of the refugee population, including a demographic analysis Assess protection environment and assistance needs in host countries
	Consultation phase		
Stabilization phase	Engage in confidence-building measures with host states in the region, countries of origin and major donor states	Targeted development assistance to support local populations (linked with humanitarian actors) Implement programmes to meet basic needs in country of origin	Stabilize the protection environment in the region Stabilize the nutritional and health status of refugees Targeted development assistance to support local populations (linked with development actors) Capacity-building of structures and systems in host country
	Consultation phase		
Consolidation phase	Convene a peace conference with the engagement of donors and principal actors Prepare for necessary peacekeeping deployment	Implement rehabilitation programmes in refugee-populated areas in host countries to encourage local integration	Develop preconditions for the three durable solutions: repatriation, local integration and resettlement

Table 16.1 (cont.)

Implementation phase	Implement peace agreement Implement disarmament, demobilization, reintegration and rehabilitation (DDRR) programmes Implement peacebuilding and institution-building programmes to support transitional government	Implement rehabilitation and reconstruction programmes in the country of origin Work with host countries to ensure the transition from relief to development	Implement comprehensive plan of action through the complementary use of the three durable solutions

gees in the countries of asylum, while supporting the development actors in their delivery of targeted assistance projects and facilitating the necessary capacity-building of requisite structures and systems in the country of origin and host countries in the region.

Once the benchmarks for the stabilization phase have been accomplished, a second consultation process should be undertaken. This process would allow for the three sets of actors to report on their progress to the major stakeholders, including donors, host states and the country of origin. On the basis of this reporting process, the political will should be secured to proceed to the medium- and long-term phases of the CPA. Crucially, the major donors should also be engaged at this stage in the process to ensure that the necessary support and funding are forthcoming for the next two phases.

The objective of the medium-term phase of the CPA is consolidation. Building on the foundations laid in the short-term phase of the CPA, the peace and security actors should then convene the necessary peace conferences with the core group of actors and a broader group of backers to formulate a resolution to the conflict and the terms of the transition. Provisions should also be made at this stage for the necessary peacebuilding, peacekeeping, demobilization and transitional activities. At the same time, development actors would be engaged in the necessary rehabilitation and reconstruction in the country of origin in preparation for large-scale repatriation, in addition to the rehabilitation of refugee-populated areas in the countries of asylum. The humanitarian actors would concurrently be engaged in developing the preconditions for enacting the appropriate durable solutions for the various groups and sub-groups within the refugee population. In the case of repatriation, this would involve assessing the needs in the country of origin and disseminating information to the refugee population on the possibilities of return. In the case of those refugees who are unable to repatriate, humanitarian actors, supported by political actors, should engage in negotiations on the number and legal status of refugees to be locally integrated. Finally, resettlement opportunities should be secured for those refugees, or groups of refugees, who cannot repatriate and cannot locally integrate.

The final phase of the CPA is the implementation phase. The peace and security actors would be responsible for supporting the transitional government and measures to ensure that the solution is sustainable over the longer term. Development actors would ensure the transition between relief and development by supporting the new government in the design and implementation of its longer-term development objectives. Crucially, it is at this stage that the humanitarian actors implement the comprehensive solution for the refugee situation through the complementary use of the three durable solutions.

While this solutions framework may provide a useful point of departure for further research and policy development, it is important to emphasize how the dynamics of such an approach need to be understood within a broader context. First, there is a long history of competition between the range of actors incorporated in this framework. In fact, it has been argued that competition between UNHCR and other UN agencies, especially development actors, has frustrated innovations in the search for solutions for refugee situations for the past 15 years.[29] More generally, there is a significant gap in our understanding of how security, development and humanitarian actors have been able to cooperate in so-called 'joined-up' or 'whole of government' responses to peacebuilding in fragile states. Recent work has outlined the challenges of such a response in cases of comprehensive responses from individual countries[30] and within the UN system.[31] While more work is required in this area, discussions on more comprehensive responses to protracted refugee situations should learn the early lessons from this research on the challenges of cooperative action by a range of national and international actors.

Second, more work is required on the differentiation of responses to different types of protracted refugee situations. As suggested by Morris and Stedman, different types of situations in countries of origin may result in different kinds of protracted refugee situations. For example, there may be important differences between the case of the Bhutanese refugees in Nepal, who have fled systematic human rights violations, as outlined in the chapter by Lama, and the case of the South Sudanese refugees in Uganda and Kenya, who remain in exile following the end of conflict, as outlined in the chapter by Kaiser. In fact, responses to particular refugee situations must be mindful of these differences and proceed accordingly. As suggested in the Introduction, these differences point to the utility of developing a typology of protracted refugee situations, including those resulting from ongoing conflict, those involving people fleeing following a military victory (so-called 'losers in exile'), and those refugee situations that remain, pending the consolidation of peace in the country of origin. Such a typology may facilitate better comparative analysis and lessons learned from historical and contemporary situations.

Preconditions for success

These differences notwithstanding, this chapter argues that the range of activities outlined in the solutions framework must, at minimum, be considered as part of a comprehensive solution. The success of such an approach will, however, depend entirely on the commitment of a number of state actors to see it succeed. This solutions-oriented approach will be successful only if it is able to engage with and address the concerns and interests of host states, donor states and important states in the region. To this end, such an engagement must be fashioned as part of an approach that recognizes the links between the issue of refugees and the broader range of state interests at play. Increased external engagement in regions of refugee origin and comprehensive solutions to protracted refugee situations are the best way to simultaneously address the concerns of Western states, meet the protection needs of refugees and respond to the concerns of countries of first asylum. Such an approach would ensure effective protection in the region of origin, thereby diminishing the need for individuals to migrate to the West to seek such protection, would be structured around managed comprehensive responses, thereby ensuring the predictability sought by Western states, and would work towards the comprehensive solution of protracted refugee situations, thereby addressing both the protection needs of refugees and the concerns of host states.

In order to engender political support for such an international approach, it will be necessary to build on state interests in resolving protracted refugee situations, as argued by Betts in this volume. As with successful efforts to resolve the situations in Indo-China and Central America, a resolution of today's protracted refugee situations will involve looking beyond humanitarian interests and engaging with the broader political, economic and strategic interests of all stakeholders. At the same time, the importance of including the country of origin, and addressing its wider interests, cannot be overlooked. While conditions in the countries of origin of many of today's protracted refugee situations seem intractable, any response which excludes a response to the root causes of refugee movements will be limited in its impact. While there remains a significant role for UNHCR to play as a catalyst for bringing together key stakeholders and for ensuring that the process is sustained, this type of broader engagement cannot occur without the sustained engagement of all branches of the UN system. In this way, the establishment of the UN Peacebuilding Commission (PBC) provides both a timely opportunity and a possible institutional context for this type of cross-sectoral approach.

Protracted refugee situations and the UN Peacebuilding Commission

In his 1992 report, *An Agenda for Peace*, UN Secretary-General Boutros Boutros-Ghali argued that the end of the Cold War presented new challenges and opportunities for both the international community and international institutions mandated with the preservation of peace and security. In considering the various tools at the disposal of the United Nations in responding to the new security environment, the Secretary-General added 'peacebuilding' to the more established activities of preventive diplomacy, peacemaking and peacekeeping. He argued that such an innovation was required as the United Nations system needed to develop the capacity to 'stand ready to assist in peacebuilding in its differing contexts: rebuilding the institutions and infrastructures of nations torn by civil war and strife; and building bonds of peaceful mutual benefit among nations formerly at war'.[32]

While few of these activities were new, it became increasingly recognized that these longer-term undertakings were essential elements in preventing a return to conflict. The importance of peacebuilding was clearly illustrated by several cases through the 1990s, including Liberia, Rwanda and Sudan. However, numerous gaps remained in the conceptual and practical understandings of peacebuilding. In particular, there has been significant debate on the scope of peacebuilding activities and who should undertake them. While there is growing empirical evidence to suggest that effective peacebuilding strategies should involve long-term activities designed to support the security, political, economic, judicial and reconciliation needs of a country emerging from conflict,[33] no single international organization had the mandate to undertake this full range of activities. While the UN system contained a number of specialized agencies with mandates to undertake some of these activities, and while these agencies have been involved with peacebuilding activities around the world for some time, it became increasingly clear that stronger leadership and institutional coherence were required to ensure that peacebuilding was more effectively and systematically undertaken.

The establishment of a UN Peacebuilding Commission was subsequently proposed as a means of ensuring better leadership and coordination of peacebuilding activities within the UN system. The initial proposal was included in the 2004 report of the UN Secretary-General's High-Level Panel on Threats, Challenges and Change. In his 2005 memo, 'In Larger Freedom', UN Secretary-General Kofi Annan endorsed the creation of a Peacebuilding Commission as an intergovernmental advisory body, which could ensure long-term political support and funding for post-conflict recovery programmes, in addition to advising on thematic issues and specific cases.

The UN Peacebuilding Commission was subsequently established by the UN General Assembly (UNGA) in December 2005. In establishing the PBC, the UNGA recognized the 'interlinked and mutually reinforcing' nature of peace and security, development and human rights, and the benefits of 'a coordinated, coherent and integrated approach to post-conflict peacebuilding'.[34] To this end, the PBC was established to serve three functions:

- to bring together all relevant actors to marshal resources and to advise on and propose integrated strategies for post-conflict peacebuilding and recovery;
- to focus attention on the reconstruction and institution-building efforts necessary for recovery from conflict and to support the development of integrated strategies in order to lay the foundation for sustainable development;
- to provide recommendations and information to improve the coordination of all relevant actors within and outside the United Nations, to develop best practices, to help to ensure predictable financing for early recovery activities and to extend the period of attention given by the international community to post-conflict recovery.

Important decisions were then taken in the first half of 2006 on the size and composition of the PBC. By mid-2006, the PBC comprised 31 member states, including members of the Security Council, members from the UN Economic and Social Council (ECOSOC), representatives of the major donor countries, troop-contributing countries, and other members of the UNGA with experience in post-conflict reconstruction, in addition to those states directly implicated in the specific peacebuilding operation under consideration. Finally, meetings of the PBC during its first session invited contributions from senior UN representatives in the field, representatives of other UN agencies, representatives of major development institutions, including the World Bank, and representatives of civil society. In this way, the PBC brings together a wide range of institutional stakeholders implicated in peacebuilding initiatives.

At the same time, the UNGA resolution created the Peacebuilding Support Office (PSO) to facilitate the ongoing work of the PBC, to gather expert opinions on thematic issues and country-specific plans, and to collect examples of 'best practices' from previous and ongoing post-conflict recovery programmes that could be replicated elsewhere. In May 2006, Carolyn McAskie, a senior Canadian diplomat who was previously the UNSG's Special Representative for Burundi, was named Assistant Secretary-General for Peacebuilding Support and Head of the PSO.

The first formal meeting of the PBC was convened in New York on 23 June 2006. As detailed in its report to the UN General Assembly in July 2007,[35] the first year of the PBC's work was largely devoted to developing a clearer understanding of the scope and nature of the commission's work and to country-specific work on Burundi and Sierra Leone. Through its country-specific work, the PBC adopted work plans, sent several missions to both Burundi and Sierra Leone, and identified key priority areas for peacebuilding in both cases. In the case of Burundi, the PBC focused on promoting good governance, strengthening the rule of law, security sector reform and ensuring community recovery. In Sierra Leone, the PBC focused on youth employment and youth empowerment, justice and security sector reform, democracy consolidation and good governance, and capacity-building, especially the capacity of government institutions. In addition, the engagement of the PBC coincided with important developments in both countries, including parliamentary elections in Sierra Leone and the development of a Strategic Framework for Burundi.

While these are important developments for peacebuilding in both cases, it is important to note the limited scope of the early work of the PBC.[36] Specifically, the early work of the PBC has focused exclusively on activities within the country in question, with little or no attention paid to either the regional nature of conflict or the significant refugee populations associated with these conflicts. This narrow approach is particularly striking given that conflicts in both Burundi and Sierra Leone were largely tied to broader regional dynamics and neighbouring conflicts – the African Great Lakes for Burundi and the Mano River Union for Sierra Leone. Given the regional dynamics of conflict and the role that refugee populations play, not only as a consequence of conflict but as a source of its perpetuation in both cases, the importance of situating peacebuilding efforts in Burundi and Sierra Leone within a broader regional context would seem logical. The PBC has not, however, adopted such an approach, and its discussions have remained country specific, with very limited discussion of the regional dynamics.

The treatment of Burundi, Sierra Leone and similar cases by the PBC, and the sustained political and donor interest this is hoped to generate, could provide a unique opportunity for engaging the full spectrum of stakeholders required to formulate and implement a comprehensive solution, not only for peacebuilding and post-conflict recovery in the country of origin but also to resolve the related refugee situations. The emerging approach of the PBC, however, does not appear to make this link. Instead, the members of the commission seem to be adopting a myopic, country-specific approach. Such an approach does not allow for a full consideration of factors outside the country that could upset post-conflict recovery. It also adopts a limited understanding of the links between long-term displacement and peacebuilding, incorporating the issue of refugees only insofar as the return and reintegration of refugees is taken to be a barometer of the success of peacebuilding efforts.

While this is an important dimension of the issue, such a limited approach not only risks missing an important opportunity to resolve protracted refugee situations, but also excludes from the work of the PBC a range of factors that could potentially undermine peacebuilding efforts. Refugee-populated areas in neighbouring states may harbour elements that seek to undermine peacebuilding in the region, especially when underlying political tensions still exist and reconciliation has not been fully achieved, and refugee populations may be drawn into a campaign of destabilization. It would therefore be problematic to assume that refugees remain passively in neighbouring countries, awaiting the opportunity to return. Instead, there are many instances where large and protracted refugee situations, left unaddressed, have the potential to undermine the consolidation of a peace process.

Likewise, the concerns of host countries and the limitations on their willingness to host refugees must be taken into account. If the concerns of host states relating to the potentially negative impact of the prolonged presence of refugees on their territory are not addressed, host states may pursue early and coerced repatriation, placing fragile institutions in the country of origin under significant strain and further undermining peacebuilding efforts. For example, Tanzania has frequently claimed that the prolonged presence of Burundian refugees has a negative economic, environmental and security impact. In response to what it sees as a limited and unpredictable donor response to address these concerns, the Tanzanian government has, in recent years, pressed for the repatriation of refugees to Burundi. Many UN and NGO officials in both Dar es Salaam and Bujumbura are concerned about the coerced nature of this repatriation, feeling that refugees are being returned to areas that are unable to adequately receive them, and that the scale of the repatriation risks undermining peacebuilding efforts in Burundi.

The composition and mandate of the PBC do not preclude it from considering these wider linkages. In fact, the UNGA specifically provided that country-specific meetings of the PBC shall include as additional members the country under consideration (i.e. the country of origin), countries in the region (i.e. host countries) and senior UN representatives in the field and other relevant UN representatives (including UNHCR).[37] More generally, the PBC now represents a primary forum for the coordination of peace and security, development and humanitarian activities. This composition would facilitate the cooperation required to implement the solutions framework presented in this chapter, and include both the range of actors and a recognition of the linked interests of states, as argued by Betts. While additional research is clearly required to further develop our understanding of the relationship between protracted refugee situations and the regional dynamics of peacebuilding, it is clear from this initial survey that the PBC should not exclude a fuller consideration of the refugee component from its work. This point was highlighted during the PBC's only consideration of regional approaches to peacebuilding in June 2007,[38] and should clearly be more comprehensively mainstreamed into its country-specific work.

Conclusion

This chapter has presented the elements of a framework that can be used both to respond to the current challenges posed by protracted refugee situations and to develop and implement truly comprehensive solutions for the long term. By drawing on the cases of Kenya, Tanzania and Guinea, the chapter considered a number of specific instances where humanitarian actors have worked with development and security actors to address specific, short-term challenges. Recognizing that a response to these challenges does not constitute a solution, however, the chapter turned to a consideration of the necessary elements of a longer-term solutions framework. The solutions framework presented in this chapter outlines how three clusters of actors – peace and security, development, and humanitarian actors – can cooperate through a series of linked activities in the short, medium and long term to respond to protracted refugee situations.

The chapter concluded by considering the potential of the UN Peacebuilding Commission to provide a forum and an institutional context where this type of cooperation could be advanced. For the PBC to fully play this role, however, the chapter argued that a greater understanding is required of the diverse ways in which peacebuilding efforts and protracted refugee situations are linked. A closer consideration of the links between protracted refugee situations and peacebuilding could provide an important basis for ensuring an effective international response to both issues. In particular, the establishment of the PBC draws together the full range of actors required to formulate and implement truly comprehensive solutions for protracted refugee situations, and therefore represents a unique opportunity to articulate a system-wide response to a long-standing challenge to the international community. More generally, however, these dynamics highlight the importance of engaging more than humanitarian actors in the quest to find solutions for protracted refugee situations. As the cases of this volume make clear, protracted refugee situations are more than humanitarian challenges that require more than humanitarian responses. In fact, the evidence from historical and contemporary cases makes it clear that protracted refugee situations will only be comprehensively resolved through the sustained engagement of a broad range of actors from all sectors of the international community.

Notes

1. This chapter draws from earlier works by the authors. See Gil Loescher and James Milner, *Protracted Refugee Situations: Domestic and International Security Implications*, Adelphi Paper 375, Abingdon: Routledge for the International Institute for Strategic Studies, July 2005; and chapter 5, 'Protracted Refugee Situations: The Search for Practical Solutions', in UNHCR, *The State of the World's Refugees: Human Displacement in the New Millennium*, Oxford: Oxford University Press, 2006, pp. 105–128.
2. See Karen Jacobsen, *The Economic Life of Refugees*, Bloomfield, CT: Kumarian Press, 2005; and Cindy Horst, 'Introduction: Refugee Livelihoods: Continuity and Transformations', *Refugee Survey Quarterly* 25, no. 2, 2006.
3. See, for example, Stephen J. Stedman, Donald Rothchild and Elizabeth Cousens, eds., *Ending Civil Wars: The Implementation of Peace Agreements*, Boulder, CO: Lynne Rienner, 2002; Elizabeth Cousins and Chetan Kumar with Karin Wermester, eds., *Peacebuilding as Politics: Cultivating Peace in Fragile Societies*, Boulder, CO: Lynne Rienner, 2001; and Taisier M. Ali and Robert O. Matthews, *Durable Peace: Challenges for Peacebuilding in Africa*, Toronto: University of Toronto Press, 2004.
4. A notable exception would be B. S. Chimni, 'Refugees and Post-Conflict Reconstruction: A Critical Perspective', *International Peacekeeping* 9, no. 2, Summer 2002, pp. 163–180.
5. For more on the links between protracted refugee situations and peacebuilding, see Gil Loescher, James Milner, Edward Newman and Gary Troeller, 'Protracted Refugee Situations and the Regional Dynamics of Peacebuilding', *Conflict, Security and Development* 7, no. 3, 2007.
6. This section is based on interviews in Dadaab and Nairobi in 2001 and 2004.
7. This section is based on interviews in Kibondo and Dar es Salaam in 2004.
8. See Bonaventure Rutinwa and Khoti Kamanga, 'The Impact of the Presence of Refugees in Northwestern Tanzania', report by the Centre for the Study of Forced Migration, University of Dar es Salaam, September 2003.
9. See Karen Jacobsen, 'Can Refugees Benefit the State? Refugee Resources and African Statebuilding', *Journal of Modern African Studies* 40, no. 4, 2002.
10. This section is based on interviews in Conakry, Kissidougou and N'Zérékoré in 2001 and 2004.
11. See James Milner, *Refugees, the State, and the Politics of Asylum in Africa*, Basingstoke: Palgrave Macmillan, forthcoming.

12. See Human Rights Watch (HRW), 'Liberian Refugees in Guinea: Refoulement, Militarization of Camps and Other Protection Concerns', New York: HRW 14, no. 8 (A), November 2002.
13. See Roy Herrmann, 'Mid-term Review of a Canadian Security Deployment to the UNHCR Programme in Guinea', Geneva: UNHCR, Evaluation and Policy Analysis Unit, EPAU/2003/04, October 2003.
14. See James Milner with Astrid Christoffersen-Deb, 'The Militarization and Demilitarization of Refugee Camps and Settlements in Guinea, 1999–2004', in Robert Muggah, ed., *No Refuge: The Crisis of Refugee Militarization in Africa*, London: Zed Books, 2006.
15. See, for example, William O'Neill, 'Conflict in West Africa: Dealing with Exclusion and Separation', *International Journal of Refugee Law* 12, special supplementary issue, 2000; and Bonaventure Rutinwa, 'Screening in Mass Influxes: The Challenge of Exclusion and Separation', *Forced Migration Review*, no. 13, June 2002.
16. Jeff Crisp, 'Lessons Learned from the Implementation of the Tanzania Security Package', Geneva: UNHCR, Evaluation and Policy Analysis Unit, EPAU/2001/05, May 2001.
17. See Lisa Yu, 'Separating Ex-Combatants and Refugees in Zongo, DRC: Peacekeepers and UNHCR's "Ladder of Options"', *New Issues in Refugee Research*, Working Paper no. 60, Geneva: UNHCR, August 2002.
18. The Refugee Policy Group in Washington, DC, produced reports on protracted refugee settlements in Africa outlining many of the problems confronting long-staying refugees at that time. T. Betts, Robert Chambers and Art Hansen, among others, conducted research on some of these groups in Africa and assessed the international community's policy responses, particularly programmes aimed to promote local integration.
19. Individual studies conducted for the research are posted on the web page of UNHCR's Evaluation and Policy Analysis Unit: http://www.unhcr.org/research/46adfe822.html. For a summary of the research findings, see Jeff Crisp, 'No Solutions in Sight: The Problem of Protracted Refugee Situations in Africa', paper prepared for a symposium on the multi-dimensionality of displacement in Africa, Kyoto, Japan, November 2002.
20. See UNHCR Africa Bureau, 'Discussion Paper on Protracted Refugee Situations in the African Region', background paper prepared for the 52nd Session of UNHCR's ExCom, October 2001; UNHCR's Africa Bureau, 'Informal Consultations: New Approaches and Partnerships for Protection and Solutions in Africa', December 2001; and UNHCR, 'Chairman's Summary: Informal Consultations on New Approaches and Partnerships for Protection and Solutions in Africa', December 2001. Papers available online at: http://www.unhcr.org/research/46adfe822.html.
21. See Ruud Lubbers, UN High Commissioner for Refugees, 'Opening Statement to the 53rd Session of the Executive Committee of the High Commissioner's Programme', Geneva, 30 September 2002.
22. See UNHCR, Executive Committee of the High Commissioner's Programme, 'Agenda for Protection', Standing Committee, 24th Meeting, UN Doc. EC/52/SC/CRP.9, 11 June 2002.
23. See UNHCR, 'Briefing Notes: High Commissioner's Forum', 27 June 2003; Ruud Lubbers, UN High Commissioner for Refugees, 'Opening Statement at the First Meeting of the High Commissioner's Forum', Geneva, 27 June 2003; UNHCR, 'Background Document: Initiatives That Could Benefit from Convention Plus', High Commissioner's Forum, Forum/2003/03, 18 June 2003.
24. See Karen Jacobsen, 'Can Refugees Benefit the State? Refugee Resources and African Statebuilding', *Journal of Modern African Studies* 40, no. 4, 2002.
25. See Oliver Bakewell, 'Repatriation and Self-Settled Refugees in Zambia: Bringing Solutions to the Wrong Problems', *Journal of Refugee Studies* 13, no. 4, December 2000.

26. See Alexander Betts, 'International Cooperation and the Targeting of Development Assistance for Refugee Solutions: Lessons from the 1980s', *New Issues in Refugee Research*, Working Paper no. 107, Geneva: UNHCR, September 2004.
27. UNHCR, Executive Committee of the High Commissioner's Programme, 'Protracted Refugee Situations', Standing Committee, 30th Meeting, UN Doc. EC/54/SC/CRP.14, 10 June 2004, p. 4.
28. See UNHCR, High Commissioner's Forum, 'Making Comprehensive Approaches to Resolving Refugee Problems More Systematic', UN Doc. FORUM/2004/7, 16 September 2004.
29. See Gil Loescher, Alexander Betts, and James Milner, *UNHCR: The Politics and Practice of Refugee Protection into the Twenty-first Century*, New York: Routledge, 2008.
30. See Stewart Patrick and Kaysie Brown, *Greater Than the Sum of Its Parts? Assessing 'Whole of Government' Approaches to Fragile States*, Boulder, CO: Lynne Rienner Press, 2007.
31. United Nations, 'Delivering as One: Report of the High-Level Panel on United Nations System-Wide Coherence', UN Doc. A/61/583, 20 November 2006.
32. UN Secretary-General (UNSG), 'An Agenda for Peace: Preventive Diplomacy, Peacemaking and Peace-keeping', report of the Secretary-General pursuant to the statement adopted by the Summit Meeting of the Security Council on 31 January 1992, UN Doc. A/47/227, 17 June 1992, paragraph 15.
33. See Ali and Matthews, *Durable Peace*, 2004, pp. 409–422.
34. UN General Assembly Resolution 60/180, 30 December 2005.
35. UN General Assembly, 'Report of the Peacebuilding Commission on its First Session', UN Doc. A/62/137-S/2007/458, 25 July 2007.
36. This section is based on interviews conducted in New York in May 2006, December 2006 and March 2007.
37. UN General Assembly Resolution 60/180, 30 December 2005.
38. See UN General Assembly, 'Report of the Peacebuilding Commission on Its First Session', UN Doc. A/62/137-S/2007/458, 25 July 2007, Annex VIII, 'Chair's Summary of Working Group on Lessons Learned Meeting on Regional Approaches to Peacebuilding'.

17

Resolving protracted refugee situations: Conclusion and policy implications

Edward Newman and Gary Troeller

The number of refugees trapped in long-running refugee situations – in effect 'warehoused' in poorer countries in the Middle East, Asia and Africa – totals almost 8 million. This figure translates to one out of approximately every 700 persons on the earth living in this condition, below the radar of policy consideration and security analysis. Moreover, available statistics are primarily based upon camp-based populations (usually focusing on populations of 10,000 or more) and do not include lower accumulations of refugees or refugees based in urban areas who are, given their dispersal, harder to track and enumerate. Therefore, the cumulative total of victims in long-standing refugee situations is probably much higher than existing statistics indicate and such situations can truly be termed 'forgotten emergencies'. As this volume demonstrates, PRS are a key issue in security, sustainable development and peacebuilding as well as a fundamental issue of human rights and humanitarian endeavours, and should be seen and addressed holistically.

The preceding chapters have examined in detail the broad range of themes and issues that are interwoven in the diverse fabric of protracted refugee situations and illustrated, through the presentation of case studies in Africa, the Middle East, and south-west and south Asia, the multifaceted importance of addressing this issue. It should be clear from the foregoing chapters that refugee situations are not 'simply' compelling human rights or humanitarian matters but have a direct bearing on, and indeed play a central role in, security and peacebuilding. In its origin as well as resolution the refugee issue is a multifaceted political phenomenon and

Protracted refugee situations: Political, human rights and security implications,
Loescher, Milner, Newman and Troeller (eds),
United Nations University Press, 2008, ISBN 978-92-808-1158-2

requires the collective political will and multilateral engagement of a wide range of actors to find durable solutions. Its resolution cannot be left to humanitarian organizations to handle largely alone.

By their very nature, long-standing refugee situations are a product of the absence or breakdown of good governance by sending and/or receiving states. These situations create despair, exacerbate regional insecurity, contribute to instability and state failure, and can be a source of political extremism and even terrorism. Resolving the refugee issue in general, and protracted refuge situations in particular, is a necessary building block for peace. The chapters in this volume illustrate the importance and multi-dimensionality of PRS, and thus the need for a concerted multilateral approach to this challenge. However, bringing together the diverse players relevant to the resolution of PRS – with their different interests, institutional cultures and capacities – is a far greater challenge.

As contributors to this volume have shown, a variety of factors contribute to the evolution and traditional handling of PRS. Conceptually, the issue has been seen, when it has been recognized at all, as a marginal humanitarian issue overwhelmed and obscured by the importance of new and thus newsworthy refugee emergency situations that capture media and policymaker attention. The challenge has been further complicated by the general tendency for asylum matters to be conflated with, subsumed in and blurred by immigration concerns. In academic circles, conventional disciplinary boundaries in security and peace studies have excluded proper consideration of PRS as actors and important players in security and conflict, instead treating them, if at all, as objects or symptoms rather than actual or potential sources of conflict. In many industrialized countries, intensifying security concerns since 9/11 and issues of 'identity politics' and 'societal sovereignty', reinforced by rising tensions between multiculturalism and integration, have led to ever more restrictive asylum policies. These developments have obscured the fact that the overwhelming majority of victims of forced displacement remain trapped in long-running refugee situations in the developing world, and indeed have contributed to the phenomenon of warehousing in poorer countries.

In addition to conceptual compartmentalization in academic and policymaking circles and restrictive asylum practices in many industrialized states, the challenge has been exacerbated by a fragmented and compartmentalized institutional approach by donor and host governments, UN agencies, NGOs and other actors involved. The result has been managing misery rather than effectively addressing it.

UNHCR – generally underfunded, understaffed, under pressure by countervailing political forces and jealously guarding its position as The Refugee Agency – has often inadvertently contributed to PRS by implementing long-running care and maintenance programmes in poor countries which involve placing refugees in camps. Poor host countries in many instances wish to keep refugee populations visible and in this way attract financial assistance from donors. In a number of cases UNHCR has fallen into an uncomfortable role, stuck between host governments attempting to meet their responsibilities to refugees and donors with obligations to appropriate burden-sharing. In particular, host governments in poor countries feel overburdened as rich countries strive to keep refugees out; there is the perception that there is not adequate burden-sharing. Donor countries in turn want durable solutions in the region – local integration or repatriation – but integration is not generally attractive to the hosting state and repatriation is not politically feasible. UNHCR is caught in the middle, dependent on the resources of donors and their selective strategic interests. Donors want field/region-based solutions, and hosting countries want more burden-sharing rather than local solutions. UNHCR does not sufficiently enjoin hosting states to meet their own state responsibilities to refugees – to integrate them – and the result is extended camp-based care and maintenance programmes.

The situation was exacerbated in the 1990s when UNHCR became involved in major repatriation movements as the Cold War ended, coupled with the political turbulence of a decade-long succession of high-profile emergencies. This led to the prioritization of major assistance operations and diverted attention from the development of long-range strategies for protracted refugee situations. While UNHCR's Indo-Chinese CPA and the International Conference on Central American Refugees have worked because of the political engagement and convergence of the interests of all concerned, purely programmatic approaches spearheaded by UNHCR in, for example, the International Conference on Assistance to Refugees in Africa failed to resolve PRS. Inability to learn from past mistakes, long a hallmark of UNHCR, led to a recent failure of the CPA for Somali Refugees where real political will and engagement was replaced by a flawed programmatic approach. The foregoing illustrates the slippery slope of descent from protection to protraction and administering misery.

UNHCR, dependent on donor funding and long used by major governments as a fig leaf for the absence of concerted and cumulative political will to resolve such problems, has not been unique in its institutional problems. Collaboration within the UN among all actors that should be involved, although beginning to emerge, has been hindered by bureaucratic turf battles and artificial divisions between the development, humanitarian and political organs of the UN. Donor governments, host governments and regional bodies such as the EU are not immune to this problem. Most donor governments, as well as regional bodies, make a clear bureaucratic demarcation between development aid and humanitarian assistance when considering the inter-relationship between refugee issues and sustainable development. Few donor countries have combined the portfolios of asylum and development in their ministerial structures in one ministry, as Sweden once did but no longer does.

In poorer host countries, finance ministries which are the recipients of multilateral or bilateral assistance see their primary responsibility as attracting and administering development funds for their own nationals, not non-national refugees who are often treated as extra-territorial populations. The connection or continuum between relief and development has yet to be effectively bridged at the strategic and planning level in many agencies, both national and international. The role of development actors encompasses supporting overall efforts at conflict prevention, mitigation and post-conflict reconstruction. With reference to protracted refugee situations, this translates into ensuring that the specific needs of refugees are met and that they have access to the range of resources made available to local populations across development portfolios. Thus far this has not been the case.

More effective communication, collaboration and coordination with a broader range of civil society are required. For UNHCR, less emphasis on mandate purity and institutional independence and a move beyond a one-dimensional focus is necessary. There is a gap between the humanitarian domain and the political domain, with actors in the former sometimes endowing their principles with a 'sacred' aura in opposition to the 'profane' arts of negotiation and compromise practised by actors in the latter. States create refugees by failing to protect citizens, while asylum countries, donors and UNHCR perpetuate protracted refugee situations by failing to offer adequate comprehensive responses. All observers agree that, since the causes of protracted refugees are inherently political, realistic durable solutions must ultimately be sought in the political sphere. Moreover, sufficient development aid must be allocated to make things happen.

Tackling the problem

Actors within the international protection regime and in development, political affairs and peacebuilding must recognize that the resolution of PRS is not a marginal but an integral part of security and state stability, and goes well beyond the humanitarian realm. Similarly it must be recognized that resolving PRS cannot be country specific but must be addressed in various regional contexts in a multilateral, comprehensive manner, as recent events in Lebanon and Gaza and the unfolding and growing tragedy in Iraq amply illustrate.

This project has sought to translate normative wishes into policy recommendations and prescriptions for comprehensive actions, underpinned by the collective political will of all concerned. The issues involved in confronting the challenge are formidable. It is important to understand the origin, evolution and multifaceted nature of long-term refugee situations and the humanitarian, political and security implications of inaction. PRS are rooted in the dynamics of fragile states and a response to this challenge is therefore closely linked to security and effective peacebuilding. Arguments that some policymakers have made about globalization, the interconnection of threats, and failed states as facilitators of terrorism and lawlessness have already made the most powerful states take notice. The central message of this volume is that there are very important political and strategic consequences of unresolved refugee situations. Prolonged and unresolved refugee crises almost universally result in the politicization and militancy of refugee communities, with dire consequences for host state and regional security. While a one-size-fits-all approach is not possible given the differing nature of PRS, it is possible to make general observations about the regional dynamics of refugee situations in terms of the diffusion of conflict and contagion.

Moving from administering misery to multilateral solutions

Resolving PRS must involve peace, security and development actors, as well as the humanitarian community. The issue should be at the top of the international political agenda. A number of conclusions and policy implications follow.

Framing policy

- In terms of evaluating the nature and needs of PRS, this volume suggests that arbitrary and fixed definitions of PRS may be unhelpful as a starting point for new policy ideas. Indeed, it may even be difficult to think in terms of 'PRS' as a single, coherent phenomenon. Aside from the common feature of protracted displacement, the situations which are regarded as PRS can vary enormously in terms of causes, impact and nature. Demographic and geographical features can also vary widely. An emphasis upon geographically concentrated or 'warehoused' PRS – whilst a reality for many – should not be allowed to dominate policy analysis on this subject. Dispersed or urban protracted

communities, whilst less conspicuous, have their own dynamics, challenges and needs. A revision, and in particular a broadening, of the PRS definition would allow international actors more flexibility in addressing PRS and would bring attention to the needs and rights of long-term refugees who exist outside the existing definitions of PRS.
- It is a truism – or even a tautology – to suggest that PRS would not exist if the conflicts or conditions which gave rise to forced migration could be prevented or resolved, or if durable solutions could be found. Does this imply that a resolution of PRS requires that that we resolve the causes of state failure, oppression and violent conflict? Truly durable solutions do require that we consider the underlying causes of PRS, in addition to the policy limitations which explain why PRS are often managed poorly or in a way that does not achieve a solution. Simultaneously, however, policy implications need to have a focus. Therefore, in considering PRS policies, it is necessary to identify different levels of abstraction: addressing the structural issues which give rise to PRS and addressing PRS challenges in ways that find durable, rights-based solutions. Whilst taking a comprehensive or holistic approach – which involves preventing forced displacement – is essential, the immediate and most viable challenge lies in finding durable, rights-based solutions for existing PRS and ensuring that, when refugee flows occur in the future, the knowledge and policy tools exist to prevent prolonged situations arising. Thus, the emphasis is not on preventing refugee flows – a fundamentally important but more ambitious goal – but on preventing refugee situations becoming protracted. In order to gain policy traction, it is necessary to understand that some progress can be made in resolving PRS and preventing their reoccurrence without necessarily resolving the underlying causes of state failure, oppression and violent conflict.
- A better understanding of the political, security and human rights dimensions of PRS can help to address these challenges in a number of ways. If the full range of implications and impacts of PRS is fully understood and policy prescriptions are couched within this broader context, it is more likely to encourage donors to devote the resources and imagination necessary to address the challenges. In addition, a broader, multifaceted approach will better equip the different actors working on refugee issues to understand their roles in relation to the roles of other actors. As this volume amply demonstrates, when actors approach these challenges independently, they find that their efforts are often inadequate or even in conflict with each other.
- The chapters in this volume suggest that a broad range of actors is required to resolve PRS and that a broader range of issues needs to be

incorporated into policy planning. Does this suggest a broader operational role for UNHCR? This should not be the case. Experiences of UNHCR expanding its role in response to donor requests and changing circumstances have not always proven to be satisfactory and have embroiled the agency in situations in which it had difficulty performing, or even which were at odds with its original mandate. However, a broader coordination role needs to be invested in the refugee architecture – within UNHCR – and in the international peace and security architecture. The latter could take the form of a special office of the UN Secretary-General – a UN Secretary-General's Special Representative for Protracted Refugee Situations, to complement the UN Secretary-General's Special Representative for Internally Displaced Persons, for example. The value of this – even though some observers might think that this is a duplication of the work of UNHCR – would be that an office independent of UNHCR could raise issues and suggest solutions which transcended the institutional agenda of UNHCR and the interests of its donors, and transcended the competition which has existed between different humanitarian actors working on these issues. Alternative ideas could be a unit – or specific portfolio – within the Peacebuilding Commission, or a separate committee under the Security Council. Whatever the shape of such a bureaucratic solution, the objective would be to facilitate a creative response to PRS which allows a comprehensive institutional approach and which reflects the enormous significance of these challenges.

- This points to an urgent need for a 'one UN approach'. The developing 'cluster approach' should be rigorously pursued. The cluster – or coordinated – integrated approach now used for IDPs should be used for refugees as well. The UN's increasing commitment to the establishment of 'integrated missions' in war-affected and post-conflict situations, bringing together the humanitarian, human rights, development, peacekeeping and political functions of the world body under the overall authority of the Secretary-General, is relevant and should be applied to PRS. These developments have an evident relevance to the task of resolving the problem of protracted refugee situations, both in supporting countries of asylum that have large numbers of refugees on their territory, and in supporting countries of origin from which those people have fled and to which many will eventually return.
- Once violence and political crises have created forced displacement, refugees continue to act in their own right, with observable effects on war and peace in their home and host countries. They are susceptible to militancy and constitute a potential source of regional security problems. While refugee movements do not necessarily result in violence, in

a number of cases refugees from neighbouring states significantly increase the risk of civil conflict for receiving states and regional conflict. As the chapters in this volume illustrate, responses to PRS must be designed and implemented according to countries of origin and host countries, and – where necessary – on a regional basis.
- Necessary multilateral action is hampered by the divergent institutional interests of the communities involved. There is a disconnect between those who analyse and advocate, policymakers in governments and regional bodies, and practitioners who implement and manage on the front-lines. This divide must be bridged to end institutional conflicts.
- More communication could bridge the gap in agency cultures. More cooperation at the strategic level is required, with NGOs, the World Bank, UN development agencies, the UN Department of Peacekeeping Operations and the Peacebuilding Commission (PBC) working together. All concerned may have to be prepared to give up some of their independence to coordinate more closely.
- PRS in some instances can be resolved only case by case (or war by war). In this respect it should be noted that, if international and local actors succeed in implementing peace after the North–South war in the Sudan and the war in Burundi, and also manage to consolidate peace in Liberia, the number of protracted refugee situations in Africa will be reduced by almost half.
- The PBC should function as the central umbrella and overall coordinating body at the political level in bringing together all actors and should consider the resolution of PRS as a central and not a marginal component of its action plans. The composition of the PBC places it in a unique position to take a broad approach to the challenges of PRS. However, the approach must be thematically comprehensive and must address the regional dimensions of long-term displacement, including challenges that fall outside the country – including PRS – which could upset recovery.
- In many instances stabilization and peacebuilding are a necessary first condition in resolving PRS. Some long-standing refugee situations are particularly challenging, given their duration and politicization. The Palestinian case illustrates this. Despite the setbacks to the Arab–Israeli peace process, the application of the CPA framework to the Palestinian case continues to have some merit. It draws out more clearly the points of similarity to and difference from other protracted refugee situations, and points to the need for nuance and flexibility in the construction of solutions. The paradigm offered by Loescher and Milner can also serve as a tool for identifying key benchmarks in the evolution of the case and thus assist policymakers and the donor community in targeting their efforts and funds.

The linkages between PRS, security and peacebuilding are clear. In an increasingly interconnected world, effective peacebuilding is in the interests of all states. Resolving PRS, in turn, is an integral part of peacebuilding. While the challenges are considerable, the human and security costs of inaction only increase with time. Only collective political will, underpinning multilateral approaches, can solve these problems.

Index

"t" tables; "f" figures and "n" notes.

Abrar, C. R., 314, 331n12, 331n24
Adelman, Howard, 77, 84n14
Afghan
 asylum seekers, 47
 urban refugees in India, 25, 94
Afghan refugees in Iran and Pakistan
 background and origins of situation, 333–35
 civil war and its aftermath (1993–present), 342–45
 demography, displacement and population mobility, 335–38
 policy responses, 345–50
 refuge experience, 338–41
 refugees, 26, 37, 333, 336, 338–59
 Soviet occupation and withdrawal, 341–42
 UNHCR's role with, 336, 343, 346–49, 350n1
Afghanistan, 3, 6, 15–16, 22t2.1, 39, 55, 57, 115, 117, 122n25, 133, 142, 144, 148, 160n27, 161n43, 222, 309, 333–40, 342–50, 350n1
Afghanistan Compact, 348
Afghanistan National Development Strategy, 348
African migrants, 64

African Union (AU), 232–33, 238, 243, 246n56, 247n60, 363
AFSC. *See* American Friends Service Committee (AFSC)
An Agenda for Peace, 369, 376n32
Agenda for Protection, 56–57, 67n30, 234–35, 247n61, 375n22
al-Barakaat, 222
al-Haramain Islamic Foundation, 222–23
al-Qaeda, 53, 198, 219, 222, 249, 253
Algeria, 22t2.1, 54–55, 156
Alliance for the Restoration of Peace and Counter-Terrorism (ARPCT), 218
American Friends Service Committee (AFSC), 193
Amnesty International, 94–95, 105n20, 106n28, 106n32, 300n31, 331n17
Angola, 22t2.1, 88, 110–11, 116, 121n8, 133, 165, 232
Annan, Secretary-General Kofi, 93, 370
Arab League, 196–97, 204, 208
Armenia, 22t2.1
ARPCT. *See* Alliance for the Restoration of Peace and Counter-Terrorism (ARPCT)
Arusha Agreement for Rwanda, 78

ASEAN. *See* Association of Southeast Asian Nations (ASEAN)
ASEAN Inter-Parliamentary Myanmar Caucus, 317, 332n28
Asian Centre for Human Rights, 93–94, 106n24, 106n26
Asian Development Bank, 324
Association of Southeast Asian Nations (ASEAN), 76, 172–74, 179–80, 185n37, 307–9, 315–17, 323–25, 330, 332n28, 332n30, 363
asylum trends in industrialized countries and PRS. *See also* protracted refugee situations (PRS)
  applications for asylum, 45–46
  asylum seekers, 4, 44–51
  background, 43–44
  conclusion, 64–65
  country of asylum, 22t2.1, 27, 30, 32, 37
  election results, 50–53
  EU asylum policy, 46–48
  EU harmonization process, 47–50, 66n8
  government responses in 1990s, 45–47
  immigration-asylum-security nexus, 53–56
  immigration *vs.* asylum, 43, 45–46
  legislation, restrictive asylum, 50
  migration and asylum, 6, 47
  multiculturalism *vs.* integration/identity politics, 60–64
  'New International Approaches to Asylum Processing and Protection,' 50
  offshore processing procedure, 49
  'Pacific Solution,' 50
  policy, 5, 38, 44, 46
  politics of, 41n30, 42n43–45, 50–53
  pressure of numbers, 44–45
  restrictions on asylum, 37–38, 46
  rights and freedoms, 10, 46
  security threats, 37
  UNHCR's response, 56–59
  Western systems of asylum, 45
AU. *See* African Union (AU)
Aung San Suu Kyi, 304, 306, 323
Australia, 47, 50, 52, 58–59, 91, 105n19, 125, 127, 214, 253, 332n33
Australian Human Rights and Equal Opportunity Commission, 93
Austria, 46, 54, 166
Azerbaijan, 22t2.1, 107n52

Balfour Declaration, 192
Balkans, 25, 29, 56, 79, 129
Bangladesh, 15, 22–23, 292, 303–4, 310–15, 318–21, 327, 330n1, 331n13, 331n15, 331n21–22, 332n36
Barre, Siyaad, 214, 216–18, 220
BCPR. *See* Bureau for Crisis Prevention and Recovery (BCPR)
Beirut Declaration, 204, 206
Betts, Alexander, 11, 57, 137, 144, 162–64, 328, 360, 368, 373, 375n18, ix
  'Comprehensive Plans of Action: Insights from CIREFCA and the Indo-Chinese CPA,' 183n1, 183n11
  'Conference Report: The Politics, Human Rights and Security Implications of Protracted Refugee Situations,' 42n46
  'International Cooperation and Targeting Development Assistance for Refugee Solutions: Lessons from the 1980s,' 183n1, 376n26
  UNHCR: The Politics and Practice of Refugee Protection into the Twenty-First Century, 42n47, 376n29
Bhutan, King of, 14, 280–82, 286, 293, 298–99
Bhutan, Kingdom of, 14, 22t2.1, 71, 277–99, 299n1
Bhutan Press Union (BPU), 296, 301n41, 302n64
Bhutan State Congress (BSC), 278
Bhutanese Refugee Durable Solution Coordination Committee, 298
Bhutanese Refugee Women's Forum, 290
Bhutanese refugees in Nepal
  background, 277
  Bhutan Citizenship Act (1977), 279–80
  Bhutan Marriage Act (1980), 280
  Bhutan Marriage Act (1985), 280–81
  bilateral process, finalization of, 296
  Census of 1988, 281–82
  conflict prevention and, 71
  democracy, attraction of, 285
  feudal interests, clashing, 284
  final uprooting, 292–93
  His Majesty's Government (HMG), 287–88
  human rights violations, systematic, 367
  illegal immigration, 284–85

Bhutanese refugees in Nepal (cont.)
  India-Nepal Peace and Friendship
    Treaty, 284
  India's pivotal role, 297
  India's role, 291–92
  Joint Ministerial Level Committee
    (JMLC), 288, 301n40
  Joint Verification Team (JVT), 289–90
  King of Bhutan and the royal
    prerogative, 298–99
  Lhotsampa refugees, 14, 277–78, 280–84,
    286–87, 291, 296, 299n1
  mass exodus from Bhutan, explaining
    the, 277–85
  Ministry of Home Affairs, 281
  nationality, beyond the issue of, 282–83
  nationality, questions of, 278–79
  Nepal camps, categories of refugees, 286
  Nepal camps, major concerns with,
    287–88
  Nepal camps, phases of exodus, 286–87
  Nepal camps, refugee management in,
    286–90
  Nepal camps and repatriation
    negotiations, 288–90
  open border, costs of an, 283–84
  political reforms, 293–94
  protracted refugee situation (PRS), 7, 14,
    22t2.1, 71, 76, 92, 149, 277, 294
  recent developments, 290–95
  Refugee Coordination Unit (RCU),
    287–88
  regional tensions and, 33
  resettlement, third country, 297–98
  SAARC-level commission for
    determination and repatriation,
    296–97
  security ramifications, 294–95
  sociocultural degeneration, 285
  Special Rapporteur mission, 92
  Statement on Protracted Refugee
    Situations, 94
  'strategic resettlement' of, 59
  UNHCR's role with, 149, 287–88,
    290–91, 294–97, 299n1, 301n35,
    301n39, 301n42, 301n48, 301n58,
    302n61
'Black Hawk Down' incident, 217
BMS. *See* Mixed Security Brigade (BMS)
'boat people,' 172–74, 315
Bonn Agreement, 335, 347

Bosnia and Herzegovina, 22t2.1, 78, 82, 96,
  108, 128
Botswana, 88
Boutros-Ghali, Secretary-General Boutros,
  369
BPU. *See* Bhutan Press Union (BPU)
Brazil, 88
British Mandate for Palestine, 192
Brookings Process, 114
Brynen, Rex, 208, 213n34
BSC. *See* Bhutan State Congress (BSC)
Bureau for Crisis Prevention and Recovery
  (BCPR), 100, 107n51, 112, 115
Burmese refugees in South and Southeast
    Asia
  background, 303–4
  causes of refugee flows and internal
    displacement, 304–5
  host state response, Bangladesh, 318–21
  host state response, Thailand, 318–19
  host states, regional dynamics, 315–17
  host states in region, Bangladesh, 310–15
  host states in region, impact of refugees
    on, 305–17
  host states in region, Thailand, 306–10
  Myanmar, addressing the political
    situation in, 322–25
  Myanmar, map of, 329f14.1
  refugee situation, addressing the, 326–30
  refugees, 33, 303–4, 307–8, 310, 315,
    318–19, 322, 326–28, 330, 330n1,
    331n10, 331n15, 331n22, 332n28,
    332n34
  solution, towards a comprehensive,
    321–30
  UNHCR's role with, 307, 310–14,
    318–21, 326–28, 331n18, 331n20
Burundi
  AU Peace force, 232
  civil war, 72, 132, 384
  Peacebuilding Commission, 81
  Peacebuilding Support Office (PSO),
    370–71
  protracted refugee situations (PRS), 3,
    22t2.1, 24–26, 33, 36, 76, 94, 132,
    140n7, 356, 372
  Refugee Convention and employment,
    88
  refugees, 24, 33, 36, 76, 132, 140n7, 372
Burundi
  Strategic Framework for, 371

INDEX 389

Cambodia, 25, 76, 129, 306
Cambodian refugees, 25, 76
Cameroon, 22t2.1
Camp David Summit, 198, 204–5, 209
Canada, 46, 48, 54, 58, 125, 127, 204, 291, 321, 332n33
Caucasus, 25–26
Central African Republic, 116, 253
Central America, 25, 102, 167, 169, 179–81, 368
Central American refugees, 167, 176, 231n23, 360, 368
Central Asia, 25–26, 49, 160n37
Central Emergency Revolving Fund (CERF), 157
Centre for the Study of Forced Migration, 357, 374n8
CERF. *See* Central Emergency Revolving Fund (CERF)
Chad, 22t2.1, 165, 253
Chambers, Robert, 110, 121n5, 375n18
Chile, 88
China, 15, 22t2.1, 25, 76, 173, 292, 295, 303, 306–7, 324
Chinese CPA, 163
Church World Service, 229
CIREFCA. *See* International Conference on Central American Refugees (CIREFCA)
civil society actors, 8–9, 85, 95–101, 104, 203, 205, 209, 229, 240, 242, 244n3, 261–63, 322, 332n39, 340, 370, 380
The Clash of Civilizations (Huntington), 63
Clinton, President Bill, 46, 204, 211
'Coalition of the Right of Return Groups,' 205
Cold War, 25–26, 34, 44–45, 72, 83n1, 125–26, 128–30, 162, 167, 172–73, 203, 225, 250, 306, 334–35, 362, 369, 379
Collier, Paul, 70, 83n5
Commissions of Venezuela and Ecuador, 93
Committee on the Rights of the Child, 96–97
Common Country Assessments, 100
Commonwealth of Independent States, 97, 102
Comprehensive Peace Agreement (CPA), 250, 252–53, 273n1

comprehensive plan of action (CPA). *See also* comprehensive plans of action (CPAs)
as basic political agreement, 173
Central America, 167
Chinese, 163
CIREFCA and, 167–68, 180–81
concept, 163–64, 175
country of origin, 180
historic examples of, 163
human rights criticisms, 175
ICARA and, 164
implementation phase. of, 367
Indo-Chinese refugees, 120, 163–65, 171–76, 176t9.1, 177–78, 181–82, 183n1, 328, 330, 379
JNA *vs.*, 238
overview of, 184–85n28
for Palestinian case, 384
preconditions for, 178–83
'Principles and Criteria for Protection and Assistance,' 167
references, 185n34, 185n38, 185n41, 185n43
short-, medium- and long-term activities of, 364, 366
Somali refugees, 144, 149, 159n15, 163–64, 176t9.1, 177–78, 182, 234–36, 247n65–66, 379
Sudan, 264, 268, 270
UNHCR and, 164, 167–68, 173, 175, 179, 181, 235, 268
UNHCR's Convention Plus initiative, 182
comprehensive plans of action (CPAs)
archetypal models for, 176t9.1
CIREFCA and, 182
historical examples of, 163
political engagement model, 158
programmatic model, 176t9.1
UNHCR and, 164, 175–76, 179, 236
UNHCR's Convention Plus initiative, 162
'Concerted Plan of Action in Favour of Central American Refugees, Returnees and Displaced Persons' (CPA), 167
Conference on the Commonwealth of Independent States, 102
Congo. *See* Democratic Republic of Congo (DRC)
Congolese urban refugees in Burundi, 25, 94, 249

Convention Against Torture and Other Cruel, Inhuman, or Degrading Treatment or Punishment, 54, 86
Convention Guiding Specific Aspects of Refugee Problems in Africa, 232
Convention on the Combating and Prevention of Terrorism, 232
Convention on the Elimination of All Forms of Discrimination Against Women, 86
Convention on the Prevention and Punishment of the Crime of Genocide, 86
Convention on the Reduction of Statelessness, 86
Convention on the Rights of the Child, 86–87, 89, 96–97, 105n4
Convention Plus, 57, 115, 134, 143–44, 162, 177, 182, 234–35, 247n62–63, 261, 375n23
Convention relating to the Status of Refugees (1951), 105n6, 212n18
Convention relating to the Status of Stateless Persons, 86
Côte d'Ivoire, 22t2.1, 25, 28, 94, 110, 159n4, 160n40
CPA. *See* Comprehensive Peace Agreement (CPA); comprehensive plan of action (CPA)
CPAs. *See* comprehensive plans of action (CPAs)
Crisp, Jeff, 10–11, 22, 57, 94, 123–39, 149, 152, 160n39, 224, 256, 262, 265, 362, ix
  'Africa's Refugees: Patterns, Problems and Policy Challenges,' 41n43
  EPAU Evaluation Report, 184n23
  'Forced Displacement in Africa: Dimensions, Difficulties and Policy Directions,' 121n3, 122n19
  'Forms and Sources of Violence in Kenya's Refugee Camps,' 245n20, 245n31
  'Lessons Learned from the Implementation of the Tanzania Security Package,' 375n16
  'Mind the Gap! UNHCR, Humanitarian Assistance and the Development Process,' 122n20, 140n11
  'A New Asylum Paradigm? Globalisation, Migration and the Uncertain Future of the International Refugee Regime,' 67n20
  'No Solutions in Sight: The Problem of Protracted Refugee Situations in Africa,' 140n3–4, 159n4, 161n46, 275n27, 375n19
  'Refugees and the Global Politics of Asylum,' 140n3
  'A State of Insecurity: The Political Economy of Violence in Kenya's Refugee Camps,' 245n21, 274n15, 275n34, 275n36
  '"Who Has Counted the Refugees?": UNHCR and the Politics of Numbers,' 40n6
  'Who Has Counted the Refugees? UNHCR and the Politics of Numbers,' 122n9
Croatia, 22t2.1, 57
Cuban asylum seekers, 46
Cyprus, 88

Dadaab refugee camp (Kenya), 142, 159n4, 219, 221–24, 227–30, 239–41, 244n5, 245n23, 245n27–28, 245n32, 258, 355–56, 374n6
Danish cartoon controversy, 44, 62–63
DAR. *See* Development Assistance to Refugees (DAR)
Darfur, 82, 133, 156, 248–49, 266, 273n1
Dayton Agreement, 78
DDR. *See* Disarmament, Demobilization and Reintegration (DDR)
DDRR. *See* Disarmament, Demobilization, Reintegration and Rehabilitation (DDRR)
de Mello, Sergio Vieira, 172–74, 179, 185n31, 185n35
Declaration on the Elimination of Violence Against Women, 87
Democratic Karen Buddhist Army (DKBN), 309
Democratic Republic of Congo (DRC), 5, 22t2.1, 91, 249, 359
Denmark, 50, 53, 63, 88, 235, 332n33
Department for International Development (DFID), 363
Department of Peacekeeping Operations (DPKO), 359, 363, 365t16.1
Development Assistance to Refugees (DAR), 134, 140n9, 153, 261, 275n28

Development Programme for Displaced Persons, Refugees and Returnees in Central America' (PRO-DERE), 168
Development through Local Integration (DLI), 115, 134, 153, 159n6
DFID. *See* Department for International Development (DFID)
Disarmament, Demobilization, Reintegration and Rehabilitation (DDRR), 366t16.1
Disarmament, Demobilization and Reintegration (DDR), 116
DKBN. *See* Democratic Karen Buddhist Army (DKBN)
DLI. *See* Development through Local Integration (DLI)
Doyle, Michael, 70, 83n5
DPKO. *See* Department of Peacekeeping Operations (DPKO)
DRC. *See* Democratic Republic of Congo (DRC)
Dublin Regulation (II), 47–48
Dumper, Michael, 12, 189–213, ix
  The Future of Palestinian Refugees: Toward Equity and Justice, 212n6, 213n32
  Palestinian Refugee Repatriation: Global Perspectives, 211n2, 212n7, 212n31, 213n34

East Africa, 26–27, 35, 66n12, 163, 245n17, 246n52, 247n58, 271, 273n4
East African Community (EAC), 230–31, 246n52
East Timor, 128
Economic and Social Council (ECOSOC), 370
Economic Community of West African States (ECOWAS), 363
'Educating Refugees Around the World,' 90, 105n15
Egypt, 22t2.1, 25, 54–55, 94, 156, 190, 193–94, 204
11 September 2001. *See* post-9/11
Eritrea, 22t2.1, 140n10, 231
Ethiopia, 22t2.1, 24, 88, 165, 177, 214–16, 218–19, 231–33, 235–36, 242–43, 245n16, 251–54
Ethiopian refugees, 24, 37
ethnic Albanians, 79–80

ethnic cleansing, 56, 286
ethnic Macedonians, 80
Euro-Mediterranean Partnership, 206
European Refugee Fund, 47
European Union (EU)
  African migrants, undocumented, 64
  asylum and burden-sharing arrangements, 46, 60
  asylum policy, 46–48
  Balkans conflict, 79
  harmonization process, 47–50, 66n8
  immigration and asylum politics, 50
  immigration ministers, 46
  immigration policy lacking, 45
  India and Bhutanese refugees, 291–92
  International Contact Group (ICG), 238
  migration to industrialized states, 127
  minimum standards of, 47
  Myanmar and targeted sanctions, 323
  Palestinian refugee problem, 205
  resettlement places, 59
  resettlement schemes, 236
  safe country lists, 49
  Somalia asylum applications, 244n7
  sovereignties among members of, 60
ExCom. *See* Executive Committee (ExCom)
Executive Committee (ExCom), 40n2, 41n24, 94–95, 106n30, 110, 121n4, 142, 361, 375n22, 376n27

FAFO. *See* Norwegian Institute for Applied Social Science (FAFO)
Fearon, James D., 70, 83n5
Ferris, Elizabeth, 8–9, 31, 85–107, 363, ix
  'Chairs' Summary,' Enhancing the Effectiveness of Humanitarian Action: A Dialogue between UN and Non-UN Humanitarian Organizations, 31
Finland, 47–48, 169, 332n33
First International Conference on Assistance to Refugees in Africa (ICARA I), 163–67, 170, 182, 183n2
1st Gulf War (1990), 55, 198, 203
Fortna, Page, 70, 83n5
'Fourth World,' 146–48
France, 44, 50–51, 53–54, 60–61, 63, 67n21, 88
Franco, Leonardo, 179, 184n12, 238

392   INDEX

Gaza Strip, 190–91, 193–94, 196, 199–202, 207, 209, 381
GDP. *See* Gross Domestic Product (GDP)
General Assembly (GA). *See* United Nations General Assembly (UNGA)
Geneva Convention Relative to the Protection of Civilian Persons in Time of War, 87, 105n7
genocide, 77, 86, 106n33, 134
Germany, 45, 50, 54, 57, 60, 62–64, 68n47, 211n3
Gleditsch, Kristian Skrede, 70, 83n6
Global Consultations, 56–57, 234, 361
Global Survey on Education in Emergencies, 90, 105n17
Gogh, Theo van, 61
Great Lakes region of Central Africa, 6, 34, 81, 129, 343
Gross Domestic Product (GDP), 334
Guatemala, 25, 88, 142, 150, 161n42, 167, 170, 176, 179–80, 184n15, 184n18, 184n21
Guatemala City Conference, 167
Guatemalan refugees, 167
Guiding Principles on Internal Displacement, 91
Guinea, 22t2.1, 27, 32, 88–89, 105n10, 159n4, 357–59, 373, 375n12–14
Gulf War, 55, 198, 203
Guterres, High Commissioner António, 134, 143, 151, 155, 159n8, 160n26, 161n44

Haitians, 46, 49
Handicap International, 321
Harrell-Bond, Barbara, 146, 148, 160n23, 160n36, 265, 275n39
His Majesty's Government (HMG), 287–88
Hispanic immigrants, 52, 54, 63
HIV/AIDS, 98, 126, 305, 315, 317, 324
HMG. *See* His Majesty's Government (HMG)
Hmong refugees, 306
Hoeffler, Anke, 70, 83n5
Homeland Security Presidential Directive 2, 53
Honduras, 88, 167
Horst, Cindy, 121n7, 147, 160n33, 222, 245n26, 245n32, 374n2
human rights
 abuses, 69, 86, 92, 96–97, 104, 306, 321–22
 actors, 9, 85, 90–95, 97, 101–2, 104, 363
 institutions, 92–93, 117
 violations, 27, 31, 85–86, 90–92, 94–95, 101–2, 105n3, 128, 223, 293, 367
Human Rights Council, 92
Human Rights Watch, 41n29, 95, 274n19, 331n10, 331n19, 332n28, 375n12
Human Security Report 2005, 72, 84n9
humanitarian aid, 34, 83n1, 235, 270, 322, 346, 359
Humanitarian Charter, 98
Humanitarian Response Review, 99, 106n46
Huntington, Samuel P., 63, 68n51
Hutu refugees, 71, 75, 78, 84n12, 132, 140n6, 244n2

IASC Task Force on Human Rights and Humanitarian Action, 99, 104n2
IASC Working Group, 103–4, 107n54
ICARA I. *See* First International Conference on Assistance to Refugees in Africa (ICARA I)
ICARA II. *See* Second International Conference on Assistance to Refugees in Africa (ICARA II)
ICG. *See* International Contact Group (ICG)
ICRC. *See* International Committee of the Red Cross (ICRC)
ICU. *See* Islamic Courts Union (ICU)
ICVA. *See* International Council of Voluntary Agencies (ICVA)
IDPs. *See* internally displaced persons (IDPs)
IGAD. *See* Intergovernmental Authority on Development (IGAD)
ILO. *See* International Labour Organization (ILO)
India, 14–15, 22t2.1, 25, 33, 94, 211n3, 278–85, 290–92, 294–97, 299n1, 300n20–21, 300n25, 300n60, 303, 324–25, 330n1
India-Nepal Peace and Friendship Treaty, 284
Indo-China, 25, 176, 179–80, 306, 368
Indo-China refugees, 125, 164, 180, 306, 368
Indo-Chinese 'boat people,' 172
Indo-Chinese CPA, 163–65, 171–76, 176t9.1, 177–78, 181–82, 183n1, 328, 330, 379
Indo-Chinese refugees, international conference on, 171

INDEX 393

Inter-Agency Standing Committee (IASC), 92, 99, 104n2, 106n35
Inter-Parliamentary Myanmar Caucus, 317, 332n28
Intergovernmental Authority on Development (IGAD), 230–32, 243, 246n53, 252–53
internally displaced persons (IDPs), 39, 66n5, 73, 75, 78, 85–86, 96, 105n3, 105n13–14, 105n17, 107n53, 114, 122n13, 139, 159n14, 194
  Afghan refugees, 344
  Burmese refugees, 305–6
  Hutu, 71
  Palestinians, 194, 200–201, 211n1, 211n5, 212n26
  Somali refugees, 216, 218–19, 235–36
  Sudanese refugees, 248, 273n2, 273n4
  UNHCR and, 274n16
International Committee of the Red Cross (ICRC), 97, 193
International Conference of the National Human Rights Institutions, 93
International Conference on Assistance to Refugees in Africa, 131, 163, 379
International Conference on Central American Refugees (CIREFCA), 102, 148, 163–65, 167–72, 175–83, 176t9.1, 183n1, 183n11–16, 183n20, 184n19, 379
international conference on Indo-Chinese refugees, 171, 185n32
International Contact Group (ICG), 230, 232, 238–39, 245n24
International Council of Voluntary Agencies (ICVA), 95, 105n22, 106n27
International Covenant on Civil and Political Rights, 86
International Covenant on Economic, Social and Cultural Rights, 86
International Covenant on the Elimination of All Forms of Racial Discrimination, 86
International Development Association (IDA), 113, 122n17
International Federation of Human Rights League, 55
international human rights institutions (NHRIs), 92–93

International Labour Organization (ILO), 65, 100, 115
International Organization for Migration (IOM), 67n30, 181, 271
Iran, 7, 15–16, 22t2.1, 26, 37, 44, 88, 144, 150, 333–49
Iraq, 22t2.1, 25–26, 36, 44–45, 54, 56, 128, 133, 140n10, 156, 160n26, 196, 203, 231, 233, 340, 381
Islamic
  extremists, 61, 231
  terrorists, 54, 219
Islamic Courts Union (ICU), 217–18, 238
Israel, state of, 13, 189–211, 212n8, 212n12–13, 212n19, 212n26–29, 384
Issaq Somali in Djibouti, 37
Italy, 50, 54, 60, 68n49, 169, 238

Jacobsen, Karen, 41n23, 41n28, 110, 121n4, 121n6, 212n21, 257, 274n21, 374n2, 374n9, 375n24
Jamaica, 88
Jamal, Arafat, 10–11, 141–61, 354, 360, ix
  'Camps and Freedoms: Long-term Refugee Situations in Africa,' 160n25
  'Minimum Standards and Essential Needs in a Protracted Refugee Situation: A Review of the UNHCR Programme in Kakuma, Kenya,' 160n28, 275n24
  'Negotiating Protection in a Central Asian Emergency,' 160n37
  'The Palestinian IDPs in Israel and the Predicament of Return: Between Imagining the Impossible and Enabling the Imaginative,' 212n27
  'Real-time Evaluation of UNHCR's Response to the Afghanistan Emergency: Bulletin no. 3,' 161n43
  'Refugee Repatriation and Reintegration in Guatemala: Lessons Learned from UNHCR's Experience,' 161n42
Jesuit Refugee Service, 95
JMLC. See Joint Ministerial Level Committee (JMLC)
JNA/RDP. See Joint Needs Assessment and the Somali Reconstruction and Development Programme (JNA/RDP)

Joint Ministerial Level Committee (JMLC), 288, 301n40
Joint Needs Assessment and the Somali Reconstruction and Development Programme (JNA/RDP), 234, 237
Joint Verification Team (JVT), 289–90
Jordan, 12, 54, 93, 156, 189–90, 193–94, 197–203, 207, 212n24
Juma, Monica, 13, 214–47, 355
  Eroding Local Capacity: International Humanitarian Action in Africa, 245n19
JVT. *See* Joint Verification Team (JVT)

Kagwanja, Peter, 13, 214–47, 274n15, 355
  'Counter-Terrorism in the Horn of Africa: New Security Frontiers, Old Strategies,' 246n54
  'Ethnicity, Gender and Violence in Kenya,' 245n20
  'Refugee Camps or Towns? The Socio-Economic Dynamics of the Dadaab and Kakuma Camps in Northern Kenya,' 244n5
  'Strengthening Local Relief Capacity in Kenya: Challenges and Prospects,' 245, 245n19
  'Unwanted in the "White Highlands": The Politics of Civil Society and the Making of a Refugee in Kenya 1902–2002,' 244n3
Kaiser, Tania, 13–14, 66n12, 137, 248–75, 367
  'Between a Camp and a Hard Place: Rights, Livelihood and Experiences of the Local Settlement System for Long-term Refugees in Uganda,' 276n44
  'Experience & Consequences of Insecurity in a Refugee Populated Area in Northern Uganda,' 275n22
  'Participating in Development? Refugee Protection, Politics and Developmental Approaches to Refugee Management in Uganda,' 275n33
  '"We Are All Stranded Here Together": The Local Settlement System, Freedom of Movement, and Livelihood Opportunities in Arua and Moyo Districts,' 273n3, 275n25, 275n35
Kakuma refugee camp (Kenya), 142, 152, 159n4, 160n28, 221, 224, 239–41, 244n5, 250, 252, 254, 258–59, 262, 264–65, 272, 274n6, 275n24, 275n34, 275n41
Kamtapur Liberation Organization (KLO), 295
Karen National Union (KNU), 307–8
Karenni National Progressive Party (KNPP), 307
Kenya
  Comprehensive Plan of Action (CPA) for Somali Refugees, 177, 235–36
  Dadaab refugee camp, 142, 159n4, 219, 221–24, 227–30, 239–41, 244n5, 245n23, 245n27–28, 245n32, 258, 355–56, 374n6
  District Development Plans, 241
  East African Community (EAC), 230
  EPAU evaluation, 159n4
  'hands-off refugee policy,' 137
  as host country, 26, 177, 248
  human rights conditions in camps, 230
  Intergovernmental Authority on Development (IGAD), 231
  Kakuma refugee camp, 142, 152, 159n4, 160n28, 221, 224, 239–41, 244n5, 250, 252, 254, 258–59, 262, 264–65, 272, 274n6, 275n24, 275n34, 275n41
  National Development Plans, 240–41
  protracted refugee situations (PRS), 7, 22t2.1, 160n28, 244n3, 244n5, 245n17–21, 245n27, 245n31, 245n33, 246n35–36, 260, 263, 266, 268, 270, 275n24, 275n27
  Refugee Bill, 225–26, 239, 243, 254
  refugee camps, 217, 227, 230
  Refugee Consortium of Kenya (RCK), 245n15, 245n28, 246n34, 246n38–41, 255, 274n19
  refugee protection system, 239
  refugees and security threats, 35
  refugees and violence, 31
  repatriation of refugees, 243, 245n10
  resettlement of refugees, 244
  resettlement programme, 228–29
  Self-Reliance Strategy for refugees, 134
  Somali refugees, 7, 13, 21, 22nt2.1, 177,

214–16, 218–24, 231, 242, 244, 249, 268
Somali urban refugees, 224–25, 242
Sudanese refugees, 7, 13, 22t2.1, 226, 249–54, 257–58, 260, 266, 268, 270–71, 367
UNHCR and, 227, 240, 246n45, 246n47–49, 247n66
UNHCR's Executive Office, 142
UNHCR's firewood project, 230
Khan, Prince Sadruddin Aga, 147, 160n34, 161n57
Khmer refugees, 306
Kiryandongo settlement (Uganda), 256–58, 260, 265
KLO. See Kamtapur Liberation Organization (KLO)
KNPP. See Karenni National Progressive Party (KNPP)
KNU. See Karen National Union (KNU)
Kosovo, 41n36, 45, 80, 97, 128
Kurdistan Workers' Party (PKK), 55

Laitin, David D., 70, 83n5
Lama, Mahendra P., 14, 277–301, 367
  'Bhutanese Refugees in Nepal and Indian Responses I and II,' 300n21
  'Experiments in Democracy in Bhutan,' 301n53
  'Managing Refugees in South Asia: Protection, Aid, State Behaviour and Regional Approach,' 300n22
  'Separatism and Armed Conflicts in North-East India,' 300n26
  Sikkim Human Development Report 2001, 299n4
Laotian refugees, 25, 306
Latvia, 88
Lebanon, 12, 149, 189–90, 193, 197–99, 201–2, 381
Lewa, Chris, 313, 331n22, 332n26, 332n38
Lhotsampa refugees, 14, 277–78, 280–84, 286–87, 291, 296, 299n1
Liberia, 3, 22t2.1, 25, 27–28, 34, 40n11, 72, 89, 91, 128, 133, 357, 369, 384
Liberian refugees, 21, 22t2.1, 25–28, 35, 39, 40n10, 94, 105n10, 160n40, 375n12
Liberian refugees from Guinea, 27
Liberian urban refugees in Côte d'Ivoire, 25, 94, 110

Libya, 50, 54
Liechtenstein, 88
Loescher, Gil, 7, 14–17, 20–41, 70–71, 73, 76–77, 80–82, 88, 143, 158, 303–31, 331n18, 331n20, 331n23, 353–79, 384, xiiin1
  'Protracted Refugee Situations and Peacebuilding,' 40n1, 83n4, 84n10, 84n15, 105n9, 159n10, 374n1, 374n5
  Refugee Movements and International Security, 41n34
  UNHCR: The Politics and Practice of Refugee Protection into the Twenty-First Century, 42n47, 67n29, 160n20, 160n38, 161n56, 376n29
London Resolutions, 46
Lord's Resistance Army (LRA), 249, 251–53, 258, 264, 266, 270, 274n13–14, 275n40
Lubbers, High Commissioner Ruud, 67n33, 115, 134, 143, 160n21, 375n21, 375n23
Lusaka peace process, 111
Luxembourg, 88

Macleod, Ewen, 15, 144, 333–50, xi
Madagascar, 88
Madrid train bombings, 57
Malawi, 37, 88, 232–33
Malkki, Liisa, 84n12, 132, 140n6, 215, 244n2
Malta, 64, 68n52
Marchal, Roland, 217, 244n8, 245n11
Mattner, Mark, 9–10, 108–21, 362, xi
Mauritanian refugees in Senegal, 24
MDRP. See Multi-Country Demobilization and Reintegration Program for the Great Lakes (MDRP)
Médecins sans Frontières (MSF), 97, 314, 321
Mexico, 54, 88, 97, 159n4, 159n14, 167, 170, 184n24
Mexico Plan of Action, 159n14
Middle East, 7, 25–26, 65, 127, 142, 156, 160n26, 198, 206, 213n32, 214, 377
  asylum applications from, 127
  conflict, 198
  Europe and Muslim states, 65
  Iraqi refugees in, 160n26
  Palestinian refugees, 25
  Peace Process, 206, 213n32, 377
  refugee situations, 7, 26

Middle East (cont.)
  refugees, 'warehousing' of, 377
  resettlement in, 142, 156
  Somali refugees in, 214
  unemployment in, 65
Millennium Development Goals, 237
Milner, James, 7, 14–17, 20–41, 70–71, 73, 76–77, 80–82, 88, 94, 143, 158, 303–31, 353–79, 384, xi, xiii
  'Golden Age? What Golden Age? A Critical History of African Asylum Policy,' 183n7
  'The Militarization and Demilitarization of Refugee Camps and Settlements in Guinea, 1999–2004,' 375n14
  'Protracted Refugee Situation in Thailand: Towards Solutions,' 105n9
  'Protracted Refugee Situations: Domestic and International Security Implications,' 40n1, 83n4, 84n10, 84n15, 159n10, 374n1
  'Protracted Refugee Situations: Human Rights, Political Implications and the Search for Practical Solutions,' 106n25
  'Protracted Refugee Situations and Peacebuilding,' xiiin1
  'Protracted Refugee Situations and the Regional Dynamics of Peacebuilding,' 374n5
  Refugees, the State, and the Politics of Asylum in Africa, 41n30, 41n42, 374n11
  UNHCR: The Politics and Practice of Refugee Protection into the Twenty-First Century, 42n47, 376n29
Minimum Standards, 98
Minimum Standards for Education in Emergencies, Chronic Crises and Early Reconstruction (INEE), 98
minimum standards of EU, 47
minimum standards of protection and assistance, 160n28, 161n45, 272, 275n24
minimum standards (Paris Principles), 93
Ministry of Immigration and Registration of Persons, 226, 239–40
Mixed Security Brigade (Brigade Mixte de Sécurité, BMS), 358–59
Moi, Daniel, 219–20, 224–25, 231, 240
Moldova, 57, 88
Moorehead, Caroline, 147, 160n32
Morris, Eric, 8, 69–83, 362, 367, xi
  'Prisons of the Stateless: A Response to New Left Review,' 140n2
  'Protection Dilemmas and UNHCR's Response: A Personal View from Within UNHCR,' 140n11
Mozambican refugees, 25, 37
Mozambique, 129, 148, 232
MSF. *See* Médecins sans Frontières (MSF)
Multi-Country Demobilization and Reintegration Program for the Great Lakes (MDRP), 116
Multilateral Framework of Understandings on Resettlement, 57
Myanmar boat people, 315
Myanmar (Burma), 3, 6, 14–15, 22t2.1, 24, 303–26, 328, 329f14.1, 330, 332n28, 332n30, 332n39–42
Myanmar refugees, 76, 305–6, 310, 325, 330

NARC. *See* National Rainbow Coalition of Kenya (NARC)
National Democratic Front of Bodoland (NDFB), 295
national human rights institutions, 9, 93, 100, 104, 105n22, 107n52
National League for Democracy (NLD), 304, 306, 316–17, 323
National Rainbow Coalition of Kenya (NARC), 219
National Resistance Movement (NRM), 255, 257–58, 264
NATO, 79
NDFB. *See* National Democratic Front of Bodoland (NDFB)
Nepal, 7, 14, 22t2.1, 33, 71, 76, 92, 94, 149, 277, 280, 284–88, 290–99, 299n1, 299n3, 300n14, 300n21, 301n38–39, 301n42, 301n46, 301n57, 302n61, 367
Netherlands, 50–51, 54, 60–61, 63–64, 68n40, 235, 332n33
'New International Approaches to Asylum Processing and Protection,' 50
New Zealand, 321, 332n33
Newman, Edward, 353–85, xi, xii
  'Protracted Refugee Situations and Peacebuilding,' 374n5, xiiin1
  Refugees and Forced Displacement: International Security, Human

Vulnerability, and the State, 19n2, 66n1
NHRIs. *See* international human rights institutions (NHRIs)
Nicaragua, 129, 167, 170, 180, 184n23
Nicaraguan refugees, 25, 167, 170–71
Nigeria, 232–33
1951 Convention relating to the Status of Refugee. *See* 1951 UN Refugee Convention
1951 UN Refugee Convention, 30, 46, 50–51, 56–57, 59, 67, 105n3, 123–25, 127, 137, 141, 146, 150, 162, 196, 225, 234, 254, 288, 298, 307, 327–28, 346
1973 War, 203
1967 Protocol, 87, 125, 234, 254, 288, 298, 346
1967 War, 194–95, 199
1967 Displaced Persons, 195
NLD. *See* National League for Democracy (NLD)
non-governmental organizations (NGOs)
 accountability of, 98
 advocacy by, 96
 Agenda for Protection, 234
 Bhutanese refugees, 287, 292, 318–20
 Burmese refugees, 313, 316
 CIREFCA and, 170
 civil society and, 95–98
 Code of Conduct, 102
 Committee on the Rights of the Child, 96–97
 competition between, 95
 Convention Plus, 235, 247n63
 gender issues and, 96, 98, 100, 106n32
 global, 9, 95
 Global Consultations process, 56
 global INEE Minimum Standards, 98
 human rights, 9, 85, 95–97, 104, 292, 312
 humanitarian, 95–96, 99, 101, 322, 363, 365t16.1
 independence of, 102
 'integrated missions' debate, 102
 international, 96, 102–3, 202, 263, 270–71, 287, 316, 320
 Islamic, 245n29
 Kenya and, 222, 228–29
 national, 9, 58, 103–4
 Palestinian, 206
 Palestinian refugees and, 193, 202

 'Protecting Refugees: A Field Guide for NGOs,' 98, 106n45
 protracted refugee situations (PRS), 95–96, 103, 106n27, 363, 365t16.1, 384
 Refugee Secretariat and, 243
 refugee-serving, 95, 101–2, 240
 regional, 9
 rights-based networks and standards, 98
 Somali refugees, 226, 234–35, 238–39, 242
 Statement on Protracted Refugee Situations, 94, 106n27, 247n61
 Thai-based, 322
 UN and, 9–11, 89, 98–101, 103–4, 163, 316, 321, 360, 372, 378
 UNHCR and, 148, 154, 161n51, 226, 228–29, 235–36, 239–41, 261–63, 313, 319, 326–27
non-*refoulement* principle, 12, 93, 124, 146, 148, 174–75, 190, 292
North Africa, 26, 65, 142, 144–45, 156
North West Frontier Province (NWFP), 339–40
Northern Iraq, 128
Norway, 50, 52, 59, 88, 169, 238, 332n33
Norwegian Institute for Applied Social Science (FAFO), 195
Norwegian Refugee Council (NRC), 97
NRC. *See* Norwegian Refugee Council (NRC)
NRM. *See* National Resistance Movement (NRM)
NWFP. *See* North West Frontier Province (NWFP)

OAU. *See* Organisation of African Unity (OAU)
Occupied Palestinian Territories (OPTs), 22t2.1, 195–98, 200–205, 207, 211
ODP. *See* Orderly Departure Procedure (ODP)
OECD's Development Assistance Committee, 114, 116–17
Ogata, High Commissioner Sadako, 58–59, 67n29, 129, 140n5, 142, 145, 159n3, 160n19
'Operation Restore Hope,' 217
Orderly Departure Procedure (ODP), 172, 181

Organisation of African Unity (OAU), 125–26, 165, 232
Organisation of African Unity Refugee Convention, 125, 225
Organization for Security and Co-operation in Europe (OSCE), 79
OSCE. *See* Organization for Security and Co-operation in Europe (OSCE)
Oslo Accords, 198, 204, 209
Oslo peace process, 195
Oxfam, 95

'The Pacific Solution,' 47, 50
Pakistan, 7, 15–16, 22t2.1, 26, 35–37, 55, 110, 144, 223, 314, 333–49
Palestinian Central Bureau of Statistics, 195, 212n25
Palestinian CPA, 384
Palestinian Liberation Organization (PLO), 196–98, 202–3, 206–9, 212n17
Palestinian National Authority (PNA), 202, 207
Palestinian refugees, 12, 21, 25–26, 211n1, 211n5, 212n6, 212n17, 212n21–24, 213n32, 213n34
  background, 189–92
  Israel, ethnic basis of, 196
  map of UNRWA's area of operation, 210f10.1
  origins of, 192–95
  Palestinian nationalism, persistence of, 196–97
  prolongation, recent contribution to, 197–98
  prolongation and central role of UNRWA, 201–2
  prolongation and impact on Jordan, 200
  prolongation and impact on Lebanon, 199
  prolongation and impact on OPTs and Israel, 200–201
  prolongation and impact on refugees, 198–99
  prolongation and impact on region, 202–4
  prolongation and impact on Syria, 199–200
  prolongation of situation, factors contributing to, 195–98
  PRS, resolution to the, 204–11
  PRS and changes needed, 206–11
  PRS and peacemaking, 204–6
  UNHCR's role with, 190–91, 193, 196, 206–7, 211n4
Papua New Guinea, 88
PBC. *See* Peacebuilding Commission (PBC)
PBSO. *See* Peacebuilding Support Office (PBSO)
Peacebuilding Commission (PBC), 8, 11, 18, 81, 119, 158, 354, 369–73, 376n35, 376n38, 383–84
Peacebuilding Support Office (PBSO), 81, 83n5, 370–71
peacekeeping, 13, 72, 82, 103, 139, 215, 230, 238, 270, 365t16.1, 366, 369, 374n4, 383
peacemaking, 204–6, 369, 376n32
People's Party of Switzerland, 51
PKK. *See* Kurdistan Workers' Party (PKK)
PLO. *See* Palestinian Liberation Organization (PLO)
PNA. *See* Palestinian National Authority (PNA)
Portugal, 88, 134
post-9/11, 13, 53, 66n8, 67n22–23, 82, 93, 198, 211, 214–15, 225, 228, 232, 248, 252, 345, 378
Post-Conflict Fund (PCF), 100, 107n53, 112
poverty reduction, 9, 108, 113, 122n16, 235, 240
Poverty Reduction Strategy Papers (PRSPs), 113
PRO-DERE. *See* Development Programme for Displaced Persons, Refugees and Returnees in Central America' (PRO-DERE)
Procedures Directive, 47–48
'Protecting Refugees: A Field Guide for NGOs,' 98, 106n45
protracted refugee situations (PRS). *See also* asylum trends in industrialized countries and PRS
  about, 3
  archetypal models, two, 175–78, 176nt9.1
  background information, 20, 69–71, 85, 108–9, 162–64
  Bhutanese refugees in Nepal, 7, 14, 22t2.1, 71, 76, 92, 149, 277, 294
  Burundi, 3, 22t2.1, 24–26, 33, 36, 76, 94, 132, 140n7, 356, 372
  causes of, 26–30
  challenges, addressing current, 355–59

INDEX 399

CIREFCA (1987–1994), 167–71
civil society and NGOs, 95–101
comprehensive response, framework for a truly, 362–68
comprehensive solutions, towards, 101–4
conclusions, 82–83, 120–21, 182–83, 373–74
conflict management policy, linkages to, 80–82
consequences of, 30–38
definitions, 21–25, 22t2.1
development activities and humanitarian, peace and security actors, 114–17
development actors, potential roles for, 117–20
development challenge, opportunities and responses, 109–14
diagnosis, better, 73–74
displacement, causes of, 74–75
exile, diagnosis and conditions of, 75–77
explained, 3–5, 8, 20–25, 22t2.1
framework for responding to, 353–55
Guinea, 357–59
human rights actors and human rights violations, 90–95
human rights/civil society actors and peace/security/development actors, 99–101
human rights implications, 30–33
human rights issue and, 85–86
ICARA I and II (1981-1984), 164–70
Indo-Chinese CPA (1988-1996), 171–75
Kenya, 7, 22t2.1, 160n28, 244n3, 244n5, 245n17–21, 245n27, 245n31, 245n33, 246n35–36, 260, 263, 266, 268, 270, 275n24, 275n27, 355–56
major situations, 22t2.1
migrants, 32–33
policy, framing, 381–85
political and security implications, 33–38
political engagement model, 178–81
problem, tackling the, 380–81
questions, remaining, 38–40
refugee-security nexus, challenge of, 71–72
refugee women, 31–32, 41n26, 56, 88, 91, 96, 105n14, 105n17, 132, 234, 242, 290, 356
refugees, adolescent, 90, 98
refugees, child, 30, 32, 41n26–27, 48, 56, 61, 86, 88–90, 96, 98, 105n10–105n12, 105n14, 105n17, 106n34, 155, 196, 199, 227, 234, 237, 242, 288–89, 309, 320, 340, 342
refugees, disabled, 98, 260
refugees, girl, 41n26, 89–90, 96, 227, 242
refugees, HIV/AIDS, 98, 126, 305
refugees, medically vulnerable, 32
refugees, older/elderly, 98
refugees, urban, 25, 32–33, 94, 110, 156, 224–25, 242, 249
refugees and civil war: better prescription, 77–80
'residual caseloads,' 33
resolving, 377–80
rights in practice, 88–90
rights in theory, 86–88
solutions, comprehensive, 359–62
solutions, multilateral, 381–85
solutions, recent interest in, 360–62
success, preconditions for, 368–69
Tanzania, examples from, 356–57
trends and dimensions of problem, 25–26
Uganda, 22t2.1, 118, 255, 260–61, 263, 266, 268, 270, 275n27
UN Peacekeeping Commission and, 369–73
PRSPs. *See* Poverty Reduction Strategy Papers (PRSPs)

Qualification Directive, 47–48
Quick Impact Projects (QIPs), 170, 184n23, 227, 245n10
Quips. *See* Quick Impact Projects (QIPs)

Rakhine Buddhists, 310, 312
RCK. *See* Refugee Consortium of Kenya (RCK)
Real ID Acts, 55
Reception Directive, 47–48
Red Crescent, 9, 99, 102, 104
Red Cross, 9, 97, 99, 102, 104, 193, 220
Refugee Appeal Board, 225
Refugee Consortium of Kenya (RCK), 225, 245n15, 245n28, 246n34, 246n36, 246n38–41, 255, 274n19
refugee status determination, 172, 228, 240, 288, 307
Refugee Status Determination Committee, 225, 228
Refugee Working Group (RWG), 204, 206

Refugees International, 95, 305, 330n3, 331n21
Registered Refugees (RR), 195, 214, 242, 274n17, 299n1
Repatriation, Reintegration, Reconciliation and Reconstruction (4Rs), 115, 122n22, 153, 159n6, 170, 267
Report of the Secretary-General's High-Level Panel on UN System-Wide Coherence, 140n13
Representative of the Secretary-General for Internally Displaced Persons, 81, 91, 275n31, 383
resettlement
   of Bhutanese refugees, 292–94, 296–98
   of Burmese refugees, 318, 321, 327, 332n33–35
   Convention Plus and, 235
   countries, 11, 17, 58–59, 101–2, 156, 163, 180, 253, 318
   CPA and, 172, 178
   denial of, 33
   IASC Working Group and, 103
   of Indo-China refugees, 125
   of Latin Americans, 159n14
   of Lhotsampa refugees, 14
   organized or voluntary, 158
   of Palestinian refugees, 209
   programmes, 50, 125, 127
   protracted refugee situations (PRS) and, 360, 365t16.1, 367
   as PRS solution, 18, 32, 55
   restrictive on direct arrivals of refugees, 59
   security and, 228–29
   Somali CPA and, 225–26
   of Somali refugees, 13, 144, 147, 215, 220–21, 228–29, 242–44, 268
   of Southeast Asian refugees, 171
   of Sudanese refugees, 251, 253, 268, 271, 273
   10-point Plan, 145
   third country, 77, 125, 133
   of Uganda Asians, 148
   UNCCP and, 194
   UNHCR and, 56–59, 67n28, 67n33, 142–43, 145, 150, 155–56, 161n55, 162, 172, 228
   UNHCR–IOM and, 181
   US commitment to, 171–73
   of Vietnamese, 59

resettlement countries, 11, 17, 58–59, 101–2, 156, 163, 180, 253, 318
RGB. *See* Royal Government of Bhutan (RGB)
Rohingya refugees (Muslims of Arakan), 22–23, 310–15, 318–22, 327, 330, 331n11–12, 331n19–21, 331n24, 332n37–38
Royal Government of Bhutan (RGB), 279–80, 283, 286, 288, 292, 300n17, 300n23, 300n28
RPF. *See* Rwandan Patriotic Front (RPF)
RR. *See* Registered Refugees (RR)
Russia, 203, 205–6, 324–25, 331n8, 332n42
Rwanda, 5, 22t2.1, 24, 34, 45, 71, 77–78, 82, 232, 369
Rwandan Patriotic Front (RPF), 34, 77–78
RWG. *See* Refugee Working Group (RWG)

SAARC. *See* South Asian Association for Regional Cooperation (SAARC)
SAFHR. *See* South Asia Forum for Human Rights (SAFHR)
Salehyan, Idean, 70, 83n6
Salvadorean refugees, 25, 167, 170–71
Sambanis, Nicholas, 70, 83n5
Saudi Arabia, 22t2.1, 156, 314
Save the Children UK, 105n10, 105n12
Schweizerische Volks Partei (SVP), 51
Second International Conference on Assistance to Refugees in Africa (ICARA II), 131, 136, 163–67, 177, 182, 183n2, 183n8–10
Sectoral Committee on Cooperation in Defence and Inter-State Security, 231
Security Sector Reform (SSR), 116–17, 371
Self-Reliance Strategy (SRS), 14, 121n8, 134, 140n9, 261–62, 269, 272
Sen, Amartya, 146–47, 150, 152, 160n29, 161n58
Sensenbrenner Bill, 53–54
Serbia and Montenegro, 22t2.1, 79
'Sexual Violence and Exploitation: The Experience of Refugee Children in Liberia, Guinea, and Sierra Leone,' 89, 105n10
Sierra Leone, 3, 81, 88–89, 97, 105n10, 115, 128, 133, 357–58, 371
Sierra Leonean refugees, 25, 35, 89, 358

INDEX 401

Slaughter, Amy, 10, 57, 94, 123–39, 149, 256, 262, 362, xi
SLORC. *See* State Law and Order Restoration Council (SLORC)
SNF. *See* Somali National Front (SNF)
SNM. *See* Somali National Movement (SNM)
Socialist Republic of Vietnam, 171–72
societal sovereignty, 59, 64, 65, 378
Somali Contact Group, 234
Somali National Front (SNF), 218
Somali National Movement (SNM), 216
Somali refugees
  Agenda for Protection (2001), 234–35
  AU stabilization force, 232–33
  in Australia, 214
  background, 214–16
  camps, security and judicial capacity in, 240–41
  conclusion, 242–44
  Convention Plus (2003), 235
  in Djibouti, 214–16, 233, 236
  EAC and IGAD, 230–32
  in East Africa, 27, 163
  in Ethiopia, 24, 214, 216, 233, 251–52, 254, 274n6
  in Europe, 214
  future, facing the, 239–42
  in Horn of Africa, 163, 214–15
  International Contact Group (ICG), 238–39
  international response to, 233–39
  in Kenya, 214–16, 218, 236, 252, 254
  in Kenya and Daniel Moi's 'abdicationist' state, 219–21
  in Kenya and enchained 'good Samaritan,' 226–28
  in Kenya and Kibaki's interventionist state (2003-2007), 225–26
  in Kenya and local Somali hosts, 223–24
  in Kenya and policy of encampment, 221–23
  in Kenya and resettlement by security lenses, 228–29
  in Kenya and Somali urban refugees, 224–25
  in Kenya and the balance sheet, 229–30
  in Middle East, 214
  in North America, 214
  protection, strengthening, 243
  PRS, causes of, 216–19

PRS and evasive peace, 217
PRS and Islamic Courts, 217–18
PRS and state collapse and civil war, 216–17
PRS and 'war on terrorism,' 218–19
refugee protection capacity, strengthening, 239–40
refugees, protecting urban, 241–42
regional responses to, 230–33
repatriation, 243
resettlement, 244
Somali CPA, 144, 149, 178, 235–36
Somali JNA-RDP, 236–38
Somaliland question, the unresolved, 233
in South Africa, 214
in Uganda, 214, 254
UNHCR's role with, 214–15, 217, 220–22, 224, 226–30, 233–36, 239–42, 244n1, 244n4, 245n10, 245n15–16, 245n18, 245n23, 245n25, 245n29, 246n42, 246n45, 246n47–48, 247n59, 247n62, 247n66
in Yemen, 25, 94, 236
Somali Youth League, 223
Somalia, 3, 13, 22t2.1, 25, 34–35, 37, 45, 57, 91, 128, 144, 178, 214–21, 223, 227–28, 230–39, 242–43, 244n4, 244n7, 245n16, 245n24, 246n57, 247n67–68
Somalia CPA, 144, 149, 177–78, 182, 235–36
Somaliland, 133, 233, 235, 237, 247n60
South Africa, 125, 129, 214, 232
South America, 125, 199
South Asia, 7, 15, 26, 300n22, 300n26, 301n46, 301n50, 377
South Asia Forum for Human Rights (SAFHR), 292, 301n50
South Asian Association for Regional Cooperation (SAARC), 14, 296–97, 363
South Sudan, 7, 148, 155, 274n6, 275n41, 367
Southeast Asia, 7, 15, 24, 26, 171, 185n28, 303–4, 306, 315, 324–28, 331n9–10
Spain, 48, 50, 54, 60, 64, 88
Special Programmes for Refugee Affected Areas (SPRAAs), 356–57
Sphere Project, 98
SPLM/A. *See* Sudan People's Liberation Movement/Army (SPLM/A)
Sri Lanka, 22t2.1, 44, 71, 115, 292, 300n26
SRS. *See* Self-Reliance Strategy (SRS)

SSR. *See* Security Sector Reform (SSR)
State Law and Order Restoration Council (SLORC), 306, 331n6
Stedman, Stephen John, 8, 69–83, 362, 367, xi
  'Conclusions and Policy Recommendations,' 84n11
  Ending Civil Wars: The Implementation of Peace Agreement, 84n14, 84n16, 374n3
  'Evaluation Issues in Peace Implementation,' 84n16
  Refugee Manipulation: War, Politics, and the Abuse of Human Suffering, 83n1, 84n11
Stevens, Jacob, 123, 128–29, 140n1–2
Strategic Framework for Burundi, 371
sub-Saharan Africa, 26, 65, 128, 156
Sudan, 3, 7, 13, 22t2.1, 24, 37, 72, 91, 110, 128, 133, 148, 155, 221, 231, 246n56, 248–60, 263–73, 273n1, 273n5, 274n6, 274n8–9, 274n12, 275n41, 369, 384
Sudan People's Liberation Movement/Army (SPLM/A), 249–50, 252, 255, 264
Sudanese Lost Boys, 156
Sudanese refugees in Uganda and Kenya
  asylum, legal basis and refugee conditions, 254–55
  asylum in Uganda and Kenya, 253–55
  background, 248–50
  conflict and forced migration, 250–51
  conflict context and failure to find peace, 252–53
  DAR/SRS in Uganda, 261
  flight, causes of, 251–52
  human rights and security challenges, 270–71
  policy lessons for the region, 271–73
  protracted exile, impact of, 255–65
  protracted exile and camp/settlement system, 258–60
  PRS, impact on refugees and states, 255–57
  refugee security, 258
  refugees, who are they?, 253–54
  refugees in Kenya, 7, 13, 22t2.1, 226, 249–54, 257–58, 260, 266, 268, 270–71, 367
  refugees in Uganda, 7, 13–14, 22t2.1, 23, 37, 249–58, 260–61, 266, 268, 270–71, 367

security challenges, local/national and regional, 263–65
solution, other dimensions of a, 268–70
Sudanese, future of, 265–70
Sudanese, immediate prospects, 265–68
UNHCR/NGOs/civil society organizations, 261–63
Sudanese urban refugees in Egypt, 25, 94, 156
Sudan's CPA, 264, 268, 270
Suppression of Terrorism Bill, 226
sustainable development., 380
SVP. *See* Schweizerische Volks Partei (SVP)
Sweden, 50, 55, 57, 88, 169, 232n33, 238, 380
Switzerland, 51, 53, 66n2
Syria, 12, 156, 160n26, 189–90, 193–94, 197–202, 207

Tajikistan, 22t2.1
Taliban, 35, 222, 335, 337, 339, 343–45, 347
Tamil Tigers, 71
Tanzania, 21, 22nt2.1, 26, 29, 32–33, 35–36, 41n19, 75–76, 84n12, 110, 132, 134, 140n6–7, 159n4, 230, 232, 238, 244n2, 356–59, 372–73, 374n8, 375n16
Temporary Protection Directive, 47
TFG. *See* Transitional Federal Government (TFG)
TFI. *See* Transitional Federal Institutions (TFI)
Thailand, 15, 22t2.1, 25, 31–32, 35, 76, 105n9, 134, 303–10, 315, 318–19, 321, 326–27, 330n1, 330n4, 331n5, 331n9–10, 331n13
Three Pronged Approach, 57
Tibet, 292, 295
Tibetan refugees, 287
Timor-Leste conflict, 108
TNG. *See* Transitional National Government (TNG)
trade sanctions, 15, 304
transit camps, 57, 250, 258, 274n10
Transitional Federal Government (TFG), 217–19, 230–31, 236–38, 243
Transitional Federal Institutions (TFI), 233, 238, 243
Transitional National Government (TNG), 217
Treaty of Maastricht, 46–47

Troeller, Gary, 7, 38, 43–67, 377–85, xii, xiii
  'Protracted Refugee Situations and Peacebuilding,' xiiin1
  'Protracted Refugee Situations and the Regional Dynamics of Peacebuilding,' 374n5
  'Refugees and Displaced Persons in Contemporary International Relations; Reconciling State and Individual Sovereignty,' 66n1
  'UNHCR Resettlement: Evolution and Future Direction,' 67n28
  'UNHCR Resettlement as an Instrument of Protection,' 67n33
Tunisia, 232
Turkey, 36–37, 44, 54, 60, 65
Turkmenistan, 149, 160n27
Turner, Simon, 132, 140n7
Tutsi refugees, 34, 71, 77–78, 132

Uganda, 66n12, 160n23, 161n53
  AU force and Somali protest, 232–33
  East African Community (EAC), 230
  EPAU evaluations of, 159n4
  as host country, 26, 248, 253, 257
  identity cards for refugees, 225
  IGAD and Somali dispute, 231
  Kiryandongo refugee settlement, 265
  peacekeeping force, 230, 232
  protracted refugee situations (PRS), 22t2.1, 118, 255, 260–61, 263, 266, 268, 270, 275n27
  Refugee Convention and employment, 88
  refugees in, 165, 263–64, 268, 273n2, 273n4, 276n44
  resettlement operations, 148
  Rwandan Patriotic Front (RPF), 34, 77–78
  Rwandan refugees, 24
  security and UNHCR, 154
  Self-Reliance Strategy for refugees, 121–22n8, 134, 140n9
  Somali refugees, 214
  Sudanese refugees, 7, 13–14, 22t2.1, 23, 37, 249–58, 260–61, 266, 268, 270–71, 367
  Tutsi refugees, 71
  Zairian refugees, 249
Uganda People's Defence Force, 266
Uganda Self-Reliance Strategy, 121n8, 140n9

ULFA. *See* United Liberation Front of Assam (ULFA)
UN. *See* United Nations (UN)
UN/IASC Country Teams, 92, 100, 104
UNCCP. *See* United Nations Conciliation Committee for Palestine (UNCCP)
UNDP. *See* United Nations Development Programme (UNDP)
unemployment, 37, 60–61, 65, 200
UNHCR. *See* United Nations High Commissioner for Refugees (UNHCR)
UNHCR and PRS
  achievement, determining scope for, 151–52
  actors, working with other, 138–41
  agenda and recognizing the phenomena, 142–43
  background, 123–24, 141–42
  black spots, camps and deprived capabilities, 145–46
  capability deprivation in 'Fourth World,' 146–48
  capacities and limitations, communicating, 138
  competing priorities, 129–30
  conclusion, 158
  funding of UNHCR, 130
  goals in asylum, focus on attainable, 152–54
  humanitarian strategy, elements of, 135–41
  ideal role for UNHCR, 148–49
  industrialized states, 127–29
  interactions between refugees and local population, 135–36
  local integration, 154–55
  more funds, not less, 156–57
  political solutions and meagre results, 143–45
  programme objectives and design, 131–33
  recent developments, 133–35
  refugee-hosting countries, 124–27
  resettlement, 155–56
  role of UNHCR, 129–33
  safety nets, provision of, 154
  solutions, other, 158
  solutions, targeted and incremental durable, 154–56
  state role, supporting, 136–38

UNHCR and PRS (cont.)
  strategy, applying by segmenting and targeting, 151–58
  strategy: focus on present, 149–51
  time for solutions, 130–31
  vision and approach, 157–58
  voluntary repatriation, 155
UNICEF. *See* United Nations Children's Fund (UNICEF)
Union of Islamic Courts, 233, 243
United Kingdom (UK), 46, 50–52, 54, 57, 63–64, 88, 235, 238, 253, 302n68, 321, 332n33
United Liberation Front of Assam (ULFA), 295
United Nationalities Alliance, 316
United Nations (UN), 93, 99–100, 105n19, 106n23, 129, 138–39, 143–44, 148, 150, 158, 160n35, 205, 246n57, 322, 369–70, 376n31
  agencies, 9–11, 91, 98–101, 103–4, 107n48, 112, 139, 163, 183, 316, 321, 360, 367, 370, 378
  Charter, 91, 227, 232
  Children's Fund (UNICEF), 106n35, 115, 321, 365t16.1
  Commission on Human Rights, 160n21
  Conciliation Committee for Palestine (UNCCP), 193–94
  Convention Against Torture, 54–55, 86
  Convention Relating to the Status of Refugees (1951) (*See under* 1951 United Nations Refugee Convention)
  Department of Political Affairs, 81
  Development Assistance Frameworks, 100
  Development Group, 237
  Development Programme (UNDP), 107n51–52, 112, 115–16, 122n11, 131, 165–70, 176, 176t9.1, 177, 179, 181, 263, 270–71, 350n1, 363, 365t16.1 (*See also* Bureau for Crisis Prevention and Recovery (BCPR))
  Economic and Social Council (ECOSOC), 370
  Emergency Relief Coordinator, 99, 106n46, 139
  High Commissioner for Human Rights, 106n39
  Human Rights Committee, 91, 96, 106n33
  Joint Needs Assessment/Somali Reconstruction and Development Programme (JNA/RDP), 234, 237–38
  Mission In Sudan (UNMIS), 270, 275n42
  Office for the Coordination of Humanitarian Affairs (UNOCHA), 28, 40n11, 106n46, 247n58, 363, 365t16.1
  Office of Internal Oversight Services (UNOIOS), 228, 246n49
  Office on Human Rights, 9
  Peacebuilding Commission (PBC), 8, 11, 18, 81, 119, 158, 354, 369–73, 376n35, 376n38, 383–84
  Peacebuilding Support Office (PBSO), 81, 83n5, 370–71
  Population Fund, 321
  as refugee agency, 124
  Relief and Works Agency (UNRWA), 12, 21, 26, 59, 190, 193–95, 199–202, 207, 210f10.1, 212n11, 212n16, 212n21
  Relief for Palestine Refugees, 193
  Work Plan for Sudan, 267
United Nations General Assembly (UNGA), 19n1, 147, 234
  agencies for Palestinian refugees, 193
  Declaration on Territorial Asylum states, 69
  Fourth World unrepresented in, 147
  ICARA I achievements, 166
  'Paris Principles,' 93
  partition of Palestine into a Jewish state, 192
  Peacebuilding Commission (PBC), 322–23, 371, 376n35, 376n38
  Resolution 181, 212n9
  Resolution 194, 205
  Resolution 35/42, 183n3
  Resolution 36/124, 183n7
  Resolution 60/180, 376n34, 376n37
  Resolution 194 III, 212n10
  Resolution 2312 XXII, 83n3
  Resolution of December 1993, 87
  Resolution on CIREFCA, 170
  Resolution 428V, 83n2
  Third Committee of the, 160n34, 161n57

INDEX 405

United Nations High Commissioner for
  Refugees (UNHCR)
 Afghan refugees in Iran and Pakistan,
   336, 343, 346–49, 350n1
 Agenda for Protection, 234
 asylum trends in industrialized countries
   and PRS, 56–59
 Bhutanese refugees in Nepal, 149,
   287–88, 290–91, 294–97, 299n1,
   301n35, 301n39, 301n42, 301n48,
   301n58, 302n61
 Burmese refugees in South and Southeast
   Asia, 307, 310–14, 318–21, 326–28,
   331n18, 331n20
 comprehensive plan of action (CPA),
   164, 167–68, 173, 175, 179, 181, 235,
   268
 comprehensive plans of action (CPAs),
   164, 175–76, 179, 236
 Convention Plus initiative, 162, 182
 Evaluation and Policy Analysis Unit
   (EPAU), 67n20, 122n10, 134, 143,
   160n28, 246n45, 275n24, 360, 375n13,
   375n16, 375n19
 Executive Committee (ExCom), 40n2,
   41n24, 94–95, 106n30, 110, 121n4,
   361, 375n22, 376n27
 *Framework for Durable Solutions for
   Refugees and Persons of Concern*,
   114, 122n21, 153, 159n6, 184n22
 Global Consultations, 56–57, 234, 361
 Global Strategic Objective, 143, 155, 158
 Guterres, António, 134, 143, 151, 155,
   159n8, 160n26, 161n44
 Handbook for Self-Reliance, 153, 161n48
 internally displaced persons (IDPs),
   274n16
 Kenya and, 227, 240, 246n45, 246n47–49,
   247n66
 Kenya and Executive Office of, 142
 Kenya and the firewood project, 230
 Lubbers, Ruud, 67n33, 115, 134, 143,
   160n21, 375n21, 375n23
 Morjane, Kamel, 159n13
 non-governmental organizations (NGOs),
   148, 154, 161n51, 226, 228–29,
   235–36, 239–41, 261–63, 313, 319,
   326–27
 Ogata, Sadako, 58–59, 67n29, 129, 140n5,
   142, 145, 159n3, 160n19

 Palestinian refugees, 190–91, 193, 196,
   206–7, 211n4
 'Refugee Livelihoods Network,' 134, 143,
   353, 361
 resettlement and, 56–59, 67n28, 67n33,
   142–43, 145, 150, 155–56, 161n55,
   162, 172, 228
 six-point plan, 56
 Somali refugees, 214–15, 217, 220–22,
   224, 226–30, 233–36, 239–42, 244n1,
   244n4, 245n10, 245n15–16, 245n18,
   245n23, 245n25, 245n29, 246n42,
   246n45, 246n47–48, 247n59, 247n62,
   247n66
 Statute of, 69
 Strengthening Protection Capacity
   Project, 100, 134, 243
 Sudanese refugees in Uganda and Kenya,
   261–63
 10-point Plan, 144–45, 156, 160n18
 Uganda and security, 154
United Nations Secretary-General (UNSG),
  4, 11, 19n1, 106n23, 139, 158, 169,
  176, 176t9.1, 179, 181, 183, 246n44,
  323, 363, 365t16.1, 376n32, 383
 Annan, Kofi, 93, 370
 Boutros-Ghali, Boutros, 369
 Report of the Secretary-General's High-
   Level Panel on UN System-Wide
   Coherence, 140n13
 Representative of the Secretary-General
   for Internally Displaced Persons, 81,
   91, 275n31, 383
United Nations Security Council (UNSC),
  151, 161n44, 365t16.1
 AU summit, 233
 peace and security, 363
 Peacebuilding Commission (PBC), 370,
   383
 Resolution 725, 243
 Resolution 794, 197, 227, 246n44
 Resolution 1725, 246n57
 US letter on Myanmar, 323–24
United Republic of Tanzania. *See* Tanzania
United States, 35, 46, 49, 52–55, 58, 61, 63,
  76, 125, 127–28, 147, 174, 198, 202,
  227, 229, 233, 244, 246n50, 252, 290,
  306, 323, 362
Universal Declaration of Human Rights, 59,
  86, 146, 150

UNOCHA. *See* United Nations Office for the Coordination of Humanitarian Affairs (UNOCHA)
UNOIOS. *See* United Nations Office of Internal Oversight Services (UNOIOS)
UNOSOM I, 217
UNOSOM II, 217
UNRWA. *See* United Nations Relief and Works Agency (UNRWA)
UNSC. *See* United Nations Security Council (UNSC)
UNSG. *See* United Nations Secretary-General (UNSG)
US Agency for International Development (USAID), 363
US Committee for Refugees and Immigrants, 30, 41n20, 105n8, 134, 143, 146, 159n11, 160n24, 245n10, 246n43, 331n19
US Homeland Security Border Patrols, 54
US Resettlement Program, 228
USA Patriot Act, 53–54, 332n34
USAID. *See* US Agency for International Development (USAID)
Uzbekistan, 22t2.1

Vietnam, 22t2.1, 76, 172–74, 181, 185n29, 306, 317
Vietnam War, 173
Vietnamese boat people (VBP), 174
Vietnamese diaspora, 173
Vietnamese refugees, 25, 59, 306
violence against women, 86, 91

Walter, Barbara, 70, 83n5
Weiner, Myron, 41n35, 55, 67n27
West Africa, 26–27, 35, 129, 134, 236, 375n15
West Bank, 190–91, 193–94, 196, 198, 200–201, 207, 209
West Nile Bank Front, 258, 274n13
Western Sahara, 22t2.1
WFP. *See* World Food Programme (WFP)
Who Are We? The Challenges to America's National Identity (Huntington), 63, 68n51
Widows and Orphans Association, 260
Women's Commission for Refugee Women and Children, 41n26, 97, 105n14, 105n17, 105n34
Working Group on Arbitrary Detention, 91, 105n19
World Bank, 9–10, 100, 103, 107n53, 108, 112–17, 119–20, 121n2, 122n11, 122n13–18, 122n25–26, 122n28, 208, 237, 270, 324, 363, 365t16.1, 370, 384
World Food Programme (WFP), 27, 29, 40n17, 154, 161n53, 222, 254, 259, 287, 313, 340, 342

xenophobia, 37, 62, 64, 214, 223–24, 242

Yemen, 22t2.1, 25, 94, 177, 196, 235–36
Yugoslavia, 45, 56, 74, 343

Zaire, 78, 165
Zambia, 22t2.1, 26, 88, 111, 115, 134, 159n4, 375n25
Zimbabwe, 88